Deutsche Gesellschaft für Kristallographie

23. Jahrestagung

16. bis 19. März 2015

Göttingen

D G K

Abstracts

Table of content

Plenary Talks

P1
Materials Crystallography

B. Iversen[1]
[1]*Aarhus University, Center for Materials Crystallography, Department of Chemistry & iNANO, Aarhus, Denmark*

Materials science projects typically revolve around synthesis, structure investigation and property characterization. Knowledge about the 3D structure of molecules, nano-particles and crystals is essential for understanding, designing and manipulating chemical behavior or physical properties of materials, and rapid development of crystallographic methods continue to push the boundaries of materials science. In this talk I will discuss recent applications of in-situ crystallography and charge density crystallography in materials science. Using unique in-situ supercritical reactors and synchrotron radiation it is possible to follow evolving atomic structure during nanoparticle formation using total scattering methods [1, 2]. Highly accurate Bragg diffraction data provide insight into the electron density distributions in crystals and recent advances include studies of heavy atom materials and powder diffraction charge density analysis [3, 4].

[1] "Evolving atomic structure during nanoparticle formation", C. Tyrsted et al., IUCr-J 2014, 1, 165-171
[2] "In Situ Studies of Solvothermal Synthesis of Energy Materials", K. M. Ø. Jensen et al., ChemSusChem 2014, 7, 1594-1611
[3] "Contemporary X-ray Electron Density Studies using Synchrotron Radiation", M. R. V. Jørgensen et al., IUCr-J 2014, 1, 267-280
[4] "Nuclear-weighted X-ray maximum entropy method - NXMEM", S. Christensen et al., Acta Crystallogr. Sect. A 2015, 71, 9-19

P2
Optimization of Anomalous Diffraction Analysis for Biomolecules

W. Hendrickson[1]
[1]*Columbia University, Biochemistry and Molecular Biophysics, New York, United States*

Anomalous diffraction has proved to be highly effective for phase evaluation in *de novo* structure determination and for elemental identifications in macromolecular studies. Although early protein structures were all analyzed by isomorphous replacement methods, now anomalous diffraction methods predominate. Analyses from multi- and single-wavelength anomalous diffraction (MAD and SAD) experiments have produced more than 80% of all *de novo* macromolecular structures obtained so far in the 21st century, and SAD alone accounted for more than 70% of those reported in 2013 (Hendrickson, *Quart. Rev. Biophys.* **47**, 49-93). Phase evaluation from anomalous diffraction depends on the strength of anomalous scattering factors, which gives us opportunities for optimization. For conventional MAD experiments and SAD studies performed at resonance, the strength of anomalous signals depends on energy resolution in the x-ray beam. For SAD experiments conducted away from a resonance condition, the strength of f'' depends on closeness of approach to a resonance edge. Light elements that are relatively prevalent in biological macromolecules (Z = 15-20; P, S, Cl, K, Ca) can be effective scatterers for SAD phasing provided that the experiment uses x-rays of relatively low energy. We have devised robust SAD-phasing procedures to study native, light-atom-only biological structures. We enhance the signal-to-noise in diffraction measurements by increasing f'' with a lower than usual x-ray energy and by reducing noise with data combined from multiple crystals. Low energy has also facilitated elemental identitifications for Ca, Cl, S, P and Mg through f'' scattering factor refinements. In addition, we are developing a new microdiffraction beamline, NYX at NSLS-II, for optimized resonance experiments, and we have planned a companion beamline, LAX, for low-energy anomalous x-ray diffraction. I will discuss these and other developments aimed at optimizing the effectiveness of anomalous diffraction experiments for biological macromolecules.

P3
Crystallography without crystals?

A. Goodwin[1]
[1]*University of Oxford, Inorganic Chemistry Laboratory, Oxford, Great Britain*

This talk will explore the extent to which scattering techniques might allow robust structural characterisation of systems exhibiting varying degrees of correlated disorder. Drawing on examples of amorphous network solids, nanoparticles, cooperative paramagnets, and proteins, this talk will seek to establish qualitative criteria for assessing which problems of structural disorder might be solvable.

P4
Trapping liquid drugs inside crystals

A. Bacchi[1]
[1]*University of Parma , Dipartimento di Chimica, Parma, Italy*

The objective of this work is to find a systematic way to embed liquid or volatile molecules inside crystalline materials, with the multiple aims of stabilizing them, of tuning their possible ways of delivery in medicine or agrochemistry, and to explore new regulatory and intellectual properties issues. Liquid or volatile formulations of active pharmaceutical ingredients (APIs) are intrinsically less stable and durable than solid forms; in fact most drugs are formulated as solid dosage because they tend to be stable, reproducible, and amenable to purification. Most drugs and agrochemicals are manufactured and distributed as crystalline materials, and their action involves the delivery of the active molecule by a solubilization process either in the body or on the environment. The poor solubility of pharmaceutical active ingredients (API) or the reverse too high solubility of agrochemicals are problems often encountered in their formulation since these phenomena limit respectively the bioavailability of the API or the duration of the action of the agrochemical. However some important compounds for the human health or for the environment occur as liquids at room temperature. We have defined a benchmark of molecules relevant to human health and environment that have been combined with suitable partners according to the well known methods of crystal engineering in order to embed them in crystalline hosts. Our approach is twofoled, by using cocrystals and MOFs.

The formation of co-crystals has been demonstrated as a means of tuning solubility properties of solid phases, and therefore it is widely investigated by companies and by solid state scientists especially in the fields of pharmaceuticals, agrochemicals, pigments, dyestuffs, foods, and explosives. In spite of this extremely high interest towards co-crystallization as a tool to alter solubility, practically no emphasis has been paid to using it as a means to stabilize volatile or labile or liquid products. In this work we trap and stabilize volatile and liquid APIs and agrochemicals in crystalline matrices by engineering suitable co-crystals. These new materials alter the physic state of the active ingredients allowing to expand the phase space accessible to manufacturing and delivery. We also explore the possibility to include liquid APIs inside the pores of suitable designed MOFs, again with the aim of stabilizing their solid state formulation.

P5

Short-range order in silicate melts at high pressure

C. Sanloup[1]
[1] *University Pierre et Marie Curie, Paris, France*

Magmas are key objects in planetary sciences, whether that be in relation with present-day volcanism or during the early stages of planetary formation while part or most of the Earth was molten. They are indeed the most efficient carriers of heat and matter between the planetary envelopes (core, mantle, crust, atmosphere).

Understanding the macroscopical physical properties of magmas requires an accurate microscopic structural description. Despite the lack of periodicity and long-range order, melts retain a characteristic short-range order, which obeys basic crystal-chemical rules. However, until recently only a limited amount of information existed about changes in silicate melts structure with temperature (T) or pressure (P), as magmas are essentially high pressure objects, being formed at depth.

The aim of our current project is thus to determine the structural environment (interatomic distances, coordination numbers) of elements in silicate melts as a function of P-T conditions using in situ synchrotron X-rays based techniques combined with high pressure devices (diamond-anvil cells and large volume presses). Structural changes in silicate melts affect their physical properties and I will show the consequences of major element coordination changes (Si, Fe) on their density and viscosity at depth. Structural changes in magmas also affect their chemical properties, for instance the way elements partition between the melt and crystals upon partial melting of a source mantle rock. To assess the effect of melt structure on partitioning at depth, we have investigated the short range order around trace and minor elements such as Br, Xe, Lu and W. Results will be presented in relation with the pressure-evolution of element partitioning.

Talks

MS01 - General Interest

MS01-T01
Invairom based point charges - a rapid way to the electrostatic potential

C. M. Wandtke[1], J. Luebben[2], B. Dittrich[2]
[1]*Georg August University Goettingen, Inorganic Chemistry, Goettingen, Germany*
[2]*University Hamburg, Inorganic and Applied Chemistry, Hamburg, Germany*

The invariom formalism [1] has, so far, mostly been applied to assign aspherical scattering factors. However, invarioms are useful for other purposes, and we present the application of rapidly computing a molecular electrostatic potential (ESP) from tabulated parameters for a set of atomic coordinates, e.g. from a crystal structure. Automatic parameter assignment relies on an evaluation of the local-atomic bonding environment. ESP computation just requires atomic point charges and bond distances to hydrogen atoms from quantum chemistry, and both are provided in a new database. Hence charge density programs and the orientation of aspherical scattering factors are not necessary to obtain an ESP anymore. Technically a "pqr" file containing invariom point charges, bond-distance adjusted hydrogen-atom coordinates and the atomic van der Waals radius is generated by the software APD-Toolkit from a set of coordinates (e.g. a cif, pdb or SHELXL res file). The pqr file can then be directly converted into a map of the electrostatic potential onto a molecular surface by freely available programs like Jmol [2] or APBS [3], depending on the kind of potential desired; the first one yields a Coulomb potential, the second a Poisson-Boltzmann potential.

Our new database relies on RESP [4] or alternatively Stockholder point charges [5] for invarioms and contains approximately 4000 entries covering a considerable fraction of organic chemistry. Theoretical bond lengths of 250 unique covalent bonds to hydrogen can be extracted. Procedures are tested and validated on a series of biologically active organic molecules. The electrostatic potentials from the atomic point charges are compared to those obtained by experimental and theoretical charge density studies.

[1] B. Dittrich, C. B. Hübschle, K. Pröpper, F. Dietrich, T. Stolper, J. J. Holstein, *Acta Cryst.* B69(2013) 91.
[2]R. M. Hanson, *J. Appl. Cryst.* 43 (2010)1250.
[3] N. A. Baker, D. Sept, S. Joseph, M. J. Holst, J. A. McCammon, *Proc. Natl. Acad. Sci.* 98 (2001) 10037.
[4] C. I. Bayly, P. Cieplak, W. D. Cornell, P. A. Kollman, *J. Phys. Chem.* 97 (1993) 10269.
[5] F. L. Hirshfeld, *Theoret. Chim. Acta* 44 (1977) 129.

MS01-T02
Extension of empirical force fields to extreme conditions.

D. W. M. Hofmann[1]
[1]*CRS4, FlexCryst, Pula, Italy*

An extreme challenge is the crystal structure prediction and their properties at extreme conditions. The obvious approach is molecular dynamics. In this case the temperature and pressure are adjustable parameters for the simulation. However, this approach is rather theoretical than a practical way for the gigantic hypothetical calculation time. Therefore we investigated the possibility to derive expressions for the weak intermolecular interactions, which are temperature and pressure dependent.

For the analysis of inter-molecular interactions two different approaches exist: On the one hand quantum chemistry can be used, on the other hand the analysis of experimental known crystal structures. Specially temperature is hardly accessible to quantum chemistry and the only remaining possibility is the analysis of experimental data. While in the beginning this was done manually, we did begin with classification as method for the analysis. Here we would like to present another method of data mining, which keep pace with the increasing number of experimental data and allows a fine-tuned analysis of the data, which reveals effects of the crystallization conditions. In a first step the problem is linearized, i.e. the interactions are described in piece-wise linear potentials. In the second step the gradients for all degrees of freedom within the crystal are calculated (cell vectors, cell angles, translation and rotation of the molecules) . These gradients have to be zero for the free energy under the given conditions, otherwise the crystal would not be stable. We obtain in this way an overdeterimined linear equation system with millions of equations and thousands of parameters. The global minimum for the parameters of such an equations system can be determined by singular value decomposition.

A side effect of data analysis is always the finding of outliers. It helps to detect erroneous crystal structures. A second effect is the accurate determination of the van der Waals radius. These two results have been verified by plotting the obtained potentials versus the radial distribution functions. The potentials have been validated by minimization of experimental structures to check, if the experimental structures are correctly reproduced as local minimums in the potential surface and by crystal structure prediction to check, if the potential reproduce correctly the global minimums in dependence of the temperature.

References:
1. D. W.M. Hofmann and L. Kuleshova, Crystal Growth and Design 3929, 14 (2014)
2. D. W.M. Hofmann and L. Kuleshova, Data Mining in Crystallography, Springer Verlag, (2010)

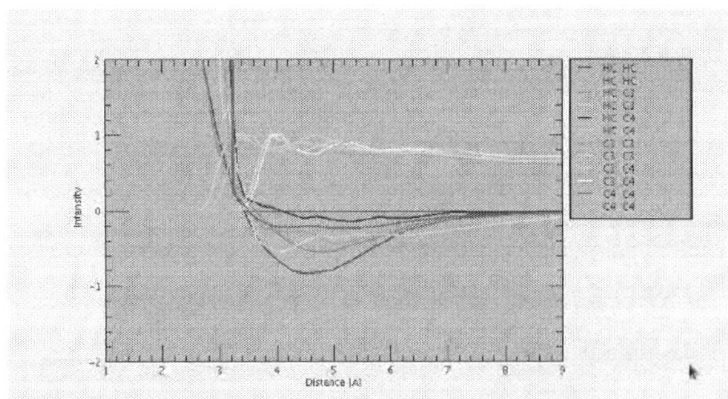

Figure 1: The derived potentials and the radial distribution functions for simple hydrocarbons at standard conditions

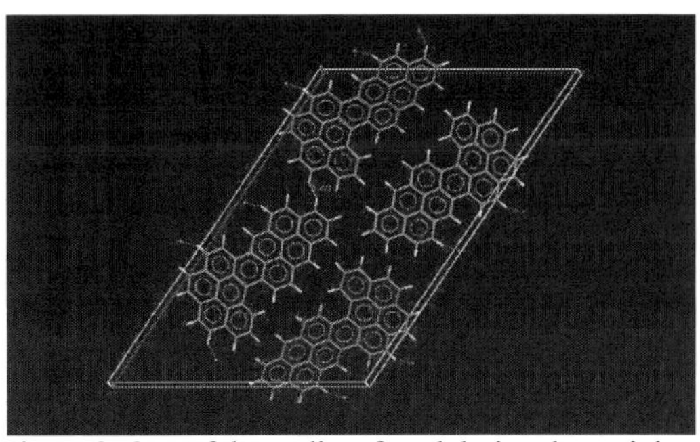

Figure 2: One of the outliers found during data mining. The short HC...HC contact is unrealistic according the potential (red curve) and the radial distribution function (pale red curve)

MS01-T03

Deuterium disturbs the molecular arrangement in the solid state

A. Kupka[1], K. Merz[1]
[1]*Ruhr-University Bochum, Inorganic Chemistry 1, Bochum, Germany*

"*For his discovery of heavy hydrogen*" the Nobel Prize was granted to Harold C. Urey in 1934. Followed by the synthesis of heavy water and after the availability of notable amounts of D_2O, deuterated compounds became an increasingly interesting topic of fundamental research and applications. H/D-exchange as the smallest possible element substitution in the molecular structure affects the physical properties that depend directly on the proton or deuterium. Spectroscopic and spin-dependent properties change in a rather predictable way and allow the use of vibrational, NMR and neutron spectroscopy.

However, the influence of H/D exchange on thermodynamic properties of compounds is not yet properly understood which is nicely shown in the case of the simple compounds H_2O and D_2O. Since the discovery of D_2 and D_2O a large number of deuterated compounds have been prepared and the crystal structures of several of these compounds were investigated. In general, it is expected that the crystal structures of compounds containing hydrogen are not affected by the substitution of hydrogen by deuterium.

A detailed survey of deuterated and non-deuterated compounds indicates that the assumption is wrong. Isotopic substitution can influence the molecular arrangement in the solid state. The aggregation of molecules in the solid state is very sensitive to small changes within the substitution pattern of the molecules. Deuterium substitution as a weak directing substituent may have quite an influence on the molecular arrangement. This phenomenon, called isotopic polymorphism, is not only restricted to compounds with absence of strong specific intermolecular interactions in crystal structures. H/D-exchange can influence hydrogen bonds, and as result the crystal structure. Particularly striking is the fact that an isotopic effect can also be observed in the use of solvents in crystallization procedures. The formation of polymorphic forms in selected cases of small organic compounds can be controlled by deuterated solvents.

[1] V. Vasylyeva, T. Kedziorski, N. Metzler-Nolte, C. Schauerte, K. Merz, *Cryst. Growth Des.* 2010; *10*, 4224.
[2] K. Merz, V. Vasylyeva, *CrystEngCom.* 2010; *12*, 3989.
[3] A. Kupka, V. Vasylyeva, D. W. M. Hofmann, K. Merz, *Cryst. Growth Des.*, 2012, *12*, 5966.
[4] O. V. Shishkin, S. V. Shishkina, A. V. Maleev, R. Zubatyuk, V. Vasylyeva, K. Merz, *ChemPhysChem*, 2013, *14*, 847.
[5] K. Merz, A. Kupka, *Cryst. Growth Des.*, 2014, in revision.

Figure 1

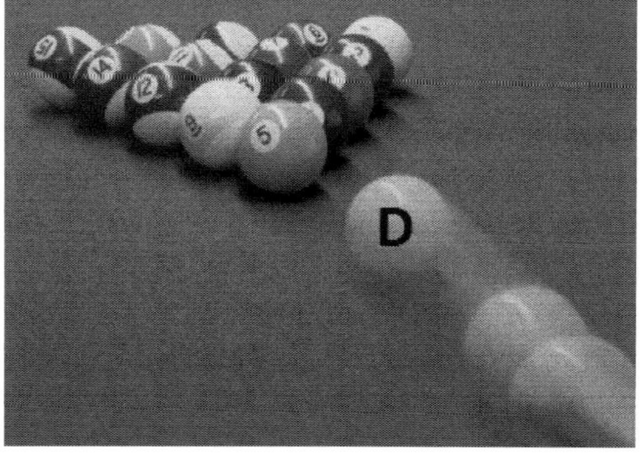

MS01-T04

Ultrafast Dynamical Study Using Time-resolved Laue Diffraction

S. Thekku Veedu[1], M. Messerschmidt[2], R. Henning[3], V. Srajer[3], I. Kosheleva[3], S. Techert[1]
[1]*Deutsches Elektronen Synchrotron, FS-SCS, Hamburg, United States*
[2]*Stanford Linear Accelerator, LCLS, Sanfrancisco, United States*
[3]*Advance Photon Source, BioCars, Chicago, United States*

Static structures for many molecules are available at high resolution but the mechanism by which these molecules function and the structures of intermediate states often remain elusive. Electron transfer reactions are fundamental processes in chemistry and also in biology. Light harvesting complexes are functional centers in plants where sunlight is converted into chemical energy. In this, optical excitation in a chromophore unit leads to the transfer of electrons within the system. However, due to the complexity of the biological photo-reaction centre, recent spectroscopic efforts have concentrated on a smaller chemical model which share characteristic with their biological counter parts. Knowledge of the geometry of molecular excited states at atomic resolution is crucial for a complete understanding of photo-induced chemical processes. Time-resolved X-ray diffraction (TR-XRD) using polychromatic synchrotron radiation allows a detailed study of the time evolution of structural intermediates and short living states of chemical systems at wide range of time-scales, drawing a complete picture of the photo-induced charge transfer process. Investigation of photo-excitation processes in molecular single crystals, where the initial photo-excitation processes occur on extremely short time-scales (femto-/picosecond time domain) and have been in the focus of scientific investigations due to their possible applications, e.g. as optical switches. We aimed to determine photo-induced structural changes in Pyrene-N,N-dimethylaniline (PyDMA)) by TR-XRD using Laser pump/X-ray probe technique and will present here the dynamics of the molecular diode system.

MS01-T05

Three-dimensional solutions of the Helmholtz equation

G. Shpenkov[1]
[1]*Military University of Technology, Warsaw, Poland*

Graphene, one-atom-thick layer of graphite, having a two-dimensional hexagonal lattice, gives us the unique possibility for the direct verification of one of the recent predictions, resulting from the comprehensive analysis of the particular solutions of the Helmholtz wave equation [1].

It is well-known that the hexagonal lattice of graphene, a two-dimensional crystal (a giant macromolecule), has a high order symmetry axis, a six fold. Hence, the physical properties of graphene, in particular, electrical conductivity, must be isotropic in a plane perpendicular to this axis, in full agreement with the basic symmetry theory [2].

In accord with the solutions of the aforesaid wave equation, it turned out that *atoms* have the *shell-nodal structure*; that is, they represent nucleon molecules. Herewith, they have the different symmetries, strictly definite inherent in each individual atom. The specific feature of the carbon atom is the fact that, as a nucleon molecule, it has the *twofold axis* of symmetry. It means that graphene, if it consists of the orderly bond atoms, apparently, is an anisotropic crystal. The laboratory tests have confirmed this prediction. The conductivity anisotropy was found on all test samples [3]. By now there are many other evidences in favour of the shell-nodal structure of the atoms.

A typical polar diagram of conductivity anisotropy of the test samples of an *unstrained pristine* graphene layer, monoatomic in thickness, has a characteristic elliptical form (Fig.1). A scheme of the measurements is shown in Fig. 2. Along the major axis of anisotropy, graphene behaves like a metal; in a direction perpendicular to the major axis graphene exhibits semiconducting properties.

[1] G. P. Shpenkov, *An Elucidation of the Nature of the Periodic Law*, Chapter 7 in "*The Mathematics of the Periodic Table*", edited by Rouvray D. H. and King R. B., Nova Science Publishers, NY, 119-160, 2006.
[2] Robert E. Newnham, *Properties of Materials: Anisotropy, Symmetry, Structure*, Oxford University Press, 2005.
[3] G. Shpenkov, *Anisotropy of Graphene*: Test Results, 2010; https://register.epo.org/application?documentId=ES2PMPUZ5321FI4&number=EP10707863&lng=en&npl=false. Supplementary data to "*Method for manufacturing nano electronic devices made from 2D carbon crystals like graphene and devices obtained with this method*". Int. Appl. No.: PCT/EP2010/052298; Publ. No.: WO/2010/097393;
https://register.epo.org/application?number=EP10707863&lng=en&tab=main

Figure 1

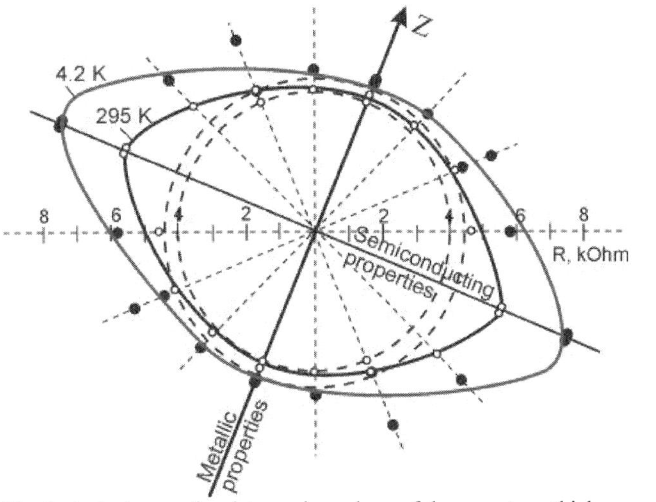

Fig. 1. Anisotropy of resistance in a plane of the one-atom-thick graphene of a round form of the diameter D=10 *mm*; a width of electrodes h=580 *nm* [3].

Figure 2

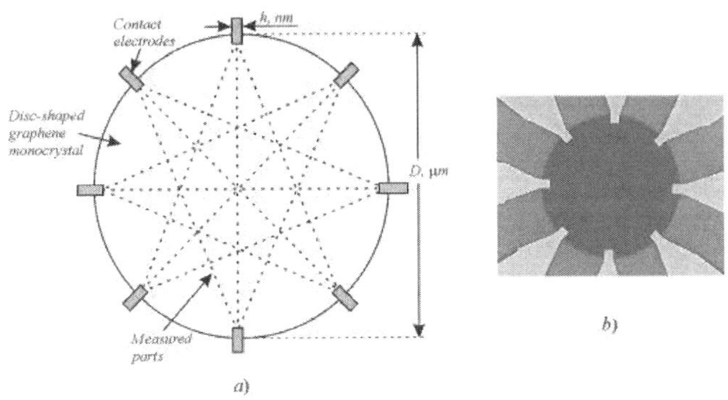

Fig. 2. A measurement scheme (*a*), and an electron microscope image of one of the test samples (*b*).

MS01-T06

Current Status of the Liquid-Metal-Jet X-ray Source Technology

E. Espes[1], B. Hansson[1], O. Hemberg[1], M. Otendal[1], T. Tuohimaa[1], P. Takman[1]
[1]*Excillum AB, KISTA, Sweden*

High-end x-ray diffraction and scattering techniques such as high-resolution XRD, protein crystallography, and SAXS rely heavily on the x-ray source brightness for resolution and exposure time. Traditional solid or rotating anode x-ray tubes are typically limited in brightness by when the e-beam power density melts the anode. The liquid-metal-jet technology has overcome this limitation by using an anode that is already in the molten state.

We have previously demonstrated prototype performance of a metal-jet anode x-ray source concept [1-3] with unprecedented brightness in the range of one order of magnitude above current state-of-the art sources. The technology has since been developed into a stable and reliable source for home-lab systems

This presentation will review the current status of the technology specifically in terms of stability, lifetime, flux and brightness. It will also discuss details of the liquid-metal-jet technology with a focus on the fundamental limitations of the technology. It will furthermore refer to some recent data from applications within x-ray diffraction and SAXS.

[1] O. Hemberg, M. Otendal, and H. M. Hertz, Appl. Phys. Lett., 2003, 83, 1483.
[2] M. Otendal, T. Tuohimaa, U. Vogt, and H. M. Hertz, Rev. Sci. Instr., 2008, 79, 016102.
[3] T. Tuohimaa, M. Otendal, and H. M. Hertz, Appl. Phys. Lett., 2007, 91, 074104

MS02 - Material Science I

MS02-T01

The Cu(2+)-mineral callaghanite Cu2Mg2(CO3)(OH)6·2H2O - a new candidate for quantum phase transitions at low pressure

S. Lebernegg[1], A. Tsirlin[2], O. Janson[3], Y. Prots[1], H. Rosner[1]
[1]*MPI CPfS, Dresden, Germany*
[2]*National Institute of Chemical Physics and Biophysics, Tallinn, Estonia*
[3]*University of Technology, Institute of Solid State Physics, Vienna, Austria*

Low-dimensional quantum magnets attracted high attention because of their exotic magnetic properties and ground states, representing an unrivaled field for challenging our understanding of collective quantum phenomena. Particularly interest are compounds exhibiting quantum phase transitions at weak magnetic fields or pressures which are experimentally still accessible. However, only few such compounds are known so far. A new promising candidate is the natural Cu(2+)-mineral callaghanite. Its crystal structure features isolated Cu2-dimers with a superweak (almost compensated) magnetic exchange coupling of J

Ref.1: Lebernegg et al., Phys. Rev. B 89, 165127 (2014)

Figure 1

Figure 1: Blue, bipyramidal callaghanite crystals (0.2 mm) on matrix from the Premiers Chemical Mine, Gabbs, Nye Co., Nevada, USA.

Figure 2: Crystal structure of callaghanite at ambient pressure. Cu2-dimers are shown in orange, CO3 groups in grey.

Figure 2

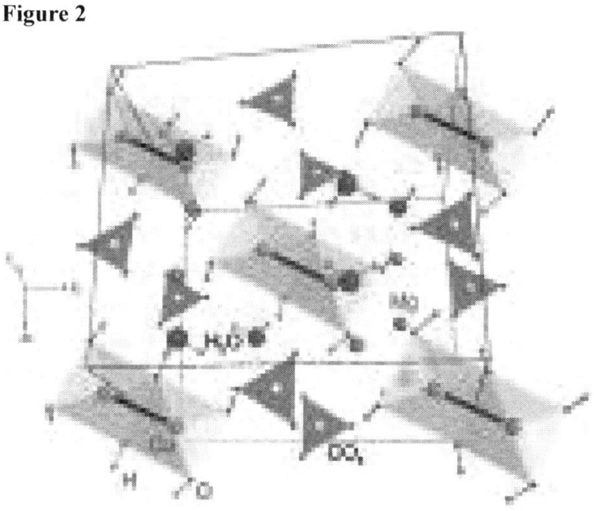

MS02-T02

Structural changes of In- and Na-doped $MnWO_4$ and their magnetic behaviours

S. Park[1], U. Gattermann[1], D. Behal[1]
[1]*LMU Mücnhen, Geo- & Umweltwissenschaften: Kristallographie, München, Germany*

Manganese tungstate ($MnWO_4$) is a relevant prototype multiferroic as: (1) $MnWO_4$ undergoes three different magnetic phase transitions in a narrow temperature range to the antiferromagnetic (AF) phases AF3, AF2, and AF1 at 13.5, 12.7, and 7.6 K, respectively [1]; (2) The AF2 state exhibits electric polarization induced by the elliptical spin spiral system [2]; (3) This magnetoelectric (ME) coupling can be manipulated readily by chemical doping and defect engineering. Recently we have investigated natural $MnWO_4$ crystals (huebnerite), featuring new nuclear and magnetic structure ordering. This surprising observation is essentially due to symmetry lowering from P2/c to P2, as shown in synchrotron and neutron single crystal diffraction studies [3]. Renninger scans of 301 reflection which is forbidden to P2/c clearly confirmed the dissymmetrization (**Figure a, -b**). In order to figure out the crucial factor influencing on the ME in the locally distorted AF2 state we have investigated In- and (In, Na)-doped $MnWO_4$ compounds. One of the unique features of the solid solution $Mn_{(1-3x)}In_{2x}\square_xWO_4$ ($\square$ = vacancy, $0 < x < 0.11$) is an abnormal lattice expansion by strong distortion around WO_6 octahedra surrounded statistically by MnO_6, InO_6, or O_6 moieties [4]. Nonetheless the basic structure of In-doped $MnWO_4$ holds the usual space group symmetry P2/c. However this basic structure cannot be preserved when In^{3+} along with Na^+ replace more than 12 % Mn^{2+}, exhibiting a super structure cell of 2a x b x c [5]. Magnetic susceptibility and specific heat capacity measurements showed that the multiferroic phase AF2 is stabilized by partial In-doping whereas the basic antiferromagnetic phase AF1 disappears, so far investigated above 2 K (**Figure c, -d**) [5]. At DGK 2015 in Göttingen we present unusual behaviour of the natural $MnWO_4$ along with interesting changes of structural and magnetic ordering in $Mn_{(1-3x)}In_{2x}\square_xWO_4$ and $Mn_{(1-3x)}In_{2x}Na_xWO_4$.

References

[1] Lautenschläger et al, H., *Phys. Rev. B*, V48 (9) (1993) 6087-6098.
[2] Urcelay-Olabarria et al, *Phys. Rev. B.*, 87 (2013) 014419.
[3] S.-H. Park, D. Behal et al, in submission.
[4] U. Gattermann, S.-H. Park, M. Kaliwoda,. J. Solid St. Chem., 219 (2014) 191-200.
[5] U. Gattermann, S.H.- Park et al, in preparation.

Figure a: Weak h0l (l= odd) reflections forbidden to P2/c (synchrotron single crystal diffractometer F1, HASYLAB); **b:** Renninger scans of the reflection 301 from a natural $MnWO_4$ crystal at 3, 10, and 13K (top) and variable temperature scan between 3 and 88K (bottom) (neutron single crystal diffractometer RESI , FRMII); **c**: Magnetic susceptibility of In:$MnWO_4$ compounds while the field sweeps at 2 K; **d**: Phase diagram of In:$MnWO_4$ compounds based on magnetic susceptibility measurements.

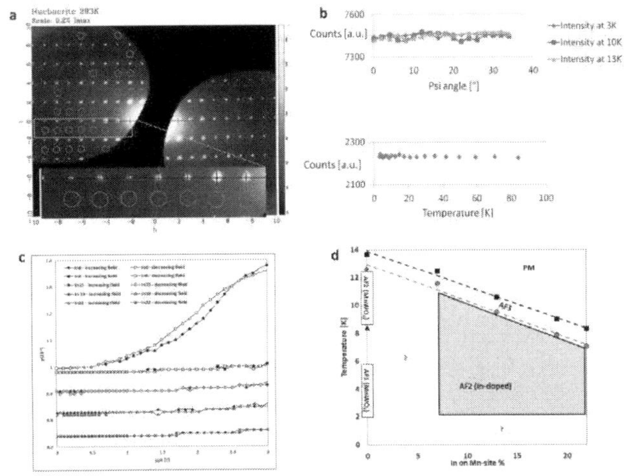

MS02-T03
Studying behavior of polar nanoregions in $(Ca,Sr)_xBa_{1-x}Nb_2O_6$ relaxor ferroelectric

C. S. Pandey[1], J. Schreuer[1], M. Burianek[2], M. Mühlberg[2]
[1]Ruhr University Bochum, Institute of Geology, Mineralogy & Geophysics, Bochum, Germany
[2]University of Cologne, Department of Crystallography, Cologne, Germany

Relaxor ferroelectrics (hereafter relaxors) are known for their frequency-dependent broad dielectric response as a function of temperature near a specific temperature T_m. For more than five decades relaxors have been extensively studied, and are still under intensive investigation because of their many peculiar features like nonlinear optical properties and applications in capacitors, actuators, transducers and electro-optical devices. Due to the disadvantage of volatility and toxicity of PbO more and more research is encouraged in lead-free relaxors. Promising candidates are members of the tetragonal tungsten bronze (TTB) structure type. A modern definition of relaxors came from one of its peculiar properties, the existence of local polar clusters with randomly oriented polarisation known as polar nanoregions (PNRs), whose first initiation takes place few hundred degrees above the Curie temperature T_C at the Burns temperature T_D. It is well accepted that relaxor properties are associated with these PNRs, however, the mechanism behind their formation is still a matter of debate. We will present the results of our studies on $(Ca,Sr)_xBa_{1-x}Nb_2O_6$ single crystals regarding relaxor properties and the behavior of PNRs using resonant ultrasound spectroscopy.

MS02-T04
New insights into the atomistic origin of chemically-enhanced response functions of $(1\text{-}x)Na_{0.5}Bi_{0.5}TiO_3\text{-}xBaTiO_3$

K. Datta[1], M. Göbbels[2], R. B. Neder[1], B. Mihailova[3]
[1]Friedrich-Alexander-Universität Erlangen-Nürnberg, Crystallography and Structural Physics, Erlangen, Germany
[2]Friedrich-Alexander-Universität Erlangen-Nürnberg, Mineralogy, Erlangen, Germany
[3]University of Hamburg, Earth Sciences, Hamburg, Germany

Perovskite-type (ABO_3) materials with transition elements on the B site are the best choice to produce multifunctional ferroelectrics with tunable properties, due to the flexibility of the perovskite framework when accommodating various chemical elements on the cationic A and/or B sites. Pb-based ferroelectric solid solutions exhibit outstanding dielectric, piezoelectric, pyroelectric, and optoelectric properties but they have become undesired for future exploration because their production and consequent waste storage are environmentally unfriendly. Recently, the solid solution $(1\text{-}x)Na_{0.5}Bi_{0.5}TiO_3\text{-}xBaTiO_3$ (NBT-xBT) has received much of attention due to its similarities with $PbZr_{1-x}Ti_xO_3$ and $(1\text{-}x)PbB'_{1/3}B''_{2/3}O_3\text{-}xPbTiO_3$ in terms of the existence of morphotropic phase boundary (MPB) with weak temperature dependence and strong enhancement of the dielectric permittivity and piezoelectric response for $x \sim x_{MPB}$. However the underlying structural phenomena in NBT-xBT are still poorly understood, which calls for further structural investigations. In this study the mesoscopic-scale structural transformations in NBT-xBT are studied by combined inelastic light scattering and synchrotron x-ray total elastic scattering in the temperature range between 80 - 870 K. The most remarkable result is that at $x \sim x_{MPB}$ the phonon mode involving displacements of both off-centred A-site Bi and B-site Ti experiences strong softening and damping near the triple-point temperature in the phase diagram, in strong contrast to the compounds with $x < x_{MPB}$ or $x > x_{MPB}$. The chemically-enhanced coupling between the Bi and Ti subsystems at x_{MPB} is facilitated by the subtle disturbance of the coupling processes within the Bi subsystem induced by Ba at x_{MPB}. These mesoscopic-scale structural and dynamic phenomena are proposed to be key factor for the enhanced response functions at the MPB in NBT-xBT rather than the complex nanodomain pattern.

MS02-T05
Investigation of the high temperature behavior of the pyroxenes $LiMGe_2O_6$ (M = Cr, V)
G. Nénert[1], M. Isobe[2], N. Dadivanyan[1],
[1]PANalytical B.V., Lelyweg 1, 7602 EA, Almelo, The Netherlands
[2]Max-Planck Institute for Solid State Research, Heisenbergstr. 1, D-70569 Stuttgart, Germany

G. Nenert[1], M. Isobe[1], N. Dadivanyan[1]
[1]PANalytical, Almelo, Netherlands

The pyroxenes have been intensively investigated for decades due to their importance in mineralogy. In the 90's, several compositions within the clinopyroxenes receive more attention due to their low dimensional magnetic properties. Finally, since 2007 a renewed interest arises for these materials after the work of Jodlauk *et al.* reporting the multiferroic and magnetoelectric properties. This was the beginning of an intensive investigation of the magnetic, magnetoelectric and multiferroic properties of these materials.

A proper understanding of these properties require a deep understanding of the crystal structures of these materials. While the silicon based clinopyroxenes have been widely investigated as function of pressure and temperature [3], the germanium based compositions remain poorly investigated. Here we report on the high temperature behavior of $LiMGe_2O_6$ (M = Ge, V) exhibiting a strong first order phase transition between P21/c and C2/c symmetries as illustrated in Figure 1. This study also clarifies the conflictual reports on the symmetry at room temperature of $LiCrGe_2O_6$ ($P2_1/c$ versus C_2/c) [4]. Our data confirm the $P2_1/c$ symmetry and a strong first order phase transition around 870°C

[1] B. Pedrini, *et al.* Phys. Rev. B 70, 024421 (2004); M. Isobe, *et al.* J. Phys. Soc. Jpn. 71, 1423-1426 (2002); Z. S. Popovic, *et al.* Phys. Rev. Lett. 93, 036401 (2004),
[2] S. Jodlauk, *et al.*, J. Phys.: Condens. Matter 19, 432201 (2007).
[3] G. J. Redhammer & G. Roth, Z. Kristallogr. 219, 585-605 (2004)
[4] G. J. Redhammer, *et al.* Acta Cryst. C 64, i97-i102 (2008), Y. Matsushita, *et al.* Solid State Sciences 12, 676-679 (2010).

Figure 1

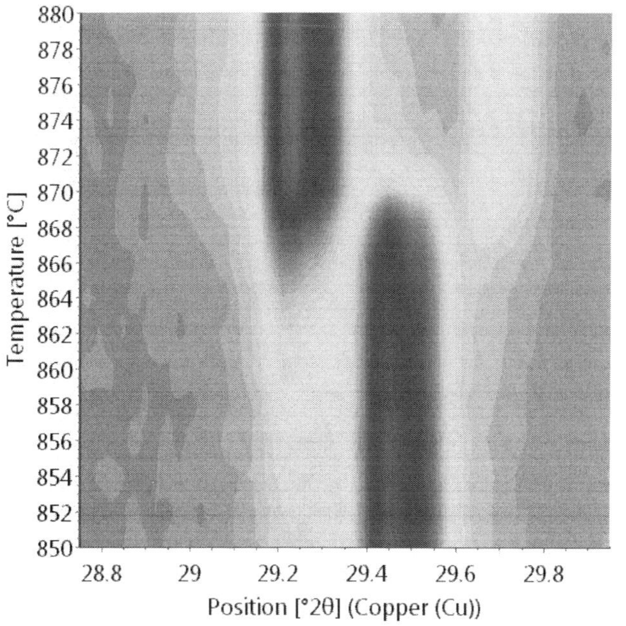

Figure 1: Powder diffraction data evidencing the phase transition in $LiCrGe_2O_6$ from $P2_1/c$ towards C2/c

MS02-T06

Ferroelectricty of Phenazine- chloranilic acid at its Incommensurate and Commensurate Phases

L. Noohinejad[1], S. van Smaalen[1], S. Mondal[1], A. Schoenleber[1], A. Woelfel[1], S. I. Ali[1], S. Dey[1]
[1]*University of Bayreuth, Lab of Crystallogarphy, Bayreuth, Germany*

Organic materials based on hydrogen-bonded supramolecular chains with a polar space group form one class of ferroelectric materials. The cocrystal of phenazine (Phz) and 2,5-dichloro-3,6-dihydroxy-p-benzoquinone (chloranilic acid, H_2ca) is one of several recently discovered hydrogen-bonded organic ferroelectrics [1].The co crystal of Phz-H_2ca becomes ferroelectric upon cooling through the loss of inversion symmetry [2]. Further cooling results in the development of an incommensurate ferroelectric phase, followed by a lock-in transition towards a twofold superstructure. Here we present the incommensurately and commensurately modulated crystal structures of Phz-H_2ca at T = 139 and 100 K, respectively, by using the superspace approach. The superspace group $P2_1(1/2\ \sigma_2\ 1/2)0$ applies for both the incommensurate (σ_2 = 0.5139) and commensurate (σ_2 = 0.5) phases [3]. The modulation mainly affects the position of protons within the part of the intermolecular hydrogen bonds which is responsible for the ferroelectricity at lower temperature. The modulation is explained as a competition between proton transfer favored for single hydrogen bonds on the basis of pKa values and avoiding unfavorable Coulomb repulsion within the lattice of the resulting ionic molecules.

[1] Horiuchi, S., Ishii, F., Kumai, R., Okimoto, Y., Tachibana, H., Nagaosa, N., Tokura, Y.: Ferroelectricity near room temperature in co-crystals of nonpolar organic molecules. Nature Materials 4 (2005) 163-166.
[2] Horiuchi, S., Kumaia R., Tokura Y.: Proton-displacive ferroelectricity in neutral cocrystals of chloroanilic acids with phenazine. J. Mater. Chem. 19 (2009) 4421-4434.
[3] Ferroelectricity of phenazine-chloranilic acid at T = 100 K, Noohinejad, L., Mondal, S., Wölfel, A., Schönleber, A., van Smaalen, S., J. Chem. Cryst. 44, (2014) 387-393.

MS03 - Challenges in macromolecular crystallography

MS03-T01
In vivo grown protein crystals - new targets for serial crystallography

L. Redecke[1], M. Klinge[1,2], R. Schönherr[3], K. Nass[4], F. Lübber[1,3], J. M. Rudolph[1,3], S. Schneegans[1,3], K. Fita[1], C. Gati[5], G. Bourenkov[6], D. Rehders[1], D. Oberthür[2,5], B. Sommer[2,7], J. Martinez-Costas[8], T. S. Schneider[6], H. N. Chapman[5,9], M. Duszenko[7], K. Hirata[10], R. Duden[3], C. Betzel[2]
[1]*University of Lübeck, Institute of Biochemistry, Lübeck, Germany*
[2]*University of Hamburg, Institute of Biochemistry, Hamburg, Germany*
[3]*University of Lübeck, Institute of Biology, Lübeck, Germany*
[4]*Max-Planck-Institute for Medical Research, Heidelberg, Greece*
[5]*Center for Free-Electron Laser Science (CFEL), Hamburg, Germany*
[6]*European Molecular Biology Laboratory (EMBL), Hamburg Outstation, Hamburg, Germany*
[7]*University of Tübingen, Interfaculty Institute of Biochemistry, Tübingen, Ghana*
[8]*University Santiago de Compostela, Department of Biochemistry and Molecular Biology, Santiago de Compostela, Spain*
[9]*University of Hamburg, Institute of Physics, Hamburg, Germany*
[10]*RIKEN/SPring-8 Center, SR Life Science Instrumentation Unit, Hyogo, Japan*

Spontaneous protein crystallization within living cells is known for more than a century, e.g. for storage proteins in seeds. However, *in vivo* crystals have not been considered for X-ray structure determination, mainly due to their small size that coincided with low diffraction capabilities and sensitivity to radiation damage. This changed with the realization of serial crystallography approaches at X-ray free-electron lasers (XFEL) and third-generation synchrotron sources. We already demonstrated that highly brilliant and ultra-short XFEL pulses are perfectly suited to collect high-resolution diffraction data (2.1 Å) from *in vivo* crystals of *Trypanosoma brucei* cathepsin B (TbCatB) grown within baculovirus-infected insect cells during gene over-expression (1,2). Comparable structural information was obtained from the same *in vivo* crystals when a micron-sized synchrotron beam was combined with high-precision diffractometry and a helical line scan approach, although the resolution was slightly reduced (3). Even if these results already indicate that *in vivo* grown protein crystals could act as suitable targets for structural biology, a more broad application of this approach requires a detailed investigation of the intracellular crystal formation process.

Applying live-cell imaging techniques we revealed that the intracellular environment displays one crucial factor for successful crystal nucleation. Depending on the protein, crystallization is restricted to specific cellular compartments, e.g. the endoplasmic reticulum or peroxisomes. The size of the protein crystals can be optimized *in vivo* by fusion of amino acids that improve the crystal contacts. However, the baculovirus infection itself is not required for *in vivo* crystal growth, as shown by direct transfection of insect cells. Inspired by these results we established an easy-to-use screening system to test spontaneous crystallization of proteins in different cellular compartments. Next to TbCatB, four other proteins formed *in vivo* crystals so far that diffracted XFEL or synchrotron radiation applying serial crystallography techniques. Diffraction data processing is currently ongoing. The identified parameters already offer exciting new possibilities to improve the success rate of this approach. However, the exceptional dynamic formation and degradation cycles observed for luciferase *in vivo* crystals highlight the need for further investigations to understand *in vivo* crystallization and to fully exploit its potential for structural biology.

(1) R. Koopmann*, K. Kupelli*, L. Redecke* *et al. Nat. Methods* 9, 259-262 (2012).
(2) L. Redecke*, K. Nass* *et al. Science* 339, 227-231 (2013).
(3) C. Gati*, G. Bourenkov* *et al. IUCrJ* 1, 87-94 (2014)

MS03-T02
Sample environment developments for serial crystallography experiments

L. Chavas[1], H. Chapman[1], S. Bajt[1]
[1]*DESY, Hamburg, Germany*

Serial femtosecond crystallography (SFX) uses x-ray pulses from free-electron laser (FEL) sources to outrun radiation damages and thereby overcome long-standing limits in the structure determination of macromolecular crystals. Intense x-ray FEL pulses allow the collection of damage-free data at room temperature, and give the opportunity to study highly time-resolved reactions including irreversible events. This method may open the way to understanding the structure of biological molecules that fail to crystallize readily into large crystals.

By taking full advantage of high repetition rate x-ray pulse delivery schemes, protein structures could be determined in just minutes of measurement time. Automation in sample delivery during SFX experiments could therefore induce a turnover of samples much higher than at today's third generation synchrotrons and automated macromolecular crystallography beamlines, as no crystal alignment or complex robotic motions are required. New challenges would emerge from this enormous amount of data that could be recorded at such facilities, at all levels of the experiment, from the beam delivery to the data management, through the sample environment and injection process.

Challenges exist at various levels of the experiments, going from the vast nature of the samples to be screened at FELs, handling a large number of samples with minimum human intervention, but also fully characterizing the overall injection and sample environment system. The Center for Free-Electron Laser science has developed a scientific unit that concentrates on breaking down these various milestones and prepares ways for the SFX scientific community to experiment in a user friendly and reliable environment. The presentation will introduce the latest tools that were engineered, emphasizing on the developments for automated replenishment of samples and sample injection devices. Our development program additionally targets serial synchrotron crystallography, by efficiently integrating the latest technologies derived from experiments performed at FELs and third generation synchrotron sources.

MS03-T03
Synchrotron based fragment screening at the HZB

F. Huschmann[1,2], R. Förster[1,3], A. Heine[2], M. Hellmig[1], G. Klebe[2], J. Linnik[1], P. Malecki[1,4], N. Radeva[2], M. Röwer[1], J. Schiebel[2], K. Sparta[1], M. Steffien[1], M. Ühlein[1], P. Wilk[1,5], M. Weiss[1], U. Mueller[1]
[1]*Helmholtz-Zentrum Berlin, Macromolecular Crystallography (HZB-MX), Berlin, Germany*
[2]*Philipps-Universität Marburg, Pharmazeutische Chemie, Marburg, Germany*
[3]*Freie Universität , Chemie-Pharmazie-Biologie, Berlin, Germany*
[4]*Max Delbrück Zentrum für Molekulare Medizin, Crystallography group, Berlin, Germany*
[5]*Humboldt-Universität zu Berlin, Institute of Biology, Berlin, Germany*

During the last 10 years fragment-based lead discovery has evolved into a reliable technique in drug development. The use of high throughput methods, high diffraction quality of the target protein and the high quality of the chosen fragment library are essential for the success of fragment screening. Thorough crystallographic analysis of protein-fragment complexes and their binding modes reveal detailed structural knowledge to develop fragments (100-200 Da) into new potential lead structures (300-500 Da).

Recently we have started to establish an experimental facility optimized for high throughput fragment screening at the BESSY II storage ring [1].

We have validated our assembled library of 96 fragments against two target proteins. These initial results revealed that this library can identify binding partners at a hit rate of about 10%.

In addition we are currently testing several novel techniques to simplify and accelerate sample preparation.

Aim is to make our library in combination with a fully automated beamline [2] available for academic and industrial users. This unique facility with automated performance of screening experiments and evaluation of bound fragments will enable efficient fragment screening on a much broader basis.

References:

[1] U. Mueller, N. Darowski, M.R. Fuchs, R. Förster, M. Hellmig, K.S. Paithankar, S. Pühringer, M. Steffien, G. Zocher, M.S. Weiss, 2012. , J. Syn. Rad., 19, 442

[2] M. Krug, M. S. Weiss, U. Heinemann and U. Mueller, 2012, J. Appl. Cryst. 45, 568-572

MS03-T04

Phasing a PH domain via caesium

W. Kölmel[1], C. Kisker[1]

[1]*Rudolf Virchow Center of Experimental Biomedicine, University of Würzburg, Structural Biology, Würzburg, Germany*

Background

The p62 protein is a subunit of the general transcription factor IIH (TFIIH). TFIIH is a ten subunit complex and fulfils a dual role in the cell. As a transcription factor it is crucial for transcription initiation as it conducts promoter opening. Furthermore it is an indispensible factor in nucleotide excision repair (NER). NER is a DNA repair pathway, which removes UV light induced DNA damages, as well as a broad range of bulky DNA adducts. p62 assumes a structural function within TFIIH. Moreover it interacts with other transcription factors during transcription and with additional NER factors during DNA repair.

Aim

p62 comprises a pleckstrin homology (PH) domain at it's very N-terminus. We wanted to attain a crystal structure of a p62 construct encompassing this domain.

Methods

We used X-ray crystallography in a single wavelength anomalous dispersion (SAD) approach.

Results

We obtained crystals diffracting to 1.8 Å in space group $C222_1$. The crystallization conditions contained among other ions also potassium. To gain a source of anomalous diffraction we substituted potassium with caesium as it is a heavier element from the same group of the periodic table. Crystals of the same space group and similar cell dimensions grew in the presence of caesium. These crystals yielded a substantial anomalous signal when collecting data at low energy (7 keV). The structure was solved using the PHENIX software package. The caesium substructure was determined via phenix.HySS. 13 low occupancy caesium sites were identified, which were used to phase the data through Phaser EP. After density modification (RESOLVE) an interpretable electron density map could be obtained. Using this map, we were able to build a complete model of the PH domain.

Conclusions

We succeeded in our primary goal to obtain a crystal structure of the PH domain of p62. Our approach to substitute an ion in the mother liquor with a heavier one from the same group of the periodic table was effective. The substitution of potassium with caesium did not hinder crystallization. Rather it provided us with sufficient anomalous signal to overcome the phase problem.

References

Compe E. & Egly J. M. TFIIH: when transcription met DNA repair. *Nature Rev. Mol. Cell Biol.* **13**, 343-354 (2012).
Gervais V. et al. TFIIH contains a PH domain involved in DNA nucleotide excision repair. *Nat Struct Mol Biol.* **7**, 616-622 (2004).
Adams P. D. et al. PHENIX: a comprehensive Python-based system for macromolecular structure solution. *Acta Cryst.* **D66**, 213-221 (2010).

MS03-T05

Overcoming challenges in macromolecular crystallography data acquisition with HPC detectors

M. Müller[1]

[1]*DECTRIS Ltd., Baden, Switzerland*

Acquisition of diffraction data is the last experimental step in the frequently challenging process of crystallographic structure determination of macromolecules. This last experimental step can pose numerous challenges depending on the properties of the macromolecular crystals. The unique advantages of Hybrid Photon Counting (HPC) detectors help to overcome many of these challenges: high frame rates and virtually dead-time-free readout allow for outrunning of radiation damage at room temperature when crystals cannot be cryo-protected; a sharp point spread function and small pixel size help to resolve reflections of long unit cell axes; the absence of readout noise and high frame rates enable fine-sliced data collection that substantially improves quality of weak high resolution data and allow for low-dose high-redundancy protocols in advanced phasing methods; curved detectors operated in vacuum open new possibilities for long-wavelength anomalous data collection from crystals that cannot be derivatized with strong anomalous scatterers.

This talk will highlight the advantageous features of HPC detectors and give examples how they are used to overcome challenges in macromolecular crystallography data acquisition.

MS03-T06

Automated high-throughput diffraction data collection at the European Synchrotron Facility.

D. von Stetten[1]

[1]*ESRF, Structural Biology, Grenoble, France*

The current Upgrade Programme for Structural Biology at the European Synchrotron Radiation Facility addresses the need to tackle the more and more challenging projects emerging in Macromolecular Crystallography (MX), for example large macromolecular assemblies, complex membrane proteins and rapid diffraction data collection on tiny crystals. Crystals from complex systems often show considerable inter- and intra- crystal variability with respect to diffraction quality. Sample evaluation of dozens to thousands of crystals to result in complete and successful data collections is thus becoming even more crucial, as will data collection facilities optimised for the collection of diffraction data from crystals that are very small and/or diffract to low resolution (i.e. $d_{min} > 5$ Å). The presentation will showcase current developments of a fully integrated and automated "hands-off" end station (MASSIF) for sample evaluation and diffraction data collection. Furthermore, an automated micro beam facility optimised for multi-crystal data collections will be presented.

MS04 - Young Crystallograhers Lightning Session

MS04-T01
Epitaxial growth of pulsed laser deposited Ge-Sb-Te thin films on (111) oriented substrates

E. Thelander[1], U. Ross[1], J. Gerlach[1], A. Lotnyk[1], B. Rauschenbach[1]
[1]*Institut für Oberflächenmodifzierung e.V., Leipzig, Germany*

Phase change materials based on the material system Ge-Sb-Te (GST) have been widely used for optical storage for decades and are gaining more interest as candidates for next generation non-volatile electronic memories. Recently, it was shown that memory devices with an ordered fiber textured structure possesses improved switching characteristics in terms of speed and stability [1]. Even more highly ordered Ge-Sb-Te-films in form of epitaxial layers have been achieved with MBE [2], however, with severe limitations regarding deposition rate. Pulsed laser deposition (PLD) employs a compound target and offers high deposition rate rendering it more interesting from an industrial point of view. Furthermore, it has been successfully used for the deposition of epitaxial GST films on cubic (100) oriented substrates[3].

In this contribution, PLD was employed to deposit epitaxial GST-layers on the (111) plane of BaF_2 and Si single crystal substrates with deposition rates between 2-200 nm/min[4]. X-ray diffraction measurements show a process temperature window for epitaxial growth between roughly 100°C and 295°C. No polycrystalline growth is observed for lower temperatures, i.e. a direct epitaxial growth is obtained, whereas higher temperatures lead to strong desorption of the film constituents. The films are of hexagonal structure with lattice parameters consistent with existing structure models. X-ray pole figure measurements reveal that the films grow with one single out-of-plane crystal orientation, but rotational twin domains are present. The out-of-plane epitaxial relationship is determined to be $Ge_2Sb_2Te_5$(0001) || BaF_2(111), whereas the in-plane relationship is characterized by two directions i.e. $Ge_2Sb_2Te_5$[-12-10] || BaF_2[1-10] and $Ge_2Sb_2Te_5$[1-210] || BaF_2[1-10] as revealed by pole figure measurements. For films deposited on Si, the epitaxial relationships are completely analogue. Aberration-corrected high-resolution scanning transmission electron microscopy was used to resolve the local atomic structure, including vacancy layer distribution, and confirm the hexagonal structure of the films.

References:

1. R. E. Simpson et al., *Interfacial phase-change memory,* Nat. Nano **6,** 501 (2011).
2. F. Katmis et al., *Insight into the Growth and Control of Single-Crystal Layers of Ge-Sb-Te Phase-Change Material,* Cryst. Growth Des. **11,** 4606 (2011).
3. E. Thelander et al. , *Epitaxial growth of Ge-Sb-Te films on KCl by high deposition rate pulsed laser deposition,* J. Appl. Phys. **115,** 213504 (2014).
4. E. Thelander et al., *Low temperature epitaxy of Ge-Sb-Te films on BaF_2(111) by pulsed laser deposition,* In Press at Appl. Phys. Lett. (2014).

Figure caption

Pole figure measurement of GST films on BaF_2(111) deposited at different substrate temperature (a). Intensity (cps) is shown on a logarithmic scale and applies for all pole figures. In (b), (c) and (d), STEM images of epitaxial GST films deposited at 180°C (b and c) and 295°C (d) are shown.

Figure 1

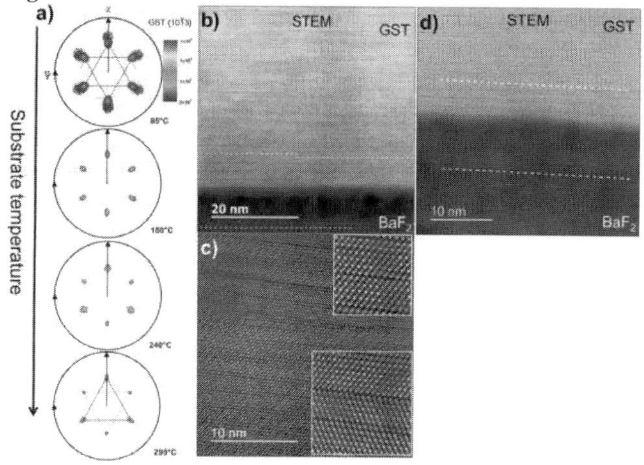

MS04-T02
Real time structural dynamics and entropy insights into the early unfolding kinetics of ubiquitin - rapid mixing micro-channel based time-resolved small angle X-ray scattering at synchrotrons

R. Jain[1], M. Petri[1], S. Sonnenkalb[1], S. Becker[1], C. Griesinger[1], A. Menzel[2], T. P. Burg[1], S. Techert[1,3,4]
[1]*Max Planck Institute of Biophysical Chemistry, Structural Dynamics of Biochemical Systems, Goettingen, Germany*
[2]*Paul Scherrer Institut, Villigen, Switzerland*
[3]*Georg-August-University, Goettingen, Germany*
[4]*Deutsches Elektronen-Synchrotron (DESY), Hamburg, Germany*

The overall size and shape of a macromolecule as a function of time, salt, or denaturant concentration can be determined with small-angle X-ray scattering (SAXS). Time-resolved SAXS (TR-SAXS) is a state of the art technique for studying protein folding-unfolding *in vitro* at all time-scales - from slow to very fast. It requires the setup of novel experimental schemes at pulsed, high flux X-ray synchrotrons. Doing so, TR-SAXS provides unique information about global structure of transient intermediates.

An experimental set-up with low protein consumption is essential for the extensive application of TR-SAXS. Together with the workgroup of T. Burg, we set up a newly developed 20-microchannel microfluidic continuous-flow mixer and combined that with SAXS at high-flux X-ray synchrotrons. With this experimental setup, I will demonstrate the kinetic ubiquitin unfolding dynamics after the rapid mixing of folded ubiquitin with unfolding agent Guanidinium-HCl. It required only ~ 40 nanoliters of protein sample per scattering image. This new platform might increase the use of TR-SAXS, TR-WAXS and neutron scattering experiments for studying milliseconds dynamics. The potential research field includes protein folding, protein misfolding, aggregation in amyloidogenic diseases, function of intrinsically disordered proteins and various protein-ligand interactions.

With the above mentioned experimental set-up, the structural dynamics investigation of ubiquitin unfolding will be presented on the diffusion limited time scales. Time-dependent variation in the X-ray scattering intensity, size and shape of ubiquitin after the complete mixing of ubiquitin and Gdn-HCl were monitored with the Fourier transformation of collected X-ray scattering signals during unfolding. The summarized results yielded a "real-time millisecond ubiquitin unfolding movie", and I will finally present that. These structural intermediate changes will be discussed as a sequential multi-state ubiquitin unfolding model along with detailed kinetic analysis with Boltzmann-fitting, time-dependent dispersive kinetics and JMAK model. The entropy and enthalpy contributions derived with the dispersive kinetics were used to describe the driving forces contributing to the ubiquitin unfolding funnel

References:

Jain et al., X-ray scattering experiments with high flux X-ray source coupled rapid mixing microchannel device...... European Physical Journal E, Sep; 36(9):109.

Real-time small-angle X-ray scattering reveals entropy-driven multistate ubiquitin unfolding reaction. (Jain et al., Submission in Nature Communications)

Visualization of millisecond ubiquitin unfolding kinetics in a microchannel device by TR-SAXS. (Jain et al., under preparation).

A glimpse of substrate induced conformational changes in human Guanylate kinase, (Jain et al., under preparation).

Fig. 1: Scheme of the TR-SAXS set-up.

Figure 1

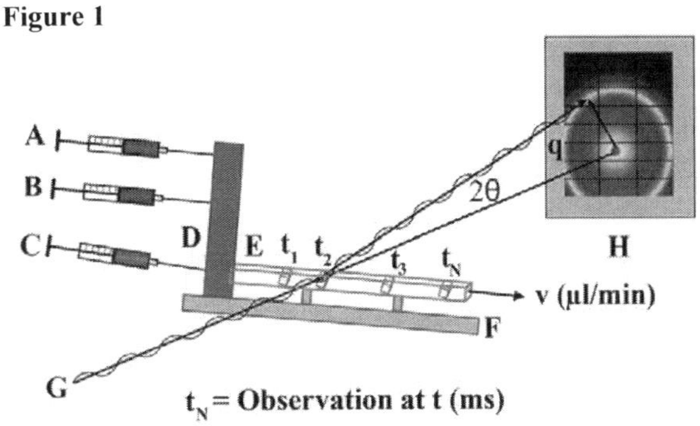

MS04-T03

Hexanuclear Metallacryptates: Cu(α-amino-carboxylate)$_2$ as Building Units for Self Assembly.

K. Lamberts[1], U. Englert[1]
[1] *RWTH Aachen, Inorganic Chemistry, Aachen, Germany*

Crown ethers and cryptands are organic molecules, introduced in the late 1960s [1], with the ability to selectively coordinate specific cations. Metalla-crowns [2] and metallacryptands [3] are their analogues in coordination-chemistry. However, their synthesis is not straightforward, as coordination chemistry is strongly dependent on weak interactions and self assembly. Cu(α-amino-carboxylate)$_2$ has proven a suitable building unit for the formation of hexanuclear metallacryptates: The square planar units occupy the equatorial positions of a distorted octahedron and bridge the axial Cu(II) cations. Perchlorate anions stabilize the four Cu(α-amino-carboxylate)$_2$-lamellas by connecting them over Cu---O Jahn-Teller contacts.

In the center of the octahedral unit the carboxylate-O form a coordination pocket in almost square prismatic geometry. Crystals with incorporated Na(I) cations were reported for glycine, L-proline, L-hydroxyproline, L-alanine, L-threonine and 1-amino-1-cyclopropane [4].

We recently discovered that such Metallacryptates are also capable of coordinating the transition metal cation Ag(I) (Fig 1). Structural and chemical properties, trends and chemical variations will be presented.

Fig. 1: Hexanuclear metallacryptate cation from Cu(II), perchlorate and L-proline with incorporated Ag(I).

[1] C.J. Pedersen, *Journal of the American Chemical Society*, 1967, 89, 7017-7036.
[2] *e.g.*: a) M.S. Lah and V.L. Pecoraro, *Comments on Inorganic Chemistry*, 1990, 11, 59-84. b) F.C.J.M. van Veggel, W. Verboom and D.N. Reinhoudt, *Chemical Reviews*, 1994, 94, 279-299.
[3] *e.g.*: a) M.S. Lah, B.R. Gibney, D.L. Tierney, J.E. Penner-Hahn and V.L. Pecoraro, *Journal of the American Chemical Society*, 1993, 115, 5857-5858. b) B.R. Gibney, A.J. Stemmler, S. Pilotek, J.W. Kampf and V.L. Pecoraro, *Inorganic Chemistry*, 1993, 32, 6008-6015.
[4] a) S. Hu, W. Du, J. Dal, L. Wu, C. Cui, Z. Fu and X. Wu, *Dalton Transactions*, 2001, 2963-2964. b) L.-Y. Wang, S. Igarashi, Y. Yukawa, Y. Hoshino, O. Roubeau, G. Aromí and R.E.P. Winpenny, *Dalton Transactions*, 2003, 2318-2324. c) S.-M. Hu, S.-C. Xiang, J.-J. Zhang, T.-L. Sheng, R.-B. Fu and X.-T. Wu, *European Journal of Inorganic Chemistry*, 2008, 1141-1146. d) S.-C. Xiang, S.-M. Hu, J.-J. Zhang, X.-T. Wu and J.-Q. Li, *European Journal of Inorganic Chemistry*, 2005, 2706-2713. e) W. Ghattas, R. Ricoux, H. Korri-Youssoufi, R. Guillot, E. Rivière and J.-P. Mahy, *Dalton Transactions*, 2014, 7708-7711.

Figure 1

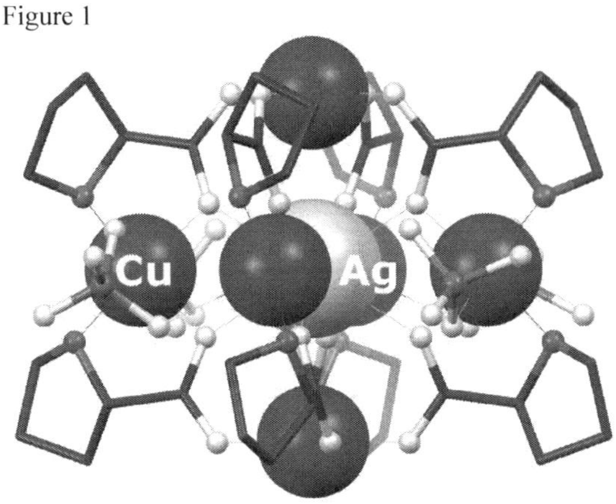

MS04-T04

Biochemical and structural studies on reaction mechanism of the acetyl-CoA synthesis

Y. Dendra[1], S. Götzl[1], J.- H. Jeoung[1], H. Dobbek[1]
[1] *Humboldt Universität zu Berlin, Institut für Biologie, Berlin, Germany*

Acetyl-CoA synthase (ACS) is one of the key enzymes of the Wood-Ljundahl pathway of anaerobic CO_2 fixation and is abundant in a number of anaerobic bacteria (sulfate reducers, methanogens, acetogens, etc) (1). ACS catalyzes the formation of acetyl-CoA by condensing the methyl group ($-CH_3$) of a methylated corrinoid iron-sulfur protein (CoFeSP) with CO and coenzyme A (1, 2). The active site of the ACS, denoted as A-cluster, contains a Fe_4S_4 cubane bridged to a nickel-nickel binuclear site. In acetogenic bacteria, ACS forms a tight tetrameric complex together with CO Dehydrogenase (CODH). The first crystal structure of ACS/CODH complex from *Moorella thermoacetica* was solved at 2.5′ Å resolution by Doukov et. al (1). The 2.2 Å resolution structure of the monomeric ACS from hydrogenogenic bacterium *Carboxydothermus hydrogenoformans* was obtained by Svetlitchnyi in 2004 (3). In methanogenic archaeae, ACS together with CODH, CoFeSP and methyltransferase are a part of a 2.4 MDa acetyl-CoA decarbonylase/synthase (ACDS) complex (4). No crystal structures of the methanogenic ACS alone or as part of the ACDS complex has been reported so far.

Despite nearly three decades of extensive spectroscopic and biochemical studies and several high-resolution structures of the ACS, many questions concerning the mechanism of acetyl-CoA formation remain open. Here we present the current state of knowledge on reaction mechanism of acetyl-CoA formation.

(1) Doukov et al., Science 298, 567 (2002);
(2) Ragsdale et al., Chem. Rev. 96, 2515 (1996);
(3) Svetlitchnyi et al., PNAS 101, 2 (2004);
(4) Terlesky et al., J. Bacteriol 168, 3 (1986).

MS04-T05

Crystallization kinetics of zeolite Li-A(BW) in dependence of the aluminosilicate educt and the reaction temperature and time

D. Zeibig[1], J.- C. Buhl[1]
[1]Leibniz Universität Hannover, Institut für Mineralogie, Hannover, Germany

Zeolites of ABW structure type like Li-A(BW) (composition: $Li_4(H_2O)_4[Al_4Si_4O_{16}]$ are microporous materials with a channel system of dimensions of 3.8 x 3.4 Å, formed of 8-rings [1]. According to the pore wide and the manyfold substitution possibilities of T-atoms and cations in this structure the A(BW) zeolites are interesting for future applications. Energy efficient crystallization procedures would thus become important. The aim of the present work was to optimize the synthesis, first performed by Barrer and White [2] and modified by Norby et al. [3] to test a complete conversion at economical conditions using different aluminosilicate educts, low temperatures and short reaction times.

Therefore the crystallization of Li-A(BW) has been examined by testing three different alumosilicate educts (Na-LTA as in [3], Na-13X and Na-LSX) in dependence of the synthesis temperature and synthesis time. The educts have been dissolved in LiCl-solution and heated under hydrothermal conditions at 200 °C, 150 °C and 100 °C for 72 h, 36 h and 18 h. All products have been analyzed using XRD. Furthermore, for some selected samples FTIR, SEM/EDX and quantitative Rietveld-refinements were carried out.

The results show that 200 °C and 72 h are suitable conditions for all starting materials to obtain pure phase zeolite Li-A(BW). 200 °C and 36 h are still adequate for Na-LTA and Na-LSX, whereas Na-13X does not crystallize completely to Li-A(BW). A further decrease of synthesis time down to 18 h leads to an incomplete conversion of Na-LSX. Na-LTA is the only reactant which produces Li-A(BW) at 200 °C and 18 h.

Even at lower temperature of 150°C the starting material Na-LTA reveals as suitable source for the crystallization of the desired product for both times (72 h and 36 h). In contrast Na-13X and Na-LSX does not show any conversion, even at 72 h. The synthesis temperature of 100 °C is generally inappropriate, because the conversion to the zeolite Li-A(BW) could not be observed for any of the three educt compounds.

Summarizing the results it could be shown that using Na-LTA as starting material is the best method for synthesizing the desired Li-A(BW) even for lower temperatures and times.

Figure 1: Conversion of the different solid educts into zeolite Li-A(BW) in dependence of the reaction time at 150°C and 200°C (results from quantitative Rietveld refinements).

Figure 2: SEM image of zeolite Li-A(BW) crystals obtained from zeolite LTA at 150°C and 72 h reaction time in 1,6 M-LiCl solution.

References:
[1] Baerlocher, Ch, Meier, W. M., Olson D. H.: "Atlas of Zeolite Framework Types". 5th ed., Elsevier, Amsterdam, New York, 2001.
[2] Barrer, R. M., White, E. A. D.:J.: "The Hydrothermal Chemistry of Silicates". Part 1. Synthetic Lithium Aluminosilicates. Chem. Soc. (1951) 1267 - 1278.
[3] Norby, P., Norlund Christensen, A., Krogh Andersen, I. G.: "Hydrothermal preparation of zeolite Li-A(BW), $LiAlSiO_4 \cdot H_2O$, and structure determination from powder diffraction data by direct methods". Acta Chem. Scand. A40 (1986) 500-506.

Figure 1

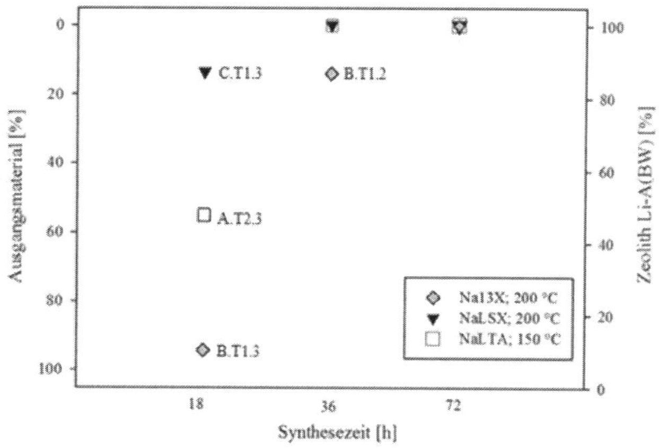

Figure 2

MS04-T06

Anisotropic Displacement Parameters from Dispersion-Corrected DFT and their Experimental Validation by X-ray Diffraction

A. Wang[1], U. Englert[1]
[1]RWTH Aachen University, Institute of Organic Chemistry, Aachen, Germany

Single crystal X-ray diffraction experiments have been conducted for the monoclinic form of pentachloropyridine[1] as a function of temperature. For the anisotropic displacement parameters (ADPs) of the five chloride atoms in the molecule experimental values have been refined based on these intensity data, and theoretical values based on dispersion-corrected density-functional theory (DFT) have been obtained by phonon computations using the PHONOPY[2] software. Dispersion correction represents an important aspect in these calculations: With the PBE+TS method[3], excellent agreement for the calculated and experimentally observed thermal motion in the crystal is found at 100, 150 and 200 K. The agreement can be conveniently visualised by thermal ellipsoids as shown in the figure. Alternative validation approaches are discussed on the poster. Our method promises reliable predictions for thermal motion in molecular crystals.

Figure: Comparion betweeen ADPs from experiment (Exp) and the most suitable theoretical method among those tested (PBE+TS).

[1] J. George, A. Wang, V. L. Deringer, R. M. Wang, R. Dronskowski and U. Englert, *CrystEngComm*, preparation.

[2]A. Togo, F. Oba and I. Tanaka, *Phys. Rev. B: Condens. Matter. Mater. Phys.*, 2008, 78, 134106.

[3]J. P. Perdew, K. Burker, M. Ernzerhof. *Phys, Rev. Lett*, 1996, **77**, 3865-3868; A. Tkatchenko and M. Scheffler, *Phys. Rev. Lett.*, 2009, 102, 073005.

Figure 1

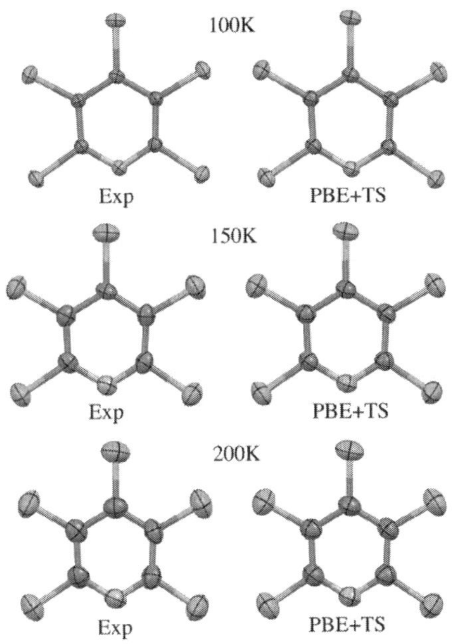

MS04-T07

Isoreticular Solvo- and Polymorphs: Cu(I) in a Stable Bimetallic Coordination Network

F. Steuber[1], U. Englert[1]
[1]RWTH Aachen, Institute of Inorganic Chemistry, Aachen, Germany

Organic-inorganic hybrid materials such as coordination polymers have gained tremendous interest in basic research as well as applied materials science. The ditopic ligand 3-cyano-2,4-pentanedione (HacacCN) has proven to be a suitable candidate for a stepwise bimetallic network construction [1, 2]. Isoreticular interpenetrated diamond networks have been synthesised from Sc(III) and Y(III) in combination with Ag(I). A similar structure from La(III) has been reported previously [3]. The preferred self-assembly of the motif allowed for the synthesis of an isomorphous Cu(I)-Y(III) coordination network. Structural relationships between five novel bimetallic networks will be discussed.

Figure 1: Structural motif found in presented compounds.

[1] A. D. Burrows, K. Cassar, M. F. Mahon and J. E. Warren, *Dalton Trans.*, 2007, 2499-2509.
[2] C. Merkens, N. Becker, K. Lamberts, U. Englert, *Dalton Trans.*, 2012, **41**, 8594-8599.
[3] C. Merkens, U. Englert, *Dalton Trans.*, 2012, **41**, 4664-4673.

Figure 1

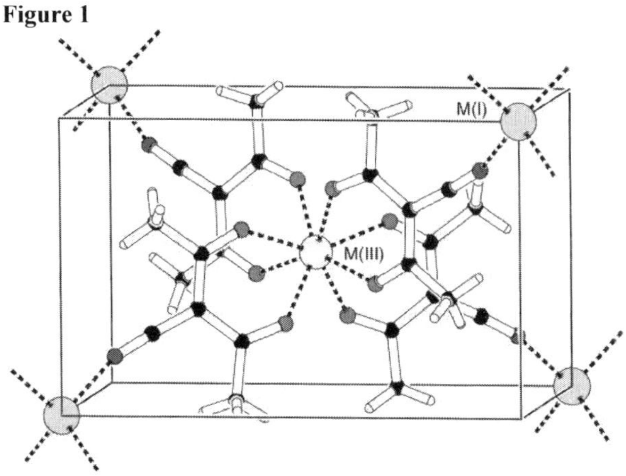

MS04-T08

Do we know the Pyroelectric Coefficient of Barium Titanate?

H. Stöcker[1], B. Störr[1], S. Jachalke[1], E. Mehner[1], J. Hanzig[1], M. Sonntag[1], R. Schmid[1], T. Leisegang[2], D. C. Meyer[1]
[1]TU Bergakademie Freiberg, Institut für Experimentelle Physik, Freiberg, Germany
[2]Fraunhofer-Technologiezentrum Halbleitermaterialien, Freiberg, Germany

Barium titanate is a ferroelectric material widely used in capacitors because of its high permittivity and as a chemically and mechanically stable pyro- and piezoelectric. For utilization in pyroelectric devices, e. g. for infrared detection or waste heat harvesting, a high pyroelectric coefficient is often desired. Literature data for barium titanate compounds vary between 4.8 $\mu C/m^2K$ and 13 C/m^2K, depending on doping, synthesis procedure, and measurement method. Hence, the pyroelectric coefficient of barium titanate seems to span more than six orders of magnitude, which renders predictions on the efficiency of devices nearly impossible.

Here, we try to give an overview on the factors influencing the pyroelectric coefficient of barium titanate to finally extract meaningful values of this fundamental material property. First, a short overview on pyroelectric measurement methods and their application criteria is given. This allows identifying reliable literature values measured with an appropriate method for further discussion. Since barium titanate can be produced in many different forms, e. g. thin films, ceramics or single crystals, polarization and measurement of the samples have to be conducted accordingly. Depending on crystallinity and morphology, different pyroelectric coefficients and Curie temperatures are reported in literature. Finally, doping of barium titanate with calcium, cobalt, iron, nickel, strontium, tin or zirconium is found to decrease the Curie temperature. Shifting the Curie temperature in turn changes the pyroelectric coefficient at a fixed temperature. Doping with calcium or nickel leads to an increased pyroelectric coefficient, whereas the substitution with strontium or tin does not affect the pyroelectric coefficient at the Curie temperature.

MS04-T09

CO Dehydrogenase II - Digging channels for small molecules

L. Domnik[1], J.- H. Jeoung[1], J. Fesseler[1], H. Dobbek[1]
[1]*Humboldt Universität zu Berlin, Berlin, Germany*

Rising of CO_2 levels in the earth's atmosphere demands the development of energy sources that are CO_2 neutral. Therefore understanding how microorganisms utilize C1-compounds will be of crucial importance in the future. The carbon monoxide dehydrogenase II (CODHII) of the thermophilic bacterium *Carboxydothermus hydrogenoformans* catalyses the reduction of CO_2 to CO and water using two protons and two electrons with a turnover rate of 15 s^{-1}.

In order to guarantee an efficient supply of substrates and release of products controlled routes are needed within the enzyme [1, 2, 3]. Biochemical and structural studies on mutants have been applied to illuminate the putative proton and water transfer pathways. Exploiting the ability of xenon to bind hydrophobic protein cavities allowed determination of gas channels in CODHII.

References:

[1] Jeoung JH and Dobbek H (2007) Science 318(5855):1461-4

[2] Kim EJ, Feng J, Bramlett MR, Lindahl PA (2004) Biochemistry 43(19):5728-34

[3] Doukov TI, Blasiak LC, Seravalli J, Ragsdale SW, Drennan CL (2008) Biochemistry 47(11): 3474-3483.

MS04-T10

Identifying new inorganic Na-ion conductors using data-mining and geometric crystallography - the Voronoi-Dirichlet approach

F. Meutzner[1], T. Nestler[1], M. Wolfram[1], N. A. Kabanova[2], T. Leisegang[3], V. A. Blatov[2], D. C. Meyer[1]
[1]*TU Bergakademie Freiberg, Institut für Experimentelle Physik, Freiberg, Germany*
[2]*Samara State University, Samara, Russian Federation*
[3]*Fraunhofer Technologiezentrum Halbleitermaterialien, Freiberg, Germany*

Due to its high achievable energy densities, stationary electrochemical energy storage devices are most important for the stabilisation of the electric grid, compensating for the capricious nature of renewable energies. Sodium is one of the most used candidates in this field and applied in stationary Na-S and Zebra cells for example in Japan. These are high-temperature cells in which Na and S react with each other, mediated by a solid Na-ion conductor such as ß-aluminate. In order to achieve reasonable conductivities, high temperatures have to be applied leaving Na and S in their liquid state. For energetic and environmental reasons, however, it would be best to lower temperatures. Therefore, new solid electrolytes have to be identified.

Crystallographic Databases, such as the inorganic crystal structure database (ICSD), store large amounts of measured structures and thus offer knowledge which is partially not systematised. Parameters of related structures can be "*data-mined*" to find their most common values or familiar structure motives. On the other hand, these databases can be used in an inductive way, using this newly created information to find compounds which show the same behaviour but have not been considered for their potential function, yet.

The Voronoi-Dirichlet tessellation offers a possibility to partition space - in this case a crystal structure. The polyhedra created by this routine depend solely on the crystal structure's geometry and metrics offering readily available parameters, such as volumes, vertex-coordinates, *etc*. These can be used to describe structures very easily and fast. These parameters also have crystallographic meanings that can be translated into functional properties as well. Through this, structure-immanent voids and channels between these voids can be identified. They are important factors in compounds showing ionic conductivity. Because of its simplicity, the partition can be easily applied on huge databases. This combination of data-mining and Voronoi-Dirichlet tessellation is used in this work to identify compounds which are new potential ionic conductors or intercalation hosts for Na-ions.

This work is funded by the German BMBF (CryPhysConcept: 03EK3029A) and the German BMU (BaSta: 0325563B).

MS04-T11

Synthesis and local structure of $Cu_xZn_{2-x}TiO_4$ spinel unveiled by Raman spectroscopy

M. He[1], J. Ruiz-Fuertes[1], T. Bernert[1], B. Winkler[1], V. L. Vinograd[2]
[1]*Goethe-Universitaet Frankfurt, Frankfurt am Main, Germany*
[2]*Forschungszentrum Juelich GmbH, Juelich, Germany*

The local structure of solid solutions is essential for understanding their physical properties. Since their structural investigations are challenging, up to now there are just a few studies addressing the local symmetry of inverse spinel-solid solutions, especially when they are affected by a Jahn-Teller distortion [1]. Here we present the solid-state synthesis of the $Cu_xZn_{2-x}TiO_4$ solid solution and the investigation of the influence of the incorporation of Cu^{2+} on the local structure of technologically important Zn_2TiO_4 [2]. X-ray diffraction shows no phase change accross the solid solution. The analysis of the lineshape and frequency of the Raman active modes reveal the incorporation of some Cu^{2+} in tetrahedral sites up to x = 0.6. The observed ordering scheme is in agreement with our *ab initio* calculations with pseudo-random generated structures, where an octahedral over tetrahedral site preference energy for Cu^{2+} of only ~-6 kJ/mol has been computed [3]. With increasing Cu^{2+} content, a non-linear change of the unit-cell volume is observed (Figure 1). The excess volume is associated to an increase of the distortion of the coordination polyhedra which is maximum for x = 0.6.

Fig. 1: (a) Lattice parameter a of $Cu_xZn_{2-x}TiO_4$ as a function of Cu concentration and (b) the oxygen thermal displacement parameter U_{iso}.

M.H. acknowledges the China Scholarship Council for a PhD

grant and J.R.-F. thanks the Alexander von Humboldt Foundation for a postdoctoral fellowship.

[1] A. Le Nestour, M. Goudon, G. Villeneuve, R. Andrissen, and A. Demourges, Inorg. Chem. **46**, 2645 (2007).
[2] T. Esaka, T. Ikebe, and M. Kamata, Solid State Ionics **76**, 237 (1995).
[3] J. Ruiz-Fuertes, T. Bernert, M. He, B. Winkler, V. L. Vinograd, and V. Milman Appl. Phys. Lett. **105**, 071911 (2014).

Figure 1

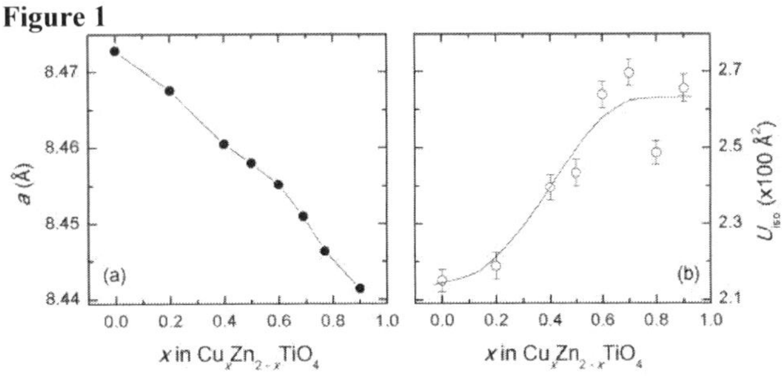

MS04-T12

High resolution polarized Raman scattering study on multiferroic $MnWO_4$

F. Ziegler[1], H. Gibhardt[1], J. Leist[1], P. Becker[2], L. Bohatý[2], G. Eckold[1]
[1]*Georg-August-Universität Göttingen, Institut für Physikalische Chemie, Göttingen, Germany*
[2]*Universität zu Köln, Institut für Kristallographie, Köln, Germany*

Multiferroic materials with magnetic and electric ordering within a single phase gained attention during the last years due to their potential for industrial applications. Materials with a cycloidal spin-arrangement are of particular interest because the electrical polarisation is directly caused by the complex magnetic order. Hence, both ordering phenomena are tightly coupled in this category of compounds. One example is $MnWO_4$, which passes successively three antiferromagnetic phase transitions on cooling. The phase AF2 between 12.5 K and 7 K exhibits a cycloidal spin arrangement with an incommensurate magnetic propagation vector. This ordering induces a spontaneous polarisation by the inverse Dzyaloshinski-Moriya-interaction. As a consequence, the interaction between spin and polarisation could possibly lead to a change in the spectrum of lattice vibrations.

Due to the tiny polarization, it was believed up to now that this effect was too small to be observed experimentally. Using high resolution polarized Raman scattering, however, we were able to detect clear signatures of this magnetoelectric interaction. The Raman spectra were taken from a $MnWO_4$ single crystal in the temperature range from 100 K to 6 K covering the three magnetic phase transitions. A careful analysis of the experimental data allowed the precise determination of both, the wave numbers and intrinsic linewidths.

Furthermore, the effect of an electric field applied along the polar axis was investigated for each of the 18 Raman active modes. Most of the phonons show a significant shift in wave number below the magnetic ordering temperature that cannot be explained merely on the basis of lattice dynamics, but should be attributed to spin-phonon-interaction. The most striking effect is observed for the A_g mode with the highest frequency (about 884 cm^{-1}), which corresponds to an oxygen stretching vibration and exhibits a pronounced softening on cooling.

MS04-T13

New $Mn_{23}Pt_{12}Ga_{11}$ phase, synthesis, crystal structure and its complex phase reaction scheme

B. Jamiyansuren[1], W. Carrillo-Cabrera[1], G. Kreiner[1], C. Felser[1]
[1]*Max-Planck-Institut für Chemische Physik fester Stoffe, Anorganische Chemie, Dresden, Germany*

Mn_2-based Heusler compounds have attracted much interest as magnetic materials because of high spin polarization at the Fermi level. In addition, tetragonal compounds can exhibit large magnetocrystalline anisotropy [1,2]. Recently the tetragonal Heusler compound Mn_2PtGa has been reported to show a large exchange-bias effect and a first order ferrimagnetic to ferromagnetic transition [3,4]. Here, we report on the crystal structure and the heterogeneous equilibria of the new intermetallic compound $Mn_{23}Pt_{12}Ga_{11}$, which forms close to the composition Mn_2PtGa. Bulk samples were prepared by inductively melting the elements in Al_2O_3 crucibles in Ar atmosphere. The resulting polycrystalline ingots were placed in Al_2O_3 crucibles. These crucibles were then encapsulated in cleaned and weld-sealed Ta containers which in turn were jacketed in fused silica ampoules. The samples were then annealed for several days at elevated temperatures and subsequently quenched in water. The samples were characterized by chemical bulk analysis (ICP-OES), metallographic examination, EDXS, WDXS, thermal analysis and by powder X-ray diffraction. The crystal structure of $Mn_{23}Pt_{12}Ga_{11}$ was solved using electron diffraction data (ADT3D, [5]) obtained by manual electron diffraction tomography as described in [6]. The structural model was validated by powder diffraction data (Rietveld technique, Jana2006, [7]). $Mn_{23}Pt_{12}Ga_{11}$ crystallizes in a new crystal structure type with 46 atoms per unit cell ($P3m1$, a = 8.649(1) Å, c = 10.848(1) Å) and shows a complex phase reaction scheme, which includes a four phase reaction of class II. The title compound exhibits heterogeneous equilibria to MnPtGa (BeZrSi type, [8]), to a structurally closely related monoclinic phase with slightly more Ga content, to a phase of $TiAl_3$ structure type close to Mn_3Pt and to a Mn-rich Heusler compound. The crystal structure of $Mn_{23}Pt_{12}Ga_{11}$ contains structural entities as found in NiAs type based MnPtGa as well as in Heusler structures.

[1] T. Graf, C. Felser, S.S.P. Parkin, Prog. Solid State Chem. 39 (2011) 1.
[2] G. Kreiner, A. Kalache, S. Hausdorf, V. Alijani, J.-F. Qian, G. Shan, U. Burkhardt, S. Ouardi, C. Felser, Z. Anorg. Allg. Chem. 640 (2014) 738.
[3] A.K. Nayak, M. Nicklas, S. Chadov, C. Shekar, Y. Skourski, J. Winterlik, C. Felser, Phys. Rev. Lett. 110 (2013) 127204.
[4] A.K. Nayak, M. Nicklas, C. Shekar, C. Felser, J. Appl. Phys. 110 (2013) 17E308.
[5] U. Kolb, T. Gorelik, C. Kübel, M.T. Otten, D. Hubert, Ultramicroscopy, 107 (2007) 507.
[6] J. Fan, W. Carrillo-Cabrera, L. Akselrud, I. Antonyshyn, L. Chen, Y. Grin, Inorg. Chem. 52 (2013) 11067.
[7] V. Petricek, M. Dusek, L. Palatinus, Jana2006. Structure Determination Software Programs. Institute of Physics, Praha, Czech Republic, 2006.
[8] P. Villars, K. Cenzual, Pearsons's Crystal Data - Crystal Structure Data Base for Inorganic Compounds , Release 2012/2013, ASM International, Materials Park, Ohio, USA.

MS04-T14

Structural chemistry of the SrO-CuO-As_2O_5 ternary system: crystal structure of $Sr_2Cu_7(AsO_4)_6$

S. Gerger[1], T. Djordjevic[1]
[1]*University of Vienna, Wien, Austria*

A new arsenate, $Sr_2Cu_7(AsO_4)_6$ was synthesised under low-temperature hydrothermal conditions during detailed structural investigations of the compounds from the *M*O-CuO-As_2O_5-H_2O system ($M = Cd^{2+}$, Sr^{2+}). $Sr_2Cu_7(AsO_4)_6$ crystallised as transparent, light-blue crystals with lengths up to 0.11 mm from a mixture of $Sr(OH)_2 \cdot 8H_2O$, $Cu(OH)_2$, As_2O_5 and distilled water (Teflon-lined steel autoclaves, 493 K, 7 d). Besides $SrCuAs_2O_7$ [1], it represents the second structurally characterized compound in the ternary SrO-CuO-As_2O_5 system.

The crystal structure of $Sr_2Cu_7(AsO_4)_6$ was refined from the single-crystal X-ray diffraction data (293 K, Mo*K*a, $2q_{max} = 70°$) starting from the atomic coordinates of the isotypic $Pb_2Cu_7(AsO_4)_6$ [2] (space group *P*-1, $a = 5.1413(10)$, $b = 8.3536(17)$, $c = 11.173(2)$ Å, $a = 90.52(3)$, $\beta = 90.62(3)$, $g = 92.76(3)°$, $V = 479.26(17)$ Å^3, $Z = 1$). The refinement (182 free parameters) yielded $R_1(F) = 0.0472$ for 4481 observed reflections with $F_o^{2\ 3}\ 4\sigma(F_o^2)$ [3].

The crystal structure of $Sr_2Cu_7(AsO_4)_6$ is built up from three crystallographically unique CuO_4 squares and two unique CuO_5 trigonal pyramids, three distinct AsO_4 tetrahedra and one SrO_8 polyhedron. CuO_x polyhedra are linked in two different kinds of chains: Cu_3O_{10} chains running along *a*-axis are formed by corner connection of one Cu(3)O_4 square and four different CuO_5 polyhedra. Cu_4O_{12} chains are parallel to [110] and are formed by edge and corner sharing of Cu(1)O_4 and Cu(2)O_4 squares and the Cu(4)O_5 polyhera. These chains are interconnected by AsO_4 tetrahedra thus forming a 3D-network, where in channels parallel to the *a*-axis slightly distorted SrO_8 tetragonal antiprisms are situated. SrO_8 polyhedra are interconnected to dimers *via* one O-O edge by a centre of symmetry. Both Cu(4)O_5 and Cu(5)O_5 polyhedra are trigonal dipyramids, which is a rare coordination figure for divalent copper atoms. However, as already observed in $Pb_2Cu_7(AsO_4)_6$ [2], the trigonal dipyramids are notably regular concerning Cu-O bond lengths and O-Cu-O bond angles: Cu-O bond legths 1.91 to 2.10 Å and O_a-Cu-O_a angles (O_a is apical ligand) are just somewhat bent and amount 174.8(2) and 173.7(2)° in Cu(4)O_5 and Cu(5)O_5 polyhedra, respectively. The O_e-Cu-O_a in equatorial plane of the polyhedra deviate only 21° from ideal value; the largest is 131.7° and the smallest is 101.7°.

Financial support of the Austrian Science Foundation (FWF) (Grant V203-N19) is gratefully acknowledged.

[1] Chen, T.-C &Wang, S.-L. (1996) J. Solid State Chem., 121, 350-355.
[2] Effenberger, H. (1995) J. Solid State Chem., 114, 413-419.
[3] Sheldrick, G.M. (2008) Acta Crystallogr., A64, 112-122.

MS04-T15

Crystal branching phenomena in natural and gel grown calcites

F. Wiethoff[1], J. Schreuer[1]
[1]*Ruhr-University Bochum, Geology, Mineralogy & Geophysics, 44780, Germany*

The morphological features of calcite crystals building up marine cements and speleothems are not easy to address due to the space filling crystallization based on competitive growth. Especially the calcite morphology depends strongly on the physico-chemical conditions during crystallization with the molar Mg/Ca ratio as the main parameter controlling the morphology [1]. Several studies pointed out that Mg^{2+} bearing solutions cause a larger growth vector parallel the c-axis resulting in elongated calcite crystals [2]. Therefore, measuring the aspect ratio in thin sections is one of the first approaches of sediment-petrologists to assess the presence of Mg^{2+} in the parental solution on a qualitative basis. Other features of such fabrics, like systematic undulous extinction and curved cleavage planes and/or crystal faces (Fiure 1a) occur quite common but can't yet employed as a tool to reveal further information due to the poorly understood underlying processes [3].

For this reason we grew calcite crystals in gels with different Mg/Ca ratios of the parental solutions. As a result of the spatially separated sites of crystallization in the porous gel media we observed plenty of different crystal morphologies some of them exhibiting subparallel branched plates with macroscopically curved adjacent faces (Figure 1b). In comparison to natural samples we found clear similarities in thin sections regarding the curved intracrystalline marks (Figure 1a), the undulous (divergent-) extinction behaviour, and in the morphology and branching phenomena of some natural, freely grown calcites.

On an atomistic level branching in calcites can be attributed to the incorporation of Mg^{2+} into the calcite lattice resulting in accumulation of internal stress-fields which can be reduced by developing slightly tilted sub-domains. Our observations suggest that crystal branching is a common feature during calcite crystallization in climate-research relevant archives and in principle influences the distribution and partitioning of trace elements [4] and isotopes through face specific branching phenomena.

Figure 1: (a)Thin-section of a marine calcitic cement showing curved intracrystalline marks (linear polarizedlight).(b)In gel grown synthetic calcite (Mg/Ca [mol/mol] = 0.5) with strong crystal branching phenomena resulting in subparallel crystal units.

References:

[1] Given, R.K. and Wilkinson, B.H. Kinetic control of morphology, composition, and mineralogy of abiotic sedimentary carbonates. J. Sediment. Petrol. **55** (1985) 109-119
[2] Folk, R.L. The natural history of crystalline calcium carbonate: Effect of magnesium content and salinity. J. Sediment. Petrol. **44** (1974) 40-53
[3] Richter, D. K., Neuser, R. D., Schreuer, J., Gies, H. and Immenhauser, A.: Radiaxial-fibrous calcites: A new look at an old problem. Sediment. Geol. **239** (2011) 23-36
[4] Rimstidt, J.D., Balog, A. and Webb, J. Distribution of trace elements between carbonate minerals and aqueous solutions. Geochim. Cosmochim Ac. **62** (1998) 1851-1863

Figure 1

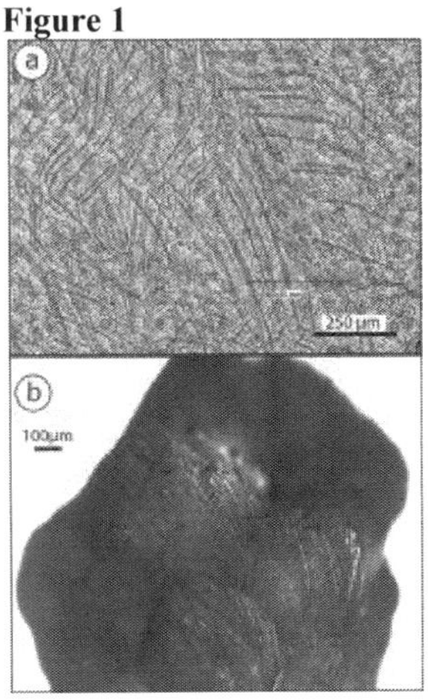

MS04-T16

Structure and Dynamics of Substituted η^3-Cycloheptatrienide-Pd Complexes

C. Jandl[1], A. Pöthig[1]
[1]*Technische Universität München, Catalysis Research Center, Garching b. München, Germany*

We report the structural characterisation of the first monomeric allyl complexes of Pd dihalides. Single crystal X-ray diffraction studies were essential not only in the identification of the monomeric structure, as both mass spectrometry and IR/Raman spectroscopy were ambiguous, but also in the further characterisation: Our approach combined crystallography and computational chemistry to gain insights into the rapid dynamic of the complexes and besides the global minimum isomer, suitable derivatisation also allowed us to characterise one structure as a snapshot of an intermediate in the fluctuation process (Figure 1). Additionally the compounds feature strong intramolecular H-bonds the exact conformation of which could only be determined *via* X-ray diffraction and in accord with DFT calculations was found to be highly solvent-dependent. This combination of molecular crystallography and computational chemistry proved to be a very valuable tool in investigations of fluxional systems that are traditionally seen as problematic for X-ray studies.[1]

[1] Christian Jandl, Karl Öfele, Fritz E. Kühn, Wolfgang A. Herrmann, and Alexander Pöthig*, *Organometallics*, 2014, *33 (22)*, 6398. DOI: 10.1021/om500738d .

Figure 1

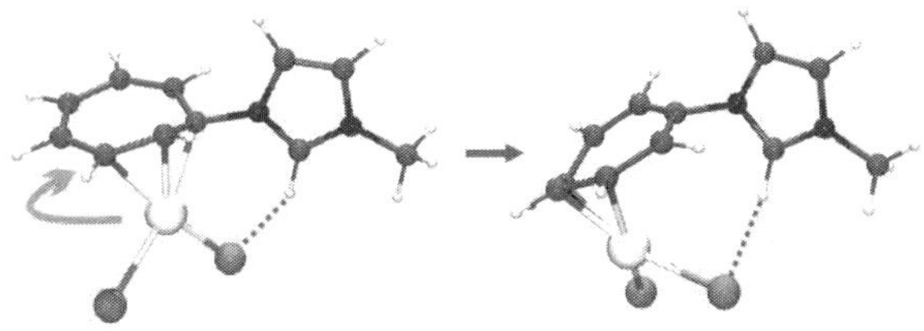

Figure 1. Fluxional coordination of a substituted cycloheptatrienide ligand towards a Pd-dihalide moiety.

MS04-T17

Synthesis, X-Ray Structure, Infrared study and anti-corrosion activity of 1,6-diammoniumhexandihydrogenodiphosphate dihydrate

A. Zaraq[1], A. Kheireddine[2], A. El Bouari, S. Belaaouad[2]
[1]*Laboratory physics and chemistry of applied materials - University of Hassan II Casablanca , Casablanca, Morocco*
[2]*Laboratory chemistry and physics of materials - University Hassan II Casablanca, Casablanca, Morocco*

The crystal structure and IR investigation of 1,6-diammoniumhexandihydrogenodiphosphate dehydrate $(C_6H_{18}N_2)H_2P_2O_7.2H_2O$ (denoted DHDP) synthesized newly are explored through measurements of X-ray four-circle diffraction. This system is triclinic with space group P-1and the lattice constants are a=9.1744(4) Å, b=13.6977(5) Å, c=13.6977(5) Å, α=115.514(4)°, β=111.940(4)°, γ=91.186(3)° and V=1501.72(10)Å^3. The main feature of the atomic arrangement of DHDP is the existence of infinite lines parallel to the c direction of the unit cell, made up of $H_2P_2O_7$ groups. These phosphoric lines are organized to create with the help of the organic cation, large rectangular channels in which water molecules are located. The compound was tested as corrosion inhibitors for mild steel in 0.1 M nitric acid medium using potentiodynamic polarization technique.

MS04-T18

High resolution structure of apo-furin reveals competitive and allosteric inhibitor mechanisms

S. O. Dahms[1], M. E. Than[1]
[1]*Leibniz Institute for Age Research - Fritz Lipmann Institute (FLI), Protein Crystallography, Jena, Germany*

Furin is a subtilisin-like endoproteinase belonging to the pro-protein convertase (PC) family. PCs are required for the activation and maturation of many secreted proteins, including peptide hormones, growth factors, matrix metalloproteases, blood coagulation factors, regulators of the cholesterol metabolism, bacterial toxins and viral capsid proteins. Therefore furin and other PCs are highly interesting targets for the treatment of many diseases, e.g. cancer as well as viral- and bacterial infections (Seidah & Prat, 2012).

Development of furin inhibitors for pharmacological use requires a detailed structural understanding of the mechanisms involved in substrate-binding, catalysis and inhibition (Dahms et al, 2014; Henrich et al, 2003). Here we present the structure of apo-furin at 1.8 Å resolution and different inhibitor-bound structures. Our crystals in absence of any inhibitor showed the protease catalytically active in the crystalline state. These crystals proved to be highly suited to investigate the mechanism of competitive and allosteric inhibitors by soaking. Allosteric mechanisms involve the reorientation of the catalytic residues and/or the structural changes proximal to the catalytic center. Crystals of apo-furin enable the investigation of inhibitory compounds with low and very low affinities, the use of novel inhibitory mechanisms and to target novel surface patches of the enzyme, e.g. in crystallography based fragment screening approaches. Thus, our work provides novel possibilities for structure-based lead discovery and the design of next-generation furin inhibitors.

Dahms SO, Hardes K, Becker GL, Steinmetzer T, Brandstetter H, Than ME (2014) X-ray structures of human furin in complex with competitive inhibitors. *ACS chemical biology* **9:** 1113-1118

Henrich S, Cameron A, Bourenkov GP, Kiefersauer R, Huber R, Lindberg I, Bode W, Than ME (2003) The crystal structure of the proprotein processing proteinase furin explains its stringent specificity. *Nature structural biology* **10:** 520-526

Seidah NG, Prat A (2012) The biology and therapeutic targeting of the proprotein convertases. *Nature reviews* **11:** 367-383

MS04-T19

In-situ X-ray studies of GaAs Nanowire growth onto silicon(111) substrate

S. M. Mostafavi Kashani[1], P. Schroth[1], H. Letzguss [2], M. Koehl[2], S. Bauer [3], T. Baumbach[2], U. Pietsch[1]
[1]*University of Siegen, Department of Physics, Siegen, Germany*
[2]*Karlsruhe Institute of Technology (KIT), Institute for Photon Science and Sychrotron Radiation (IPS)/ANKA, Karlsruhe, Germany*
[3]*Karlsruhe Institute of Technology, Synchrotron Facility ANKA, Karlsruhe, Germany*

GaAs nanowires (NWs) are promising candidates in optoelectronic applications e.g., low power-consumption light emitting diodes (LEDs) and laser diods, and solar cells with high efficiency. Naturally, NWs contain random structural defects and polytypism which may degrade electrical and optoelectrical properties of NW devices. In-situ x-ray studies of NW growth can play a key role to reveal understandings in order to control the NW structures.

We report on monitoring the evolution of NW polytypism during growth using high intensity x-rays provided at NANO beamline of ANKA synchrotron (KIT). The growth of GaAs NWs onto silicon (111) substrate was performed using a Portable Molecular Beam Epitaxy (PMBE) chamber. Before the in-situ experiment, the growth parameters such as gallium and arsenic flux, and the substrate temperature have been calibrated by ex-situ growth runs. Inspected by SEM, it turned out that at a substrate temperature of 600 °C, a V/III ratio of 2.55, and a growth time duration of 1 hour, the epitaxially grown NWs have an average length of ~1 µm, a diameter of ~40 nm, and a number density of 0.76 μm^{-2}.

During the in-situ experiment, we inspected the (220), (311), and (103) Bragg reflection which were sensitive for zinc-blende (ZB), twin ZB, and wutzite (WZ) phases, respectively, using time resolution of 5 minutes. NWs have been grown preferentially in ZB phase with a small content of WZ phase.

MS04-T20

Symmetry-mode analysis of the commensurately modulated superstructure of tetragonal tungsten bronze $KMnCrF_6$

C. Drathen[1]
[1]*ESRF - The European Synchrotron, Grenoble, France*

Multiferroic materials showing coupling of the different order parameters (ferroelectric, ferromagnetic, ferroelastic) are interesting not only from a fundamental perspective, but also from a technological point of view, e.g. for to the development of new storage technologies. However, the coexistence of (ferro)magnetism and ferroelectricity is considered a rare phenomenon. Whilst this may be true for perovskite oxides, where empty *d*-shells favor the off-centering of ions but counteract magnetism, this intrinsic limitation can be avoided by moving to different structure types, and/or away from oxides. An example of non-perovskite, non-oxide multiferroic systems are the tetragonal tungsten bronze (TTB) fluorides $K_xM^{2+}{}_xM^{3+}{}_{1-x}F_3$ (x = 0.4 - 0.6), which show coexistence of electric and magnetic ordering [1]. Here we present a detailed structural study of the TTB fluoride $K_xMn_xCr_{1-x}F_3$ (x = 0.5). $K_{0.5}Mn_{0.5}Cr_{0.5}F_3$ has been previously described as tetragonal $P4_2bc$ and orders ferrimagnetically below T ~ 30 K [2]. We used high-resolution powder diffraction techniques to reinvestigate the crystal structure as a function of temperature. Our results reveal a structural distortion to monoclinic symmetry (*P*2) at room temperature, with a $2\sqrt{2}x2\sqrt{2}x1$ supercell as compared to the tetragonal cell, containing 400 and 15 independent atoms, respectively. To handle the high number of parameters, we use the symmetry mode description in a combined neutron- and X-ray refinement. This approach leads us to a model with only ~ 60 variables describing the fractional coordinates of all atoms and allows us to identify those modes that are driving the structural transition to tetragonal symmetry at high temperatures. Although structurally subtle, this distortion may indicate a ferroelectric state, similar to K_xFeF_3, where ferroelectricity is observed only in the orthorhombic phase.

References:

[1] Scott, J.F., Blinc, R., *J. Phys. Condens. Matter*, **2011**, *23*, 113202.,
[2] Banks E., Shone, M. Hong, Y.S., Williamson, R.F., Boo, W.O.J., *Inorg Chem.* **1982**, *21* (11) 3894.

MS04-T21

Spinophilin restricts Neurexin signaling to protect from excessive seeding of new active zones

J. H. Driller[1], K. G. Muhammad[2,3], S. Reddy[2,3], U. Rey[3], M. A. Böhme[3], C. Hollmann[3], N. Ramesh[2], H. Depner[2,3], J. Lützkendorf[2,3], T. Matkovic[2], D. Bergeron[2], C. Quentin[2,3], J. Schmoranzer[4], F. Goettfert[5], M. Holt[6], M. C. Wahl[1], S. W. Hell[5], A. Walter[2,4], B. Loll[1], S. J. Sigrist[2,3]
[1]*Freie Universität Berlin, Biochemie, Berlin, Germany*
[2]*Freie Universität Berlin, Biologie, Berlin, Germany*
[3]*Charité, NeuroCure, Berlin, Germany*
[4]*Leibniz Institut für Molekulare Pharmakologie, Berlin, Germany*
[5]*Max Planck Institute, Department of NanoBiophotonics, Göttingen, Germany*
[6]*VIB Center for the Biology of Disease, Leuven, Belgium*

Cytoplasmic active zone scaffolds, composed of several conserved multi-domain proteins, decorate the presynaptic membrane. While scaffold sizes were suggested to influence synaptic vesicle release, core mechanisms to assemble functional synapses remain to be characterized. Assembly and maturation of synapses depend on trans-synaptic Neurexin/Neuroligin (Nrx/Nlg) signaling, which is promoted by the regulator Syd-1 at *Drosophila* neuromuscular synapses.Here, we show that presynaptic Spinophilin, dependent on its PDZ domain by which it binds Nrx with micromolar affinity, attenuates Nrx/Nlg signaling and antagonizes Syd-1.Loss of Spinophilin resulted in the formation of excess but atypically small active zone scaffolds. We verified these interactions *in vitro* via pull-down assays and characterized them by ITC measurements. We also determined a high resolution crystal structure of the PDZ domain of Spinophilin bound to a peptide derived of the last 10 C-terminal amino acids of Nrx-1.

MS04-T22

Crystallographic analysis of the ClpP1/2 heterocomplex from Listeria monocytogenes

M.- T. Vielberg[1], M. Dahmen[2], S. A. Sieber[2], M. Groll[1]
[1]*Department of Chemistry, Technische Universität München, Chair of Biochemistry, Garching, Germany*
[2]*Department of Chemistry, Technische Universität München, Chair of Organic Chemistry, Garching, Germany*

The virulence of many obligate as well as facultative pathogens is mediated by the caseinolytic protease P (ClpP). This multimeric serine protease was shown to degrade small peptides independently, whereas digestion of proteins requires the interaction with an AAA+-chaperone, as for example ClpX or ClpA. Listeria monocytogenes encodes not only one, but two isoforms of ClpP. Prior investigations revealed that LmClpP2 alone forms a catalytically active, tetradecameric complex. In contrast, LmClpP1 was found to exist as a heptameric ring in vitro that does not exhibit any enzymatic activity. However, labeling studies with LmClpP1-specific inhibitors proved that both isoforms interact to build up a functional degradation machinery in vivo. Recently, we could determine the crystal structure of the LmClpP1/2 heterocomplex at 2.8 Å resolution (PDB: 4RYF)[1]. It preserves all characteristic features of active ClpP proteins which is particularly reflected in the correct alignment of all catalytic centers. The substitution of Asp172 with an asparagine in LmClpP1 leads to replacement of the highly conserved triad by a functional dyad. The reactivity of this assembly is strengthened by the polarizing properties of the mutated residue. Further differences between LmClpP1 and LmClpP2 were found in both the N-terminal region and the S1-substrate pocket. The latter may be useful for isoform-specific drug design in the future.

[1] Maria Dahmen*, Marie-Theres Vielberg*, Michael Groll, Stephan A. Sieber: Structure and mechanism of the caseinolytic protease ClpP1/2 heterocomplex from Listeria monocytogenes, Angew. Chem. Int. Ed. 2015, in press

MS05 - Material Science II

MS05-T01
Pushing the detection limit of in-situ PDF - highly diluted nanoparticles in organic solvents

M. Zobel[1], S. A. J. Kimber[2], R. B. Neder[1]
[1]*University Erlangen, Erlangen, Germany*
[2]*European Synchrotron Radiation Facility, Grenoble, France*

Pair distribution function (PDF) measurements have originally been used to study the structure of liquids and glasses [1]. Almost forgotten since, the method experienced a tremendous revival as Rapid Acquisition PDF with fast readout 2D detectors at dedicated high-energy beamlines [2]. The recent interest focusses on nanoparticles and in-situ studies of chemical reactions, where common x-ray diffraction fails [3,4]. The current state-of-the-art is the application in nanoparticle nucleation in liquid media. Precursor species of e.g. SnO_2 or WO_3 were identified and particle formation observed [5, 6]. To our knowledge, all previous in-situ PDF studies have been undertaken in either aqueous or supercritical solvents, although manifold actual nanoparticle syntheses use organic solvents. The drawback to study nucleation in organic solvents are the direct implications of using them. Firstly, nanoparticle formation in organic solvents often involves non-classical nucleation pathways, challenging data interpretation. We show this for the formation of 2.8 nm ZnO nanoparticles from their primary molecular precursor Zn_4OAc_6 in ethanol. Secondly, the solubility of the reactants, which are frequently salts, are noticeably lower in organics than in water, which reduces the scattering signal. Thirdly, from the experimental and data treatment point of view, the probably largest drawback to study organic solvents at ambient conditions is the scattering contribution of the solvents themselves. They not only scatter x-rays stronger than water, but more importantly organic molecules possess several interatomic distances as well as feature pronounced intramolecular ordering within the bulk solvent. This bulk liquid structure of alcohols was, as stated earlier, the main point of interest in the very first PDF studies and now again becomes essential [1]. In highly diluted systems as often found in nanoparticle nucleation, these intra- and intermolecular distances of the solvent molecules contribute to more than 99 % of the overall signal in the experimental PDFs. The extraction of the signal from the nanoparticles hence requires understanding of the organic solvent as well as the ongoing nucleation process. With our latest state-of-the-art in-situ PDF experiments on ZnO nanoparticle formation, we ventured into this field. We could successfully extract and model difference-PDFs of ZnO nanoparticles, which undergo a gradual crystallization process from a disordered to an entirely crystalline state in an ethanolic solution with a mere 0.4 wt% zinc species.

[1] Zachariasen, W. H., J. Chem. Phys. 3 (1935), 158
[2] Chupas, P. J., et al, J. Appl. Cryst 36 (2003), 1342
[3] Korsunski, V. I., et al. J. Appl. Cryst. 36 (2003) 1389
[4] Borkiewicz, O. J., et al, Phys. Chem. Chem. Phys. 15 (2013) 8466
[5] Jensen, K. M. O., et al., J. Am. Chem. Soc. 134 (2012), 6785
[6] Saha, D., et al. Angew. Chem. Int. Ed.2014, 53, 3667

MS05-T02
Nanocrystalline composite ceramics synthesized by high-pressure crystallization of amorphous materials

N. A. Gaida[1,2], N. Nishiyama[2], A. Holzheid[1], O. Beermann[1]
[1]*University Kiel, Kiel, Germany*
[2]*Deutsches Elektronen-Synchrotron, Photon Science, Hamburg, Japan*

State-of-the-art high pressure technologies allow synthesizing of novel poly-nanocrystalline materials, in particular ceramics with excellent mechanical properties made from Earth's high abundant geomaterials. The positive correlation of grain fining and increasing hardness known as the Hall-Petch relation is widely observed for metals and was recently also found in ceramics, e.g., MgO [1] or stishovite (SiO_2) [2]. The opposite - a decreasing hardness below a certain grain size - is called the inverse Hall-Petch relation [1]. Poly-nanocrystalline stishovite is seven times tougher than single crystal stishovite [2, 3]. This points towards that in ceramics grain fining not only can enhance material hardness, but also the fracture toughness. This study focuses on combining the positive correlations of grain fining on hardness and of phase interlocking by grain boundary pinning on fracture toughness of composite ceramics [4] using components of the hardest and second hardest oxides (stishovite and corundum, respectively [5, 6]).

Poly-nanocrystalline aggregates of Al_2O_3 (corundum) and SiO_2 (stishovite) were synthesized at a pressure of 15.6 GPa and temperatures from 1300 to 1900°C. The starting materials were glasses with Al_2SiO_5 in composition produced by container-less processing in an aerodynamic levitation furnace. X-Ray diffraction high resolution powder diffraction show that the starting material (Al_2SiO_5-glass) completely decomposes during the experiments into corundum and stishovite with Si^{4+} and Al^{3+} arranged in octahedral coordination. Corresponding to the Hall-Petch relation Vickers hardness and fracture toughness of the synthesized poly-nanocrystalline materials increase with decreasing grain sizes from 500 nm to 100 nm (from 20 to 24 GPa and from 2.5 to 3.2 MPa $m^{1/2}$, respectively). The inverse Hall-Petch relation in these poly-nanocrystalline ceramics occurred at grain sizes

Our results show that poly-nanocrystalline geomaterials made from low-cost raw substances are very promising in creating materials with excellent mechanical properties exceeding those of high quality metal alloys, which is a great challenge of material science so far.

References:
[1] Ehre D, Chaim R (2008) *J. Mater. Sci.*, 43, 6139-6143.
[2] Nishiyama N, Seike S, Hamaguchi T, Irifune T, Matsushita M, Takahashi M, Ohfuji H, Kono Y (2012) *Scr. Mater.*, 67, 955-958.
[3] Nishiyama N, Wakai F, Ohfuji H, Tamenori Y, Murata H, Taniguchi T, Matsushita M, Takahashi M, Kulik E, Yoshida K, Wada K, Bednarcik J, Irifune T (2014) *Sci. Rep.*, 4.
[4] Nicholson GC (1966) *J. Am. Ceram. Soc.*, 49, 47-49.
[5] Nishiyama N, Taniguchi T, Ohfuji H, Yoshida K, Wakai F, Kim B-N, Yoshida H, Higo Y, Holzheid A, Beermann O, Irifune T, Sakka Y, Funakoshi K (2013) *Scr. Mater.*, 69, 362-365.
[6] Qin J, Nishiyama N, Ohfuji H, Shinmei T, Lei L, He D, Irifune T (2012) *Scr. Mater.*, 67, 257-260.

MS05-T03
Zinc phosphate nanoparticles for electrochemical energy storage systems

C. Hoch[1], S. Harm[1]
[1]Ludwig-Maximilians-Universität München, Chemie, München, Germany

One of the damage mechanisms limiting the life span of lithium ion batteries is ablation of the redox-active cathode material from the electrode carrier foil, caused by volume change during repetitive intercalation and deintercalation of lithium ions [1]. It was previously shown that the cycle stability of cathode materials can be drastically improved in materials with rigid framework structures incorporating channels as pathway for lithium cations, like the mineral arrojadite [2]. Following classical structure-property relationships, search algorithms could be established to identify similar crystal structures potentially suitable for electrochemical functional materials. One of the searches over chemical, mineralogical and structural data bases yielded a well-known material, hopeite ($Zn_3(PO_4)_2$ x $4H_2O$), meeting all the requirements. This orthophosphate crystallises in the orthorhombic space group *Pnma* (a = 10.629(2) Å, b = 18.339(3) Å, c = 5.040(1) Å) [3], see Figure 1. When Zn is substituted by redox-active transition metal cations as Mn^{2+} or Co^{2+}, hopeite can be applied as cathode material in lithium ion batteries in the form of the monoclinic dihydrate $Zn_3(PO_4)_2$ x $2H_2O$ (a = 15.872(4) Å, b = 5.048(1) Å, c = 10.480(2) Å, β = 91.71(2)°). To further improve the performance of this cathode material we looked for a way to prepare hopeite nanoparticles. Nanosizing electrode materials can result in improved charge time and cyclic stability [4]. By optimizing an already established synthesis route to hydroxy zinc phosphate nanoparticles [5], nanoparticles with the desired composition could be obtained (see Figure 2). These always were observed together with nanosheets of crystalline hopeite. It could be shown that these crystals directly grow from agglomerated nanoparticles. Following transmission electron microscopic investigations the nanoparticles only have small crystalline domains. Substituting Zn^{2+} with Co^{2+} was successful while retaining a nanosized material with small particle size distribution. To effectively apply the nanoparticles in cathodes for lithium ion batteries a more homogenous product is needed. Optimisations of the synthetic conditions and subsequent electrochemical characterisations are in progress.

Figure 1: Left: Crystal structure of orthorhombic $Zn_3(PO_4)_2$ x $4H_2O$ projected along the c direction. Channels are emphasised by a grey semicircle. Right: Crystal structure of monoclinic $Zn_3(PO_4)_2$ x $2H_2O$ projected along the b direction. Shape but not size of the channels have changed during the partial dehydratation.

Figure 2: Scanning electron micrograph (secondary electron mode) of a sample with nanosized zinc phophate particles. The particle size distribution is very narrow with an approximate diameter of 40 nm.

References:
[1] C. Hoch, H. Schier, C. Kallfaß, C. Tötzke. A. Hilger, I. Manke, *Micro & Nano Lett.*7, *262* (2012).
[2] C. Hoch, *Nachr. Chem.* **60**, *1181* (2012).
[3] A. Whitaker, *Acta Crytallogr. B* **31**, *2026* (1975).
[4] Y. Wang, G. Cao, *Adv. Mater.* **20**, *2251* (2008).
[5] X. Yuan, B. Zhu, X. Ma, G. Tong, Y. Su, X. Zhu, *Langmuir* **29**, *12275* (2013).

Figure 1

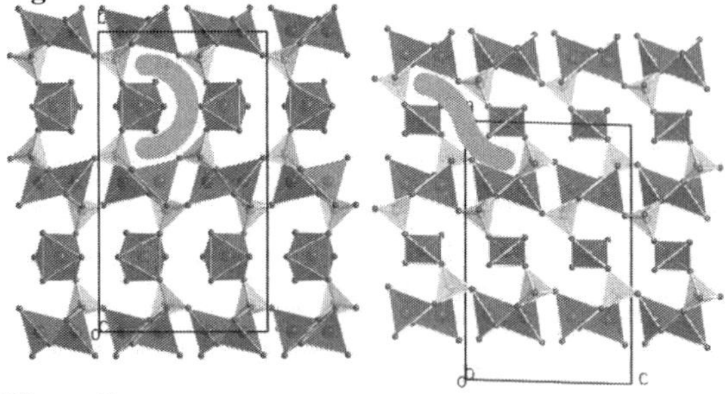

Figure 2

MS05-T04
Influence of habitat on texture: a systematic case study of sea urchin teeth

B. Maier[1], E. Griesshaber[1], F. Brümmer[2], B. Ruthensteiner[3], W. Schmahl[1]
[1]LMU München, Department für Geo- und Umweltwissenschaften, München, Germany
[2]Universität Stuttgart, Biologisches Institut, Stuttgart, Germany
[3]Zoologische Staatssammlung, München, Germany

Sea urchins form their hard body parts, e.g. shell, spines, and teeth, in a biomineralization process which produces a hybrid composite material consisting of co-oriented calcite ($CaCO_3$) micro- and nanocrystals and an organic matrix. Sea urchin teeth are composed of five major morphologic structure units, i.e. *primary plates*, *secondary plates*, *carinar process plates*, the *lamelar needle complex*, and the *stone*, which form a T-shaped girder-like structure along the curved tooth. The stone consists of Mg-calcite fibers reinforced by ultra-high-Mg-calcite nanoparticles and is encased in the softer plates and needles. The crystallographic orientation within each structural unit is uniform, despite its complex morphologic shape. However, there is a systematic difference in crystallographic orientation between different units in the order of 3-6° for the example of *Paracentrotus lividus*. This complex architecture of the sea urchin teeth features a self-sharpening mechanism which enables grinding on hard rock substrates.

In this study we show that the crystallographic misorientation between (or within) the structural units differs between different species of sea urchins. In a systematic approach we want to clarify if the degree of crystallographic co-orientation depends on the habitat (e.g. stony or sandy enviroment) and/or the species of the sea urchin. The key question is if the biologic control over crystallographic texture during the biomineralization process is subject to an adaptive evolutionary process.

MS05-T05

A new method to derive Crystallite Size Distributions (CSD) from 2D X-ray diffraction data

S. Neher[1], K. Nützmann[1], M. Chaouachi[1], A. Falenty[1], H. Klein[1], W. F. Kuhs[1]
[1]*Georg-August-Universität Göttingen, GZG, Kristallographie, Göttingen, Germany*

We present our new *"fast diffraction CSD analysis"* method for deriving CSDs of polycrystalline material or powders from 2D diffraction data. A fast and easily implemented measurement technique what can be carried out with standard lab equipment equipped with a 2D detectors.

At this point we give a insight view at the measurement technique **(1)**, the data reduction **(2)** and **(3)** the calibration, correction and scaling of the data.

(1) For the CSD analysis an image sequence of a stepwise rotation measurement (phi or omega rotation) with a "spotty" diffractions patterns is needed. The rotational range is commonly a few degrees with a small fraction of the rocking curve as step size. The occurrence of each spot needs to be resolved in a sufficient number of images to proper evaluate the rocking curve. **(2)** Data reduction is done by a computer program written in Python. Every spot is detected, if needed isolated from other overlapping spots and assigned to its occurrence in consecutive images. All integrated intensities form the intensity distribution (ID) of the sample. **(3)** For each analysis a standard material with known CSD needs to be measured under the same conditions as the sample. Both, the sample ID and the standard ID are corrected for the Lorentz factor using the general Lorentz correction by (Milch and Minor 1974). With a 1st scaling factor between the known standard CSD and the measured standard ID and a 2nd scaling factor, the ratio between the structure factors of sample and standard, the sample CSD can be calculated.

So far the method has been successfully tested with a Bruker APEX II CCD-detector using a laboratory X-ray source, at ESRF ID 15 in Grenoble using the PIXIUM 4700 and with the Perkin Elmer XRD 1622 at PETRA/DESY Hamburg. For testing and calibration we use several well defined grain size fractions of a single crystal corundum powder. The grain size fractions of the powder were separated via sedimentation in water and characterized with SEM imagery.

Milch, J. R. and T. C. Minor (1974). "The indexing of single-crystal X-ray rotation photographs." Journal of Applied Crystallography**7**(5): 502-505.

MS05-T06

Debye scattering formula: still no valid alternatives after 100 years?

M. Leoni[1]
[1]*University of Trento, DICAM, Trento, Italy*

Debye scattering formula, developed by Peter Debye in Göttingen [1], is 100 years old. It is a unique tool to obtain the powder diffraction pattern of any object described atomistically, independently of its complexity. Even if very simple and flexible, Debye scattering formula is however highly demanding from the computational point of view, as it requires the calculation of all interatomic distances of the object(s) under study. The major advantage is that the simulation can be directly compared with the measured data. During the years, several scientists have tried to improve the computational speed in order to employ the formalism for the characterization of nanostructured materials. In several practical cases, however, the question arises on the real advantage of a fully atomistic approach. The recent advances in simulation and modelling of the powder diffraction pattern using whole pattern methods are here shown [2,3]. The current formalism works with real-space models to generate the whole diffraction pattern and allows for an accuracy similar to that of the Debye scattering formula, but with computing times that are orders of magnitude smaller. A comparison of the two formalisms is proposed, with illustration of state of the art, advantages, disadvantages and caveats.

[1] Debye, P. "Zerstreuung von Röntgenstrahlen". Nachrichten von der Koeniglichen Gesellschaft der Wissenschaften zu Goettingen, Math. Klasse 1915, 70-76 (1915).

[2] Scardi, P. & Leoni, M. "Whole powder pattern modelling". Acta Crystallogr. A 58, 190-200 (2002).
[3] Koch, R., & Leoni, M. "Combined quantitative analysis of structure and nanostructure in disordered crystalline materials". Adv. Mater. Submitted (2014).

MS06 - Computational & Theoretical Crystallography: Macromolecular structures

MS06-T01

Solving macromolecular structures with shelxc/d/e

G. Sheldrick[1]
[1]*University of Goettingen, Goettingen, Germany*

The programs shelxc, shelxd and shelxe provide an approximate but robust and efficient route to macromolecular structure solution, starting with either determination of a heavy atom substructure or a small fragment that has been located by molecular replacement. For experimental phasing only one type of anomalous scatterer is assumed, so it is not necessary to know which element is responsible for the anomalous scattering. Only a poly-alanine backbone is traced, so no amino-acid sequence is required. The tracing is performed iteratively and is designed to be able to start from rather weak phase information. In general the best results are obtained when the resolution of the native data and the solvent content are high. Recent develpments will be presented, with particular emphasis on improved algorithms for tracing structures at lower resolution and for exploiting non-crystallographic symmetry.

Rodríguez, D.D., Grosse, C., Himmel, S., Gonzales, C., de Ilardya, I.M., Becker, S., Sheldrick, G.M. & Usón, I. (2009). *Nat. Methods* 6, 651-653.
Rodríguez, D.D., Sammito, M., Meindl, K., de Ilarduya. I.M., Potratz, M., Sheldrick, G.M. & Usón, I. (2012). *Acta Cryst.* D68, 651-653.
*Sammito, M., Millán, C., Rodríguez, D.D., de Ilarduya, I.M., Meindl., K., De Marino, I., Petrillo, G., Buey, R.M., de Pereda, J.M., Zeth, K., Sheldrick, G.M. & Usón, I. (2013). *Nat. Methods* 10, 1099-1101.
Sheldrick, G.M. (2010). *Acta Cryst.* D66, 479-485.
Thorn, A. & Sheldrick, G.M. (2013). *Acta Cryst.* D69, 2251-2256.

MS06-T02

Macromolecular Phasing with ARCIMBOLDO: Single workstation implementations and combination of partial solutions in reciprocal space.

I. Usón[1]
[1]*Barcelona Science Park, ICREA at Departament of Structural Biology, Instituto de Biología Molecular de Barcelona (IBMB-CSIC), Barcelona, Spain*

Macromolecular *Ab Initio* phasing, from the native intensities alone with no experimental phase information or previous particular structural knowledge has been the object of a long quest, limited by two main barriers: structure size and data resolution. Beyond the first atomic resolution successes, current approaches have been developed, exploiting alternative constraints to atomicity, through use of the Patterson function, density modification and data extrapolation [1].

Our own approach relies on the combination of locating model fragments with the program PHASER [2] and density modification with the program SHELXE [3]. Appropriate search models are polyalanine alpha-helices[4], small polypeptide folds[5], DNA-binding motifs[6] or fragments from distant homologs[7]. Given the difficulties in discriminating correctly positioned fragments, many putative hypotheses have to be tested in parallel, thus calculations are performed on a grid or supercomputer. The method has been called after the Italian painter Arcimboldo, who used to compose portraits out of fruits and vegetables. In the case of our program, most collections of fragments remain a "still-life", but some are correct enough for density modification to reveal the protein's true portrait (http://chango.ibmb.csic.es/ARCIMBOLDO). Using these methods, a number of unknown macromolecules with a few thousand atoms and resolutions around 2 Å have been solved.

Performance of the single-workstation implementation ARCIMBOLDO_LITE and combination of partial solutions in reciprocal space will be presented.

[1] Sheldrick, G.M., Gilmore, C., Hauptman, H.A., Weeks, C.M., Miller, R. and Usón, I. International Tables for Crystallography (2012). Vol. F, ch. 16.1, pp. 413-432; Burla, M.C., Carrozzini, B., Cascarano, G.L., Giacovazzo, C., Polidori,M. J. Appl. Cryst. 45, 1287-1294 (2011).
[2] McCoy, A.J. Grosse-Kunstleve RW, Adams PD, Winn MD, Storoni LC, Read RJ.. J. Appl. Crystallogr. 40, 658-674 (2007).
[3] A. Thorn and G.M. Sheldrick. Acta Crystallogr. D69, 2251-2256 (2013).
[4] Rodríguez D.D., Grosse, C., Himmel, S., González, C., de Ilarduya, I. M., Becker, S., Sheldrick, G. M. and Usón, I. Nature Methods 6: 651-653 (2009).
[5] Sammito, M., Millán, C., Rodríguez, D.D., de Ilarduya, I.M., Meindl, K., De Marino, I., Petrillo, G., Buey, R.M., de Pereda, J.M., Zeth, K, Sheldrick, G. M. and Usón, I. Nature Methods 10: 1099-1101 (2013).
[6] Pröpper, K., Meindl, K., Sammito, M., Dittrich, B., Sheldrick, G.M., Pohl, E. and Usón, I. Acta Cryst. D70, 1743-1757 (2014).
[7] Sammito, M., Meindl, K., de Ilarduya, I.M., Millán, C., Artola-Recolons, C., Hermoso, J.A. and Usón, I. FEBS Journal 281: 4029-4045 (2014)

MS06-T03

Recent crystallographic software developments at webapps.embl-hamburg.de

F. Dall'Antonia[1], T. R. Schneider[1]
[1]*EMBL, Hamburg Outstation, Hamburg, Germany*

Crystallographic software tools developed in our group at EMBL Hamburg cover the fields of experimental macromolecular phasing and comparative structure analysis. HKL2MAP [1] is a graphical user interface for the SHELXC/D/E phasing programs [2], supporting recently added features such as multi-CPU-based substructure determination (SHELXD), and improved phasing (SHELXE) thanks to heavy atom site refinement and density modification aided by iterative backbone auto-tracing. The power of these new functions in combination with HKL2MAP has been demonstrated on several difficult cases of SAD-phasing.

We are further enhancing the applicability of our tools to complicated phasing scenarios. SITCOM [3] was developed to compare substructures for site agreement. The features of solution similarity clustering and NCS detection help to identify correct solutions when the usual SHELXD criteria such as CC scatter plots and peak height profiles do not provide sufficiently clear guidance. HKL2MAP now features 3d-graphs for the poly-ala models traced by SHELXE. In cases where electron density contrast and sheer residue numbers can hardly discriminate the two maps stemming from enantiomorph substructures, the on-the-fly look at the geometry of the built protein models can often answer the question whether one hand represents a true solution.

The RAPIDO web service [4] is an online tool for the alignment of protein structures in the presence of domain motion. In addition to the 3d-alignment of the protein sequences, the structurally invariant parts ("rigid bodies") of the distinct conformations are identified based on the analysis of error-scaled difference distance matrices.

The server can, for example, be used to identify common fragments between homologues serving as molecular replacement models.

All of our tools are free for academics and can be easily accessed for download or online use from http://webapps.embl-hamburg.de.

References
[1] Pape, T. & Schneider, T. R. (2004). J. Appl. Cryst. 37, 843-4.
[2] Sheldrick, G. M. (2010). Acta Cryst D66, 479-85.
[3] Dall'Antonia, F. & Schneider, T. R. (2006). J Appl. Cryst. 39, 618-9.
[4] Mosca, R. & Schneider T. R. (2008). Nucl. Acids Res. 36, W32-6.

MS06-T04

From data collection to ligand hits: XDSAPP and automated refinement pipeline

K. Sparta[1], U. Mueller[1], M. S. Weiss[1], M. Ühlein[1], F. Huschmann[1,2], J. Schiebel[2], A. Heine[2], G. Klebe[2], U. Heinemann[3]
[1]*Helmholtz-Zentrum Berlin, Macromolecular Crystallography, Berlin, Germany*
[2]*University of Marburg, Department of Pharmaceutical Chemistry, Marburg, Germany*
[3]*Max-Delbrück Center for Molecular Medicine, Macromolecular Structure & Interaction, Berlin, Germany*

Hardware improvements at today's third generation synchrotrons have enabled the fast collection of diffraction images from protein crystals with exposure times well below one second per image. This makes the availability of automated solutions for data reduction and structure refinement highly needed, especially in the case of fragment screening experiments where the same molecule is tested against hundreds of ligands.

We have developed the expert software XDSAPP [1] to support synchrotron users in the analysis of their diffraction data during measurements at the beamline with minimal effort and time. It mainly uses the diffraction data processing program XDS [2], along with additional software like POINTLESS from the CCP4 suite [3], XDSSTAT [4], SFCHECK [5] and PHENIX.XTRIAGE [6] for automatic decision making. An independent refinement pipeline for automated ligand search based on PHENIX is currently under development for use within XDSAPP at the macromolecular crystallography beamlines BL14.1-3 at Helmholtz-Zentrum Berlin [7].

XDSAPP is available free of charge for academic users from www.helmholtz-berlin.de/bessy-mx.

References
[1] Krug, M. et al. (2012). *J. Appl. Cryst.* **45**, 568-572.
[2] Kabsch, W. (2012). *Acta Cryst.* D**66**, 125-132.
[3] Collaborative Computational Project (1994). *Acta Cryst.* D**50**, 760-763.
[4] Diederichs, K. (2006). *Acta Cryst.* D**62**, 96-101.
[5] Vaguine, A et al. (1999). *Acta Cryst.* D**55**, 191-205.
[6] Adams, P.D. et al. (2010). *Acta Cryst.* D**66**, 213-221.
[7] Mueller, U. et al. (2012). *J. Synchrotron Rad.* **19**, 442-449.

MS06-T05

Enzyme mechanisms at ultrahigh resolution

S. Lüdtke[1], P. Neumann[1], D. Meyer[1], A. Lehwess-Litzmann[1], R. Ficner[1], K. Tittmann[1]
[1]*Goettingen University , Goettingen, Germany*

Although general principles of enzyme catalysis are fairly well understood nowadays, many important details of how exactly the substrate is bound and processed in an enzyme remain often invisible and as such elusive. In fortunate cases, structural analysis of enzymes can be accomplished at true atomic resolution (≤1 Å) thus making possible to shed light on otherwise concealed fine-structural traits of bound substrates, intermediates, cofactors and protein groups. I will present recent structural studies of several enzymes using ultrahigh-resolution X-ray protein crystallography showcasing its enormous potential as a tool in the elucidation of enzymatic mechanisms and in unveiling fundamental principles of enzyme catalysis. This includes the observation of seemingly hyper-reactive, physically distorted cofactors and intermediates with elongated scissile substrate bonds, the detection of "hidden" conformational and chemical equilibria and the analysis of protonation states with surprising findings. In addition to the pivotal structural findings and the implications for our understanding of enzyme catalysis, I will further provide a practical framework for resolution enhancement through optimized data acquisition and processing.

MS06-T06

Pseudo-crystallographic refinement of the *E. coli* ribosome structure against cryo-EM map

P. Neumann[1], N. Fischer[2], H. Stark[2], R. Ficner[1]
[1]*Georg-August-University Göttingen, Molecular Structural Biology, Göttingen, Germany*
[2]*Max Planck Institute for Biophysical Chemistry, 3D Electron Cryo-Microscopy, Göttingen, Germany*

Recent progress in high-resolution structure determination of large macromolecular complexes by single particle electron cryo-microscopy (cryo-EM) made it possible to obtain cryo-EM reconstructions with the overall quality comparable to the crystal structure derived electron density maps of aforementioned complexes, like ribosomes. Although local resolution of the cryo-EM reconstructions can be non-uniform and structures available to date are usually reported to an overall resolution lower than 3 Å, the level of visible structural details legitimates building of atomic models, which can not be modeled by simple rigid body fit of known 3D structures or their fragments/domains. In order to be able to build a model observed in cryo-EM map conformational state of the macromolecular complex, the initial atomic model needs to be fitted and refined against the target cryo-EM density. This raises the need to develop new tools for cryo-EM based 3D modeling or customize the usage of existing powerful programs already available for X-ray crystallography.

Here we present an atomic model of the 70S ribosome from *Escherichia coli* in complex with elongation factor Tu, aminoacyl-tRNA and the antibiotic kirromycin at 2.65 - 2.9 Å resolution obtained by cryo-EM reconstruction and a pseudo-crystallographic refinement approach. The superior quality of the cryo-EM reconstruction allowed for the first time to build all 35 RNA modifications in the bacterial ribosome explaining their roles in fine-tuning ribosome structure and function. The starting model of the crystal structure of *Escherichia coli* ribosome has been refined against reciprocal structure factors obtained from EM maps using a combination of efficient conformational sampling with "Deformable Elastic Network" restraints, manual modeling and standard crystallography-like refinement. Application of available crystallographic refinement programs (CNS, PHENIX) and newly established protocols to yield a reliable atomic model based on cryo-EM maps will be discussed.

Figure 1

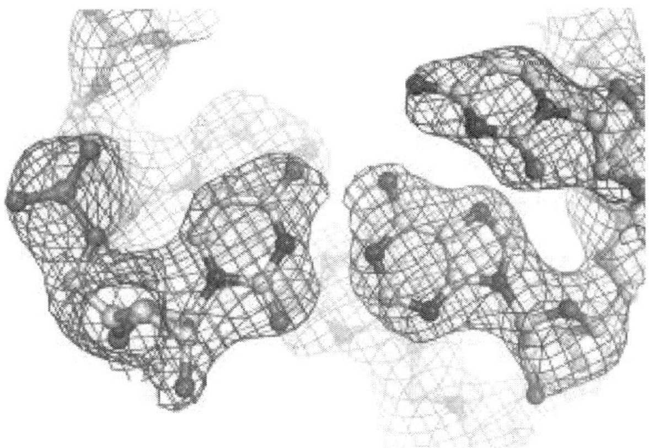

Figure 2

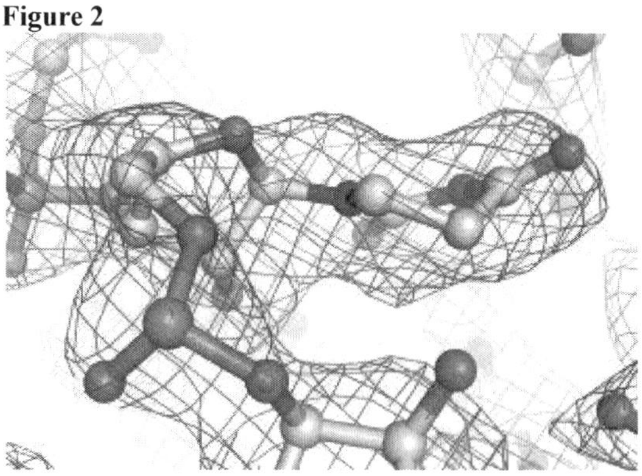

MS07 – Inorganic Structural Chemistry

MS07-T01
Enhanced methods of crystallization: The autothermal reaction, a case study of sodalite formation

J.- C. Buhl[1]
[1]*Institut für Mineralogie, Leibniz Universität Hannover, Kristallographie, Hannover, Germany*

The present investigation is a case study of autothermal synthesis of sodalites as a new model for enhanced crystallization methods saving any external heating. The novelty thereby is an allocation of the whole process energy by the enthalpy of a tailored reaction system itself. The heat of the exothermic reaction of aluminum in aqueous alkaline solution was revealed as the only thermal energy source of the crystallization process.

In [1-2] it was shown that synthesis temperature and reaction time both can be remarkably reduced under insertion of high Na_2O concentrations of 250 g/l for zeolite synthesis from aluminosilicate gels. A substantial progress would be further reached, if the thermal energy for those synthesis procedures could be produced chemically within the reaction batch itself as an autothermal process. The development of suitable autothermal conditions efforts exothermic reactions with high degree of energy transfer of the reaction enthalpy. In this manner the hydrolysis of Al (ΔH= -277 kJ/mol H_2; ΔS= 26.2 J/K und ΔG= -284 KJ [3]) was taken into account and the idea of the present study was to exploit the aqueous alkaline conditions necessary for zeolite synthesis twice, namely: (1) as reaction component and (2) as energy source for heat production to run the whole synthesis process. As the aluminate component of zeolite gel formation is usually $NaAlO_2$ in aqueous solution, this aluminate was therefore substituted by alumina swarf and sodium hydroxide and the reaction starts as follows:

$Al + NaOH + H_2O \rightarrow NaAlO_2 + 1{,}5H_2\uparrow$ [4].

The autothermal process was tested using the hydrosodalite, the tetrahydroborate and the nitrate systems as model cases for sodalites with interesting properties. It could be shown that 8-16 M NaOH, aluminum, sodium silicate solution and the template salt ($NaBH_4$ or $NaNO_3$) yield to the products after only 2-4 hours reaction time. A thermos beaker with sufficient thermal isolation was therefore created to prevent energy loss by heat conduction and minimized convection and heat radiation. After synthesis and washing all products were only dried on air over night to prevent any effect of drying in a cabinet dryer at elevated temperature. Analyses results according to SEM and FTIR spectroscopy are given in Fig. 1.

The experimental investigation offers simple possibilities for creation of enhanced less energy consuming and less cost-intensive syntheses procedures.

Fig. 1: SEM photographs and FTIR spectroscopy of autothermal synthesized sodalites

References:
[1] Hadan, M., Fischer, F.: Synthesis of fine grained NaA-type Zeolites from superalkaline solutions. Cryst. Res. Technol. 27 (1992) 343 - 350.
[2] Fischer, F., Hadan, M., Horn, A.: Investigations to the synthesis of zeolite Na A for using in detergents from superalkaline solutions. Chem. Tech. 43 (1991) 191 - 195.
[3] Petrovic, J., Thomas, G.: Reaction of Aluminum with Water to Produce Hydrogen". U.S. Dept. of Energy (2008) 1-26.
[4] Belitskus, D.: Reaction of Aluminum with sodium hydroxide solution as a source of hydrogen. J. Electrochem. Soc. 117 (1970) 1097-1099.

Figure 1

MS07-T02
On the Concept of Hemilability: Structural Insights into a Donor-Functionalized Iridium(I) NHC Motif and its Impact on Reactivity

A. Pöthig[1]
[1]*Technische Universität München, Catalysis Research Center, Garching b. München, Germany*

Although hemilabile coordination of attached donor ligands is considered beneficial for molecular catalysts, no detailed study of this phenomenon in these systems has been reported earlier. The presented report provides insight into the hemilabile bonding properties of an *N,N'*-bispyridyl-imidazolylidene (NCN) ligand motif on iridium(I), mainly based on detailed structural studies by means of single-crystal XRD. In most cases, the presented compounds exhibit rare fluxional hemilabile coordination of the *N*-donor and a remarkable performance in catalytic transfer hydrogenation is observed.[1]

Figure 1. X-ray crystallographic snapshot of Bis(pyrid-2-yl)-NHC ligands in the mono- (left) and bidentate coordination mode in Ir(I) complexes.

[1] Korbinian Riener , Mario J. Bitzer , Alexander Pöthig*, Andreas Raba , Mirza Cokoja , Wolfgang A. Herrmann , and Fritz E. Kühn*, *Inorg. Chem.*, **2014**, asap. **DOI:** 10.1021/ic5016324

Figure 1

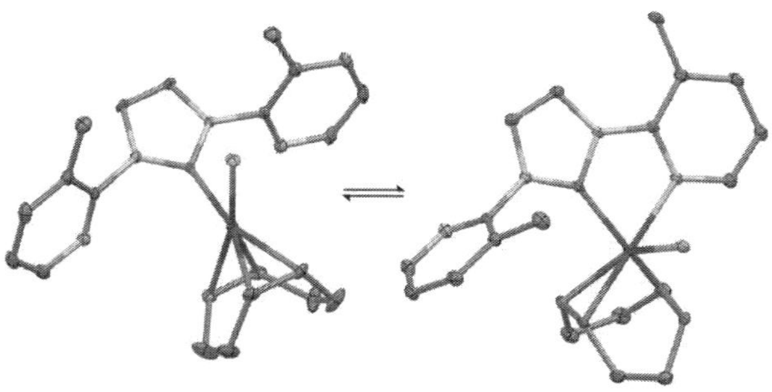

MS07-T03
On the ambient pressure polymorph of $K_2Ca_3Si_3O_{10}$ - structural and spectroscopic characterization of an unusual mixed-anion silicate

D. Schmidmair[1], V. Kahlenberg[1], L. Perfler[1], M. Tribus[1], J. Hildebrandt[1], D. M. Többens[2]
[1]*University of Innsbruck, Institute of Mineralogy and Petrography, Innsbruck, Austria*
[2]*Helmholtz-Zentrum Berlin für Materialien und Energie GmbH, Department of Crystallography, Berlin, Germany*

As a part of an ongoing research project on the phases belonging to the ternary system K_2O-CaO-SiO_2, an ambient pressure polymorph of $K_2Ca_3Si_3O_{10}$ has been synthesized via solid state reactions. Single crystal X-ray diffraction experiments show, that this new modification crystallizes in the triclinic space group *P*-1 with the following lattice parameters: a = 5.6699(6) Å, b = 7.3754(12) Å, c = 11.8310(13) Å, α = 86.199(11)°, β = 80.625(9)°, γ = 88.700(11)°. A special feature of the structure is the coexistence of two different types of silicate anions. Isolated $[SiO_4]$- tetrahedra as well as $[Si_4O_{12}]$- *vierer* single rings occur in the ratio 2:1, resulting in the crystallochemical formula $K_4Ca_6[SiO_4]_2[Si_4O_{12}]$. To the best of our knowledge, this is the first example of an oxo-silicate where insular and cyclic silicate anions appear concomitantly. Both building units are arranged in layers parallel to (100). Charge compensation is provided by Ca and K cations. All calcium atoms are coordinated by 6 oxygen atoms, forming distorted octahedra. By sharing common corners, edges and faces, these $[CaO_6]$- polyhedra build up octahedral layer-like motifs parallel to (010). Potassium ions are located in voids between the silicate anions and $[CaO_6]$- octahedra and are coordinated by 8-9 oxygen atoms.

Further characterization of this new compound was carried out by electron microprobe analysis and Raman spectroscopy. DFT calculations were employed (i) to assign Raman bands to certain vibrational modes and (ii) to determine the relative stabilities of the monoclinic high pressure and the triclinic ambient pressure polymorph of $K_2Ca_3Si_3O_{10}$.

MS07-T04
DSR- A tool for disorder modelling and refinement with SHELXL

D. Kratzert[1], I. Krossing[1]
[1]*Albert-Ludwigs-Universität Freiburg, Freiburg im Breisgau, Germany*

X-ray crystallography as a method used for identifying the atomic and molecular structure of a crystal has led to a better understanding of chemical bonds. It is an essential tool to determine the absolute configuration of molecules, has been important for the characterization of coordination complexes, as well as identifying supramolecular assemblies in biology and material science.

Being so successful, one of the remaining problems in practical crystallography is the description of disorder in crystal structures. The Cambridge Structural Database includes 23% of disordered structures in its collection of nearly 700000.[1] The program described here is able to simplify many aspects of disorder modelling.

SHELXL is able to treat almost every possible kind of disorder but with a lot of manual work. For this purpose, a new program called DSR (Disordered Structure Refinement) was developed. It needs a free variable and a part number in combination of displacement parameters and bond length restraints. DSR transfers a molecular fragment from a database of molecular fragments to the desired position in the unit cell automatically and generates restraints to stabilize the model (Figure 1).

In practice, the user writes a command line into the SHELXL .res file, which subsequently is interpreted by DSR. The special commands main purpose is to tell DSR, how to orient a molecular fragment from the database in the unit cell. Further opions define the occupancy, residue or part number of the fitting fragment. Molecular fragments can be either imported directly from the GRADE server of Globalphasing Ltd.[4], from existing crystal structures or from ab initio calculations. DSR offers several more options available to make disorder modelling a convenient process. DSR can be obtained from http://goo.gl/BL6wP1.

References

[1]http://www.ccdc.cam.ac.uk/Lists/ResourceFileList/2014_stats_entries.pdf
[2] G. M. Sheldrick, *Acta Crystallogr.* **2008**, *A64*, 112-122.
[3] Grade Web Server, Global Phasing Ltd. http://grade.globalphasing.org.

Figure 1

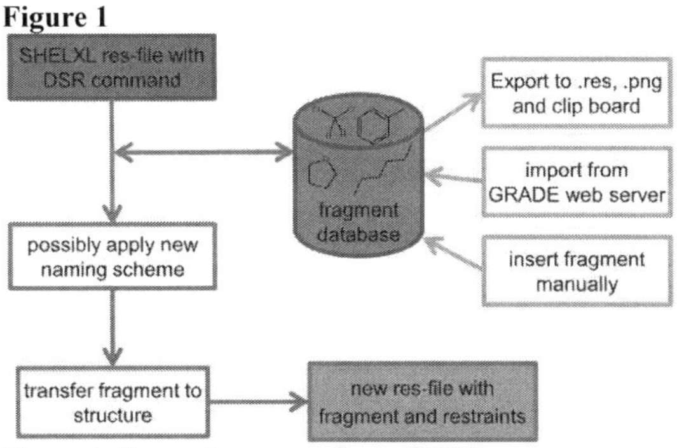

Figure 1. General program flow of DSR.

MS07-T05
The Next Level of Single Crystal Software - APEX3

H. Ott[1], C. Campana[2], J. Kaercher[3], B. Noll[2], M. Ruf[3]
[1]*Bruker AXS GmbH, Application, Karlsruhe, Germany*
[2]*Bruker AXS Inc., Application, Madison, United States*
[3]*Bruker AXS Inc., Madison, United States*

In 2004, the APEX2 single crystal suite was first launched and deemed a huge leap forward in terms of functionality and design, in comparison to previously available software. This suite allowed for the complete crystal structure determination process, from crystal quality check to the finalization of a publishable structure, to be carried out within the one program. Since its launch, the performance of the APEX2 suite has been continuously enhanced by the addition of various features. Today it is one of the most popular software suites used in chemical crystallography.

Now, the most extensive revision is available. The new version takes full advantage of important developments in computer hardware and operating systems. Improvements to the new package include a state-of-the-art graphical user interface, the modern QT4 programming environment and multi-CPU support for faster data processing, structure solution and publication.

The updated AUTOSTRUCTURE plug-in makes full use of the revolutionary intrinsic phasing structure solution engine increasing the out-of-the-box success rate of structure determinations far beyond 90 %. The new Structure Determination plug-in incorporates various structure solution modules, while ShelXle is incorporated as the new default option for fast and convenient structure refinement. Moreover, the efficient twin handling routines known from the APEX2 suite are now fully integrated into the GUI for a new level of seamless twin support.

MS08 - Material Science III

MS08-T01

Controlling defects in lithium aluminium intermetallics by hydrogenation

H. Kohlmann[1], C. Reichert[2]
[1] *University Leipzig, Leipzig, Germany*
[2] *Saarland University, Saarbrücken, Germany*

Lithium aluminium hydride, $LiAlH_4$, is well-known in preparative chemistry as hydriding and reducing agent. It has also been discussed as potential hydrogen storage material and decomposes via Li_3AlH_6 and Al to LiH and Al under hydrogen release [1]. The direct hydrogenation of lithium aluminium compounds was not reported, however, as yet. We report on the hydrogenation of Li_9Al_4, Li_2Al, Li_3Al_2 and LiAl, which were synthesized from the elements in welded niobium tubes at temperature of 723 K. The hydrogenation was studied by *in situ* thermal analysis at hydrogen pressures up to 5.0 MPa. Exothermic signals around 700 K for LiAl and around 400 K for the lithium richer compounds indicate a reaction with hydrogen. *Ex situ* X-ray and *in situ* neutron powder diffraction using a sapphire gas pressure cell [2] reveal the formation of lithium hydride, LiH, and intermetallic phases richer in aluminium in several steps, according to a simplified reaction pathway 2 Li_9Al_4 + 5 H_2 = 8 Li_2Al + 2 LiH + 4 H_2 = 4 Li_3Al_2 + 6 LiH + 2 H_2 = 8 LiAl + 10 LiH [3]. Interestingly, each of the phases shows a decrease in the unit cell volume during hydrogenation up to 1.4%, which can be interpreted as leaching out of lithium by the formation of lithium hydride. For LiAl, a detailed reaction pathway was discovered by Rietveld analysis based on *in situ* neutron powder diffraction data. Starting with a lithium rich $Li_{1+x}Al_{1-x}$ first lithium on aluminium sites reacts to lithium deuteride Li_{Al} + ½ D_2 = LiD + V_{Al} and in a second step lithium on lithium positions reacts according to Li_{Li} + ½ D_2 = LiD + V_{Li}, resulting in a lithium poor, defect rich compound $Li_{1-y}Al_{1-x}$ [3]. Hydrogen is thus a versatile tool to tailor point defects in lithium aluminium intermetallics. Thus might be of interest for potential applications as anode material in rechargeable batteries.

[1] U. Eberle, M. Felderhoff, F. Schüth, *Angew. Chem.* 2009, *121*, 6732-6757
[2] H. Kohlmann, N. Kurtzemann, T. C. Hansen, *Powder Diffr.* 2013, *28*, S242-S255
[3] C. Reichert, Ph D thesis, Saarland University, Saarbrücken, Germany, 2013

MS08-T02

Investigation of sodium borohydride $NaBH_4$ in different halide matrices (NaX): hydrogen release reactions and new compositions of the type $Na[BH_4]_xX_{(1-x)}$, (X= Cl, Br, I)

Z. Assi[1], C. H. Rüscher[1]
[1] *Institut für Mineralogie, Leibniz Universität, Hannover, Germany*

Sodium borohydride $NaBH_4$ is a promising material for hydrogen storage application. For some reasons we have been interested in the reaction of $NaBH_4$ in halide matrices under more or less dry conditions at temperatures between 20 and 450°C.

One aspect of these studies was to follow the effect of hydrogen release in temperature dependent IR (TIR) absorption spectroscopy. Using NaCl pressed pellet (1 mg $NaBH_4$ in 200 mg NaCl) the hydrogen release could be traced with $NaNO_3$ addition in the pellet . Active hydrogen release could be observed by a reduction of $NaNO_3$ to $NaNO_2$ starting smoothly above 200°C and strongly intensifies above 300°C. Above 300°C the crystallization of $Na_3B_3O_6$ was identified.

Another aspect was the observation of anion exchange between $NaBH_4$ and NaCl. The IR spectrum of $NaBH_4$ in NaCl of as received pressed pellet show peaks at 1121, 2229, 2303 and 2400 cm^{-1} assigned respectively to *v4*, *2v4*, *v3* and *v2+v4* of the $[BH_4]$-unit. TIR spectra showed between 300° and 450°C a shift to higher wavenumbers up to 32 cm^{-1} for *v4*, 74 cm^{-1} for *2v4*, 72 cm^{-1} for *v3* and 78 cm^{-1} for *v2+v4* ***(fig.1)***. The new peak positions are attributed to the incorporation of BH_4 in the NaCl matrix indicating a strong shortening of B-H bonds. For a further study of this effect different mixtures of $NaBH_4$:NaCl (50:50, 20:80, 10:90) were thermally treated under N_2 to different temperatures (300°- 400°C) for heating times varying between 20 and 48h. The end product of each isothermal was then characterized by XRD, IR and Raman methods. XRD analyses indicated the formation of two compositions $Na[BH_4]_xCl_{(1-x)}$ with x ≈ 0.1 and 0.9, principally unmodified NaCl, and some borates by-products ***(fig.2)***. IR and Raman spectra show the presence of two types of BH_4 units, too. While a substitution of 10% BH_4^- by Cl^- in $NaBH_4$ does not affect significantly the initial geometry of BH_4^-, a substitution of 10% Cl^- by BH_4^- in NaCl increases strongly the corresponding IR frequencies. Previous work[1] suggested an homogeneity range that obey to Vegard´s law between $NaBH_4$ and NaCl.

Some preliminary experiments were conducted with $NaBH_4$ (1mg) diluted in other sodium halides matrices (NaBr and NaI, 200mg).IR results of $NaBH_4$ / NaBr are 1138, 2252, 2326 and 2426 cm^{-1}; and of $NaBH_4$/NaI: 1112, 2212, 2281 and 2364 cm-1. These peak positions suggest an incorporation of NaBH4 in NaBr and NaI with estimated compositions of $NaBH4_{0.005}Br_{0.995}$ and $NaBH4_{0.005}I_{0.995}$. In these two cases, dissolution of BH_4^- was achieved without any further thermal treatment as in the case of NaCl. The B-H bonding length becomes significantly less compressed for BH_4^- diluted in NaBr compared to BH_4^- in NaCl. For BH_4^- incorporated in NaI the B-H bonding length is elongated. This is understood comparing the unit cell parameters of NaCl: a_0= 5.64 Å, NaBr: a_0= 5.97 Å, NaI: a_0= 6.47 Å with respect to NaBH4: a_0= 6.17 Å.

In conclusion, complexes of the type $Na[BH_4]_xX_{(1-x)}$ could be synthesized starting from mixture of $NaBH_4$ and NaX. IR frequencies were found to be dependent of B-H distances and were assumed to vary linearly with x.

[1] J.E. Olsen et al., *J. Alloy Compd*, 509, L228-31 (2011)

Figure 1

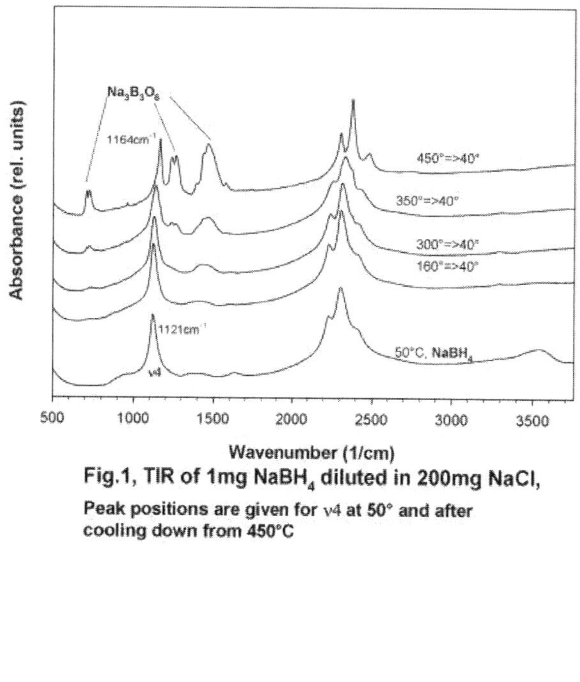

Fig.1, TIR of 1mg $NaBH_4$ diluted in 200mg NaCl,

Peak positions are given for v4 at 50° and after cooling down from 450°C

Figure 2

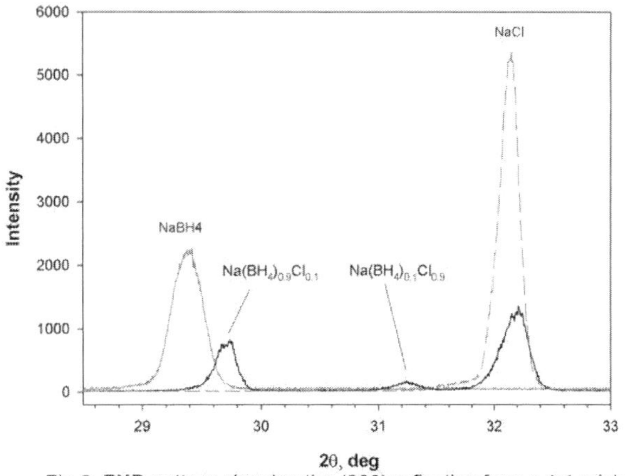

Fig.2, PXD pattern showing the (200) reflection from a 1:1 mixture of NaBH4 and NaCl after an isothermal at 350°C for 20h.
gray dashed line is pure NaCl, gray line is pure NaBH4

MS08-T03

A new crystal structure type for A-site ordered quadruple perovskites

M. Etter[1], M. Isobe[2], H. Sakurai[3], R. E. Dinnebier[1], H. Takagi[2]
[1]Max Planck Institute for Solid State Research, Röntgenographie, Stuttgart, Germany
[2]Max Planck Institute for Solid State Research, Quantum Materials, Stuttgart, Germany
[3]National Institute for Materials Science, Tsukuba, Ibaraki, Japan

A-site ordered quadruple perovskites with chemical formula $AA'_3B_4O_{12}$ (with A', B = metallic ions) recently attracted a lot of attention in various research fields as these perovskites exhibit a rich field of different temperature dependent electronic phenomena like inter-site charge transfer (ISCT), charge disproportionation (CD) and charge ordering. The investigation of quadruple perovskites with Cu on the A'-site and Fe on the B-site proofed, that the occurrence of these phenomena is usually accompanied by magnetic and/or structural phase transitions[1]. For this special class of compounds it could be shown that depending on the lanthanide cation on the A-site either an ISCT or a CD phenomenon can be found[1].

For most of the investigated perovskite compounds the ISCT is accompanied by metal-to-insulator, antiferromagnetic and isostructural phase transitions with negative-thermal-expansion-like volume changes[1], whereas the CD is accompanied by metal-to-semiconductor, ferrimagnetic and structural phase transitions without volume changes and rock-salt ordering of the B-site oxidation states[1]. Recently a similar ISCT was found for quadruple perovskites with Cr ions on the B-site[2]. Interestingly the ISCT effect in this compound shows a positive-thermal-expansion-like volume change at the transition point rather than a negative-thermal-expansion-like volume change as for the corresponding iron analogue. This behavior suggests that the Cr containing quadruple perovskites require a different theoretical description as it was given for the iron compounds[3].

A rather interesting candidate to investigate the behavior of Cr containing quadruple perovskites is $BiCu_3Cr_4O_{12}$ as it shows a phase transition to a new monoclinic crystal structure type (see figure 1) accompanied by a possible CD phenomena which is not fully understood yet.

Fig 1: Monoclinic crystal structure of $BiCu_3Cr_4O_{12}$ at low temperatures. (a) View along b-axis. (b) View along the [101] direction.

References:

[1] I. Yamada, H. Etani, K. Tsuchida, S. Marukawa, N. Hayashi, T. Kawakami, M. Mizumaki, K. Ohgushi, Y. Kusano. J. Kim, N. Tsuji, R. Takahashi, N. Nishiyama, T. Inoue, T. Irifune, M. Takano, Inorg. Chem. 52, 13751 (2013).
[2] S. Zhang, T. Saito, M. Mizumaki,Y. Shimakawa, Chem. Eur. J. 20, 1 (2014).
[3] N. Rezaei, P. Hansmann, M.S. Bahramy, R. Arita, Phys. Rev. B 89, 125125 (2014).

Figure 1

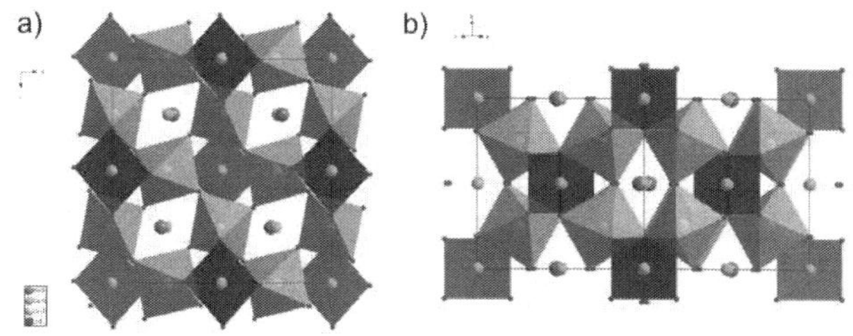

MS08-T04

Formation and properties of Ice XVI - the lowest density crystalline form of ice produced by empting sII Ne hydrate

A. Falenty[1], T. C. Hansen[2], W. F. Kuhs[1]
[1]Uni. Goettingen, Germany, Kristallographie, Göttingen, Germany
[2]Institut Laue-Langevin, Grenoble, France

Formation and properties of Ice XVI - the lowest density crystalline form of ice produced by empting sII Ne hydrate

FALENTY Andrzej[1], HANSEN Thomas C[2], KUHS Werner F[1]

1. GZG Abt.Kristallographie, Universität Göttingen, Goldschmidtstr.1, 37077 Göttingen, GERMANY, afalent@uni-goettingen.de,wkuhs1@gwdg.de; 2. Institut Laue-Langevin, B.P.156X, 38042 Grenoble Cedex, FRANCE,hansen@ill.fr

The water chemical activity of the empty clathrate structure is the reference frame for the statistical thermodynamic theory of gas hydrates on which our understanding of stability limits and composition of clathrates is based; it is a cornerstone for chemical engineers e.g. in problems related to flow assurance of gas and oil pipelines as well as in geo-engineering. As the open water-host framework is stabilized by guest molecules the empty hydrate lattice was considered experimentally difficult to access and could only be approximately calculated using thermodynamic reasoning. Recently we have managed to overcome this limitation and prepared this phase in a region of thermodynamic metastability by pumping on small particles of sII Ne-clathrate at temperatures of ~ 140 K [1]. The obtained empty water frame is the least dense of all known crystalline phases of water and it is predicted to be one of the stable low-temperature configurations of water at negative pressures [2]. According to the current nomenclature it is called ice XVI. We have established its crystal structure, thermal expansivity and metastability limit by means of neutron powder diffraction. Ice XVI is topological identical to SiO_2-, Si- and Ge-clathrates (Si_{136}, Ge_{136}) and similar open framework structures, including a hypothetical carbon clathrate. The open water framework shows a marked negative thermal expansion at low temperature, similar to ice Ih; this can be attributed to the increased low-energy framework-bending modes for the empty structure. We have also observed a considerable lattice expansion upon gas removal quantifying the importance of attractive interactions of water and small gas molecules in clathrate hydrates and beyond. The structure is stable up to temperatures of ~145K at which it transforms into a stacking-faulty ice I_{ch}[3,4].

[1] Falenty A., Hansen T.C. and Kuhs W.F., Formation and properties of ice XVI obtained by emptying a type sII clathrate hydrate, Nature 2014; 516, 231-233, doi:10.1038/nature14014
[2] Conde M.M, Vega C., Tribello G.A., Slater B. *The phase diagram of water at negative pressures: Virtual ices*, J.Chem.Phys. 2009; 131: 034510.
[3] Kuhs W.F., Sippel C., Falenty, A., Hansen T.C. *Extent and relevance of stacking disorder in "ice Ic"*. PNAS 2012; 109(52): 21259-21264.
[4] Hansen T.C., Sippel C., Kuhs W.F. *Approximations to the full description of stacking disorder in ice I for powder diffraction* Z.Krist. 2014; doi: 10.1515/zkri-2014-1780.

Figure 1

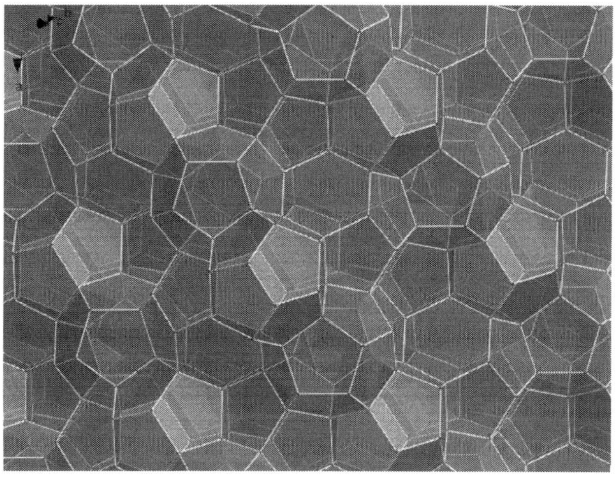

MS08-T05

A Story of Anomalous Thermal Expansion: Nuremberg Scissors, Stacked Columns and Two-Dimensional Networks

T. Runcevski[1], R. E. Dinnebier[1]
[1]*Max Planck Institute for Solid State Research, Suttgart, Germany*

Most solid state materials respond to changes in temperature with expanding their lattices upon heating (positive thermal expansion, PTE) due to population of higher energy levels of anharmonic lattice vibrations. The manifestation of a negative thermal expansion (NTE), that is lattice contraction upon heating over a certain temperature range, is a rare but technologically important phenomenon. Even less frequent is the occurrence of zero thermal expansion (ZTE) among crystalline materials. Successful engineering and manipulation of thermal expansion of solids promises plethora of novel applications. Therefore, the research on crystalline solids exhibiting anisotropic, positive, negative and/or close-to-zero thermal expansions is strongly encouraged, as it provides valuable insights into the underlying physical processes which are giving rise to the anomalous thermal response. We focus our attention to three different systems with unusual behavior of their lattices when subjected to heat. Herein, we report on a positive vs. negative axial thermal expansion of $MgBr_2{\cdot}4H_2O$, rooted in "nuremberg scissors" motions of the magnesium-centered octahedra [1]; a colossal uniaxial positive and biaxial negative thermal expansion of a molecular solid where the molecules are packed stacked in columns [2]; and near-to-zero in-layer expansion vs. colossal positive inter-layer expansion of a two-dimensional extended molecular solid [3]. A comparison between the crystal structures of these materials is made, the reasons for the peculiar thermal expansions are rationalized and a perspective of their applicability is envisaged.

MS09 - Hot new structures of biological macromolecules

MS09-T01
Insights into dynein motor function from crystal structures in two different nucleotide states

H. Schmidt[1], E. S. Gleave[1], R. Zalyte[1], L. Urnavicius[1], A. P. Carter[1]
[1] *MRC Laboratory of Molecular Biology, Structural Studies, Cambridge, United Kingdom*

Dynein motor proteins generate movement along microtubules and can be classified into several isoforms. Cytoplasmic dynein-1 (dynein-1) carries out most of the microtubule minus end directed transport in cells and moves cargoes including mitochondria, nuclei, protein and mRNA complexes [1,2]. It also plays important roles during mitosis [2] and defects in dynein-1 function are implicated in several neurodegenerative diseases [3]. Cytoplasmic dynein-2 (dynein-2) participates in intraflagellar transport in the primary cilium [4], a cellular structure often described as the cells „antenna". Mutations in dynein-2 lead to degenerated primary cilia and are associated with skeletal ciliopathies [5].

All dyneins exist as multi protein complexes with molecular weights of around 1.4 MDa. They contain a ~3000 amino-acid residue motor domain consisting of a ring of six AAA+ domains (ATPases Associated with diverse cellular Activities), the linker and an elongated coiled-coil helix (~150 Å) with the microtubule binding domain (MTBD) at its tip [2]. ATP hydrolysis causes the linker to switch between a post- and pre-powerstroke conformation to produce the necessary force for movement. The linker swing is also synchronized with cycles of microtubule binding/release in the MTBD, another important prerequisite for movement along the microtubule. Previously, it was unknown how ATP hydrolysis causes linker remodelling and how this remodelling is correlated with microtubule binding/release.

Here we present two dynein motor crystal structures: dynein-1 from *Saccharomyces cerevisiae* in the apo state [6] and dynein-2 from *Homo sapiens* in complex with ADP.vanadate [7], an ATP hydrolysis transition state analogue. These two structures reveal that ATP hydrolysis causes a transition in the AAA+ ring from an open, extended conformation to a closed, compact one. The linker switches from the post- to the pre-powerstroke conformation by a rigid-body movement of its N-terminal domain, which is induced by a steric clash with the closed AAA+ ring. AAA+ ring closure also induces a sliding movement of one of the coiled-coil helices that causes the MTBD to release from the microtubule. The transition of the AAA+ ring from open to closed is therefore crucial for the coordination of linker swing and microtubule binding/release.

References:
1. Allan VJ. 2011. *Biochem Soc Trans* 39: 1169-1178.
2. Roberts AJ, Kon T, Knight PJ, Sutoh K, et al. 2013. *Nat Rev Mol Cell Biol* 14: 713-726.
3. Chen XJ, Xu H, Cooper HM & Liu Y. 2014. *Sci China Life Sci* 57: 372-377.
4. Ishikawa H & Marshall WF. 2011. *Nat Rev Mol Cell Biol* 12: 222-234.
5. Schmidts M, Arts HH, Bongers EM, Yap Z, et al. 2013. *J Med Genet* 50: 309-323.
6. Schmidt H, Gleave ES & Carter AP. 2012. *Nat Struct Mol Biol* 19: 492-497.
7. Schmidt H, Zalyte R, Urnavicius L & Carter AP. 2014. Nature, accepted October 29th

MS09-T02
Crystal structure of the dynamin tetramer

T. F. Reubold[1], K. Faelber[2], N. Plattner[3], Y. Posor[4], K. Branz[4], U. Curth[1], J. Schlegel[2], R. Anand[1], D. J. Manstein[1], F. Noé[3], V. Haucke[4,5], O. Daumke[2,5], S. Eschenburg[1]
[1] *Medizinische Hochschule Hannover, Institute for Biophysical Chemistry, Hannover, Germany*
[2] *Max-Delbrück-Centrum for Molecular Medicine, Crystallography, Berlin, Germany*
[3] *Freie Universität Berlin, Institute for Mathematics, Berlin, Germany*
[4] *Leibniz-Institute for Molecular Pharmacology, Berlin, Germany*
[5] *Freie Universität Berlin, Institute for Chemistry and Biochemistry, Berlin, Germany*

Dynamin is the prototype of the dynamin superfamily of large GTPases. The mechano-chemical protein forms tetramers in the cytosol, which oligomerize at the neck of clathrin-coated vesicles to mediate membrane scission. Previous studies have described the architecture of dynamin dimers, but the molecular determinants for dynamin assembly and its regulation remained unclear. Here, we present the crystal structure of the dynamin 3 tetramer. Combining the structural data with mutational studies, oligomerization measurements and molecular dynamics simulations, we suggest a mechanism of how oligomerization of dynamin is linked to the release of intramolecular auto-inhibitory interactions. Moreover, we elucidate how mutations that interfere with auto-inhibition may lead to centronuclear myopathy, a congenital muscle disease. Strikingly, the bent shape of the tetramer explains dynamin's ability to assemble into a right-handed helical oligomer of defined diameter, which has direct implications for its mechanism in membrane scission.

MS09-T03
Extreme longevity of a diiron(III)-peroxo intermediate in DOHH, an oxygenase involved in hypusination

Z. Han[1], N. Sakai[1], L. Böttger[2], S. Klinke[1], A. X. Trautwein[2], R. Hilgenfeld[1]
[1] *University of Lübeck, Institute of Biochemistry, Lübeck, Germany*
[2] *University of Lübeck, Institute of Physics, Lübeck, Germany*

Deoxyhypusine hydroxylase (DOHH) is a non-heme diiron enzyme involved in the posttranslational modification of a critical lysine residue of eukaryotic translation initiation factor 5A (eIF-5A), to yield the unusual amino-acid residue hypusine. This modification is essential for the role of eIF-5A in translation and nuclear export of a group of specific mRNAs. Human DOHH (hDOHH) is a potential drug target for the treatment of HIV/AIDS, chronic myeloid leukemia, and diabetes. The diiron center of hDOHH forms an intermediate species, peroxo-diiron(III), when its reduced form is activated by exogenous O_2. The peroxo-diiron(III) intermediate in hDOHH has a lifetime exceeding that of other diiron enzymes by several orders of magnitude. Here we report the crystal structure of hDOHH, determined in two forms, the peroxo-diiron(III) intermediate and a complex with the product analogue glycerol, both of them at 1.7 Å. The structure of the intermediate reveals the presence of a μ-1,2-peroxo-diiron(III) species at the active site. The crystal structures offer explanations for the extreme longevity of the peroxo-diiron(III) intermediate in hDOHH and illustrate how the enzyme specifically recognizes its only substrate, deoxyhypusine-eIF-5A.

MS09-T04

Structural basis for organohalide respiration

M. Bommer[1], C. Kunze[1,2], J. Fesseler[1], T. Schubert[2], G. Diekert[2], H. Dobbek[1]
[1]*Humboldt Universitaet Berlin, Strukturbiologie, Berlin, Germany*
[2]*Universitaet Jena, Jena, Germany*

Surviving in the absence of oxygen, organohalide respiring microorganisms can use a variety of persistent pollutants including trichloroethene (TCE) as terminal electron acceptors. The final two-electron transfer step in organohalide respiration is catalyzed by reductive dehalogenases, a class of enzymes, whose evolution, mechanism and substrate selectivity is scarcely understood. We present the X-ray crystal structures of PceA, an archetypal dehalogenase from Sulfurospirillum multivorans, in complex with TCE, product analogs and alone. PceA's active site harbors a deeply buried norpseudo-B_{12} cofactor within a nitroreductase-fold, also found in a mammalian B_{12} chaperone. Norpseudo-B_{12} is flanked by a succession of two [4Fe-4S] clusters. The structures of PceA reveal how a cobalamin supports reductive haloelimination exploiting a conserved B_{12}-binding scaffold capped by a highly variable substrate-capturing region.

Figure 1

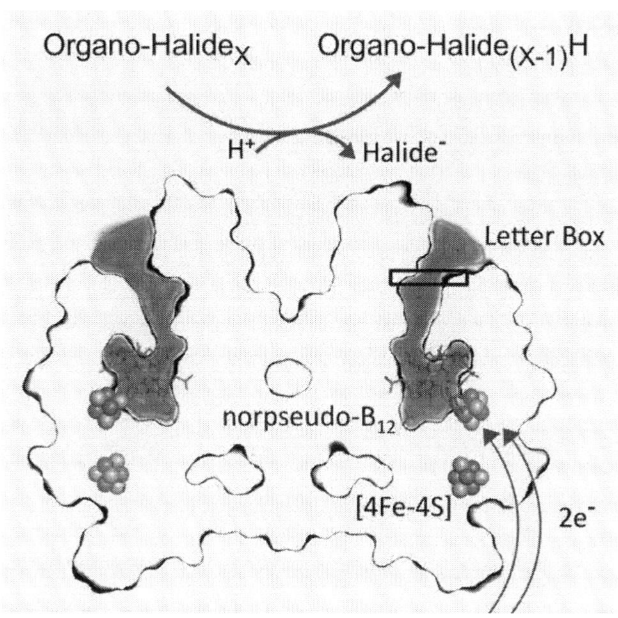

MS09-T05

Crystallographic study of 2'-5'-oligoadenylate synthase provides new insights into the function and activation of OAS/cGAS innate immune sensors

J. Lohöfener[1], N. Steinke[1,2], P. Kay-Fedorov[3], P. Baruch[4], A. Nikulin[5], S. Tishchenko[5], D. J. Manstein[1,4], R. Fedorov[1,4]
[1]*Hannover Medical School, Institute for Biophysical Chemistry, Hannover, Germany*
[2]*University of Oxford, Department of Biochemistry, Oxford, United Kingdom*
[3]*Hannover Medical School, Institute for Virology, Hannover, Germany*
[4]*Hannover Medical School, Research Division for Structural Analysis, Hannover, Germany*
[5]*Hannover Medical School, Institute of Protein Research, Hannover, Germany*

2'-5'-Oligoadenylate synthases (OASs) trigger intrinsic antiviral signaling in the interferon-inducible OAS/RNase L pathway. Upon activation by virus-derived dsRNA, OAS produces the second messenger 2'-5'-oligoadenylate (25A). 25A activates RNase L, which in turn cleaves RNA in the cells inhibiting viral replication. Despite significant progress in structural and mechanistic studies of the OAS/RNase L pathway, the molecular basis of OAS activation is still poorly understood. In this study we determined crystal structures of the fully-resolved inactive apo-form of OAS1, the OAS1 in complex with dsRNA without substrates, with dsRNA and the AMP-donor substrate, and with dsRNA and both the AMP-donor and the AMP-acceptor substrates (pre-reactive state). These structures in combination with biochemical studies specified the individual effects of dsRNA, each substrate and two Mg-cofactors on OAS1 conformation and indicated that OAS activation follows a sequential mechanism. Our structural analysis and Molecular Dynamics simulations revealed the allosteric mechanism of propagating the effect of dsRNA binding through the structure of OAS and identified the functional role of Mg-coordination and the origin of the 2'-specificity of the enzyme. The new structural data allowed a detailed comparison of the activation mechanisms in OAS/cGAS family of innate immune sensors and identified structural feature which leads to differentiation of product classes synthesized by OAS and cGAS enzymes.

MS10 - Computational & Theoretical Crystallography: Diffraction versus wave function?

MS10-T01
Experimental Charge Density Studies: Improvements in Data Processing and Refinement

R. Herbst-Irmer[1], B. Niepötter[1], L. Krause[1], F. Engelhardt[1], C. J. Schürmann[1], D. Stalke[1]
[1]*University of Göttingen, Institute of Inorganic Chemistry, Göttingen, Germany*

In experimental charge density investigation it is indispensable to use the highest possible quality of data. Therefore, the multiplicity should be as high as possible but poor data should be omitted. To decide about discarding outlier data the 'paired refinement method' [1] can be used. Models for several data sets are derived by the same refinement protocol. The data sets are compared by the fit of these models to all data sets.

The paired refinement strategy can also be used to investigate the influence of different scaling methods. In a recent version of SADABS [2] a new error model and a 3λ correction is implemented. With the paired refinement strategy the improvement in data quality gets obvious.

In several charge density studies we noticed that the residual density improved significantly after ten resolution dependent scale factors were refined. The dependency of the scale factors against resolution showed a typical pattern for all data sets, but the variation was much smaller for data collected at 15 K instead of 100 K. So this procedure seems to cure errors that are resolution and temperature dependent. Similar results could be obtained by processing data with relatively small integration boxes instead of refined box sizes. This procedure seems to emulate a rough TDS correction and can be further developed.

A further concern in charge density investigation is the question of overfitting. In macromolecular refinement this is answered by the R_{free} concept [3]. Here a refinement protocol is developed by refining against a work set of reflections, e.g. 90 % of the data. The remaining reflections are not used during the whole refinement process. An R_{free} value is calculated using only this test set of reflections. Overfitting can clearly be identified by a decrease in R_{work} but an increase in R_{free}. An optimized refinement protocol is then used for a final refinement against all data. It will be discussed how this method could support charge density studies.

[1] P. A. Karplus, K. Diederichs, *Science* 2012, *336*, 1030-1032.
[2] L. Krause, R. Herbst-Irmer, G. M. Sheldrick, D. Stalke, *J. Appl, Crystallogr.* 2015, *48*, in print.
[3] A. T. Brünger, *Nature* 1992, *355*, 472-475.

MS10-T02
"EXAFS in 3 Dimensions" with Single-Crystal X-Ray Diffraction on a Zirconium Compound

B. Dittrich[1], P. P. Samuel[2]
[1]*Universitaet Hamburg, Institut f. Anorganische und Angewandte Chemie, Hamburg, Germany*
[2]*Georg-August-Universität Göttingen, Institut fuer Anorganische Chemie, Göttingen, Germany*

We have recently synthesized and characterized a new zirconium coordination compound. Crystals of high quality - disorder is absent - were grown. We have carried out a series of single-crystal diffraction experiments to study the effect of anomalous dispersion around the K alpha edge with tunable wavelengths at the SLS synchrotron on these crystals and observe an ionization of the Zr atom. The energy around the absorption edge corresponds to a wavelength that is high enough to measure diffraction data to rather high resolution. This in turn allows to observe difference electron-density maps on a grid, using the tools developed for charge density studies. These maps show electron-density rearrangements around the zirconium atom in three dimensions and in full detail. Results are in some way analogous to EXAFS studies, but provide unprecendented detail. An important aspect in data analysis is that only accurate model phases do allow further detailed visualization of relevant remaining un-modeled electron density; features that remain after conventional independent-atom model refinement, for example using SHELXL, do not permit this. Only structure refinement with an aspherical scattering factor model [1] minimizes residual electron density enough. It is therefore mandatory for successful analysis. We have reported very recently how such refinements [2] can be carried out in a computationally efficient manner by starting from tabulated invariom [3] scattering factors using bond distances to hydrogen atoms contained in the invariom database. Comparison of the experimental results to those of theoretical computations is crucial and is currently being carried out. We are convinced to have a new analytical tool at hand that is useful and widely applicable, e.g. to small-molecule biomimetics of metal-containing proteins or to "bioinorganic" macromolecules.

References:
[1] N. Hansen, P. Coppens, Acta Crystallogr A, 1978, 34, 909
[2] B. Dittrich, C. M. Wandtke, A. Meents, K. Pröpper, K. C. Mondal, P. P. Samuel, N. A. Sk, A. P. Singh, H. W. Roesky, N.Sidhu, ChemPhysChem., 2014, 15, DOI: 10.1002/cphc.201402600R2,
[3] B. Dittrich, C. B. Hübschle, K. Pröpper, F. Dietrich, T. Stolper, J. J. Holstein, The generalized invariom database (GID), Acta Crystallogr B, 2013, 69, 91

MS10-T03
Diffraction AND wavefunction - How the use of a wavefunction allows to get more out of experimental diffraction data

S. Grabowsky[1], M. Woinska[2], D. Jayatilaka[3]
[1]*Universität Bremen, Fachbereich 2 - Biologie/Chemie, Bremen, Germany*
[2]*University of Warsaw, Warsaw, Poland*
[3]*The University of Western Australia, Perth, Australia*

In a single-crystal X-ray diffraction experiment, the observed diffraction pattern contains information about the total electron-density distribution, including the valence electrons. Since this information is what lies at the heart of chemistry, it is important to extract and interpret it. This has so far been done by extending the commonly used Independent Atom Model (IAM), which only uses spherically averaged atomic electron densities, with multipoles, which describe pseudoatoms of aspherical shape. The Hansen-Coppens formalism, which employs single-zeta atomic radial functions, is the most frequently used multipole model.[1] Various chemical applications of multipole modeling have shown the importance of the electron-density analysis for chemistry.[2]

For a better electron-density modeling and hence a better reconstruction of the observed diffraction pattern, the atomic basis sets must be improved. This can be done by using standard wavefunctions at any level of theory for the reconstruction of the experiment. It is not at all a contradiction to use more sophisticated theory in order to extract more information from the experiment. The opposite is true, the crystallographic model will be improved significantly. Whereas for multipole modeling high-resolution data are needed, the below introduced wavefunction method can work with routine low-order data sets.

The method introduced and applied in this contribution is termed X-ray wavefunction refinement (XWR). It consists of two related steps - Hirshfeld atom refinement (HAR, [3]) and X-ray constrained wavefunction fitting (XCW fitting, [4]). As there are two major advantages of the multipole model over the IAM - namely i) obtaining a more accurate geometry, and ii) the possibility of chemical electron-density analysis - there are two major advantages of the XWR over the multipole model - namely i) more accurate hydrogen-atom parameters, and ii) the possibility of going beyond standard electron-density analysis.

To illustrate i) we will show that for a large series of organic compounds already at low resolution of 0.8Å hydrogen positions, X-H bond lengths and ADPs are in statistical agreement with results from neutron diffraction. To illustrate ii) we will discuss the signficance of experimentally restrained orbitals for the determination of bond orders and lone-pair properties in sulfur-containing compounds [5] or protease inhibitors.

[1] N. K. Hansen, P. Coppens, Acta Cryst. A 1978, 34, 909.
[2] e.g. summarised in D. Stalke, Chem. Eur. J. 2011, 17, 9264.
[3] a) D. Jayatilaka, B. Dittrich, Acta Cryst A 2008, 64, 383; b) S. C. Capelli, H.-B. Bürgi, B. Dittrich, S. Grabowsky, D. Jayatilaka, IUCrJ 2014, 1, 361.
[4] D. Jayatilaka, D. J. Grimwood, Acta Cryst. A 2001, 57, 76.
[5] S. Grabowsky, P. Luger, J. Buschmann, T. Schneider, T. Schirmeister, A. N. Sobolev, D. Jayatilaka, Angew. Chem. Int. Ed. 2012, 51, 6776.

MS10-T04
On an Enhanced Hirshfeld Test

J. Luebben[1,2], G. M. Sheldrick[1], B. Dittrich[1,2]
[1]*Georg-August-University Goettingen, Institute of Inorganic Chemistry, Goettingen, Germany*
[2]*University Hamburg, Institut für Anorganische und Angewandte Chemie , Hamburg, Germany*

An improvement to the commonly used Hirshfeld test for validating crystal structures is proposed. [1] In particular bonding situations the Hirshfeld test can fail to detect clearly unphysical atomic displacement parameters (ADPs). [2] The proposed validation method is inspired by the enhanced rigid bond restraint (RIGU) [3] available in Shelxl [4] and analyses all three principal components of the involved ADPs. The method is therefore able to appropriately probe planar and linear bonding environments. A second problem of the Hirshfeld test is that atomic masses are not taken into account. This can result in unreasonable large test results for bonds between atoms with significantly different masses (e.g. carbon-hydrogen bonds and bonds involving metal atoms). Our proposed enhanced rigid body test now includes the atomic mass in the validation procedure and is able to handle all bonding situations independent of the elements involved.

Test result are given in form of a single indicator that represents how well the RIGU restraint is fulfilled and, therefore how well the bond matches an ideally rigid bond. The test values are rapidly computed and provide helpful information during the refinement process as well as useful quality indicators for structure analysis.

The impact of different refinement strategies on the test results is investigated by analysing the differences between independent atom refinements (IAM), invariom refinements [5] and multipole refinements of selected crystal structures. [6, 7]

[1] F. L. Hirshfeld, Can X-ray data distinguish bonding effects from vibrational smearing?, Acta Cryst, **1976**, A32, 239—244.
[2] R. E. Rosenfield et al., A test for rigid-body vibrations, based on a generalization of Hirshfeld's 'rigid-bond' postulate, Acta Cryst, **1978**, A34, 828—829.

[3] A. Thorn, B. Dittrich and G. M. Sheldrick, Enhanced rigid-bond restraints, Acta Cryst, **2012**, A68, 448—451.
[4] G. M. Sheldrick, Crystal structure refinement with SHELXL, Acta Cryst, 2014, C70, in the press.
[5] B. Dittrich et al., The generalized invariom database (GID), Acta Cryst, 2013, B69, 91—104.
[6] A. Volkov et al., XD2006 - A Computer Program Package for Multipole Refinement, Topological Analysis of Charge Densities and Evaluation of Intermolecular Energies from Experimental or Theoretical Structure Factors, 2006.
[7] N. Hansen and P. Coppens, Testing Aspherical Atom Refinements on Small-Molecule Data Sets, Acta Cryst, 1978, 909—921.

MS10-T05
Electron-density study of boron-carbide at 100 K

S. Mondal[1], E. Bykova[2], S. Dey[2], S. I. Ali[2], N. Dubrovinskaia[2], L. Dubrovinsky[3], G. Parakhonskiy[3], S. van Smaalen[2]
[1]*Max-Planck-Institut fuer Kohlenforschung, Chemical Crystallography and Electron Microscopy, Muelheim an der Ruhr, Germany*
[2]*University of Bayreuth, Laboratory of Crystallography, Bayreuth, Germany*
[3]*University of Bayreuth, Bayerisches Geoinstitut, Bayreuth, Germany*

With a unique combination of unusual properties, boron-carbide is one of the most versatile materials of technological importance. Nevertheless, origins of most of the physical and chemical properties of boron carbide have not been properly understood. None of the structural and bonding models available in the literature [1] are able to properly explain properties of boron-carbide. Even the exact chemical composition of boron carbide is uncertain, with a generally accepted formula of B_{12+x} C_{3-x} (0.06 > x > 1.7). The present experimental electron-density study has been undertaken in order to obtain more insight into the nature of chemical bonding in boron-carbide. Single crystals of boron-carbide have been synthesized under high pressure and high temperature. Single crystal X-ray diffraction data was collected at a temperature of 100 K at the beamline F1, Hasylab, Hamburg, Germany. Modeling of electron densities in boron-carbide has been performed by the multipole method using the software package XD2006 [2]. The final multipole model shows excellent fit to the experimental data with R_F = 0.0197. Electron densities have been analyzed with the aid of Bader's *quantum theory of atoms in molecules* (QTAIM) [3]. From the analysis of electron densities a bonding model has been proposed for boron-carbide that puts more light onto the structure-property relationship of boron-carbide.

References:
[1] V. Domnich, S. Reynaud, R. A. Haber, M. Chhowalla, *J. Am. Ceram. Soc.* 94, 3605-3628 (2011).
[2] A. Volkov, P. Macchi, L. J. Farrugia, C. Gatti, P. R. Mallinson, T. Richter, T. Koritsanszky, XD2006, *A Computer Program Package for Multipole Refinement, Topological Analysis of Charge Densities and Evaluation of Intermolecular Energies from Experimental or Theoretical Structure Factors*, (2006).
[3] R. F. W. Bader, *Atoms in Molecules*, Oxford University Press, (1990).

MS10-T06

Investigating Anharmonic Thermal Motion in Glutathion using the Maximum-Entropy-Method (MEM)

C. B. Hübschle[1], C. Ruhmlieb[2], B. Dittrich[2], S. van Smaalen[1]
[1]Universty of Bayreuth, Laboratory of Crystallography, Bayreuth, Germany
[2]University of Hamburg, Institute of Inorganic and Applied Chemistry, Hamburg, Germany

X-ray diffraction of γ-L-Glutamyl-L-cysteinyl-glycine (Glutathion) has been measured at a temperature of 100K, using Mo-Kα radiation. A structure refinement with Invarioms [1] and harmonic ADPs revealed electron density not described by the model. In one approach, the structure model was extended by anharmonic ADPs up to 3rd order for the carboxyl oxygen atoms of the glycine part of the molecule. Although this did improve the fit to the data, the resulting probability density function possessed regions with unphysical significantly negative values. In an alternative approach, the dynamic electron density [2] of the Invariom model (with harmonic ADPs) has been used as PRIOR density for Maximum-Entropy-Method calculations with the computer program BayMEM. The MEM-electron density map successfully describes all density contained in the diffraction data. The effect of the anharmonic motion on the dynamic electron density is analyzed.

Fig1: Isosurface of the MEM-electron density at 11.8 eÅ^{-3} mapped with the electron density value of the harmonic prior density. Atoms in green show almost no anharmonic effect.

[1] B. Dittrich, C. B. Hübschle, K. Pröpper, F. Dietrich, T. Stolper and J. J. Holstein Acta Cryst. B 69 (2013) 91-104.
[2] S. Mondal, S. J. Prathapa and S. van Smaalen Acta Cryst. A 68 (2012) 568-581.

Figure 1

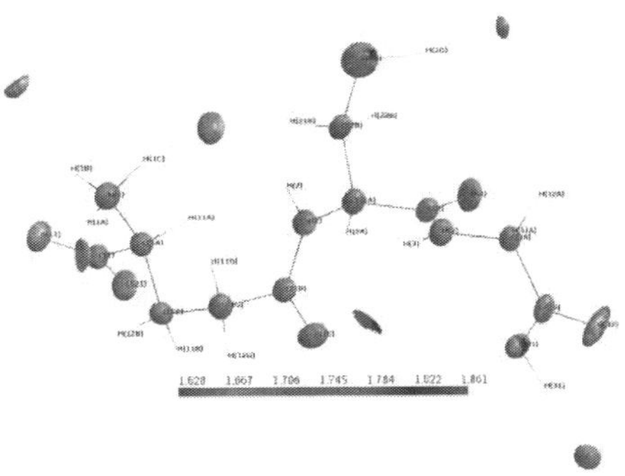

MS11 - Spectroscopy I

MS11-T01

Local ordering of the atoms in intermetallic compounds: An experimental and theoretical approach

F. Haarmann[1]
[1]*RWTH Aachen, Institut für Anorganische Chemie (IAC), Aachen, Germany*

Intermetallic compounds are a large class of materials with great impact on our daily life. Therefore, a strategy for systematic investigation and detailed knowledge about the interplay of properties, chemical bonding, and structure is desired. For electrically non-conducting materials experimental NMR-strategies are well established.[1] This is different for electrically conducting materials. The combination of NMR-spectroscopy, X-ray diffraction and density functional theory (DFT) calculations has been proven to be very useful for analyzing intermetallic compounds.[2,3,4] Especially the quadrupole coupling is well suited, because the coupling parameters can be obtained independently by NMR experiments and DFT methods. Furthermore, the orientation dependence of this coupling can be used by alignment of the conducting crystallites in the magnetic field.[5] In addition useful information about the chemical bonding of intermetallic compounds can be obtained by careful analysis of the DFT results.

Since local ordering of the atoms is frequently observed for this class of materials their characterization and structure determination is still a current issue. While NMR-spectroscopy provides access to local atomic arrangements, the long-range information is obtainable by diffraction methods. Local ordering of the atoms may also be tackled using the above mentioned combination of methods (XRD-NMR-DFT). One of the remaining challenges is the description of the crystal structures for disordered materials required in quantum mechanical solid state calculations. Super lattice structures obtainable by the *Bärnighausen* formalism using group-subgroup relations are suited to set up the required models for the calculations.[6]

The intermetallic compound $Cu_{1-x}Al_2$ was recently investigated by NMR-spectroscopy to characterize the local ordering of the atoms.[7] A QM-approach using super lattice models for the calculation of NMR parameters results in good agreement of experimental and theoretical methods.

[1] H. Eckert, Bunsen-Magazin, 2008, *10*(5), 159.
[2] F. Haarmann, Quadrupolar NMR of Intermetallic Compounds. In: R. K. Harris and R. E. Wasylishen (Eds-in-chief), *Encyclopedia of Magnetic Resonance*. John Wiley & Sons, Ltd, Chichester (2011).
[3] F. Haarmann, K. Koch, P. Jeglič, O. Pecher, H. Rosner, Yu. Grin, *Chem. Eur. J.* 2011, *17*, 7560.
[4] F. Haarmann, K. Koch, D. Grüner, W. Schnelle, O. Pecher, R. Cardoso-Gil, H. Borrmann, H. Rosner, Yu. Grin, *Chem. Eur. J.* 2009, *15*, 1673.
[5] O. Pecher, F. Haarmann, *Z. anorg. allg. Chem.* 2010, *636(11)*, 2089.
[6] H. Bärnighausen, *Group-Subgroup Relations between Space Groups: A Useful Tool in Crystal Chemistry*, "MATCH", Communication in Mathematical Chemistry 1980, 9, 139-175.
[7] F. Haarmann, M. Armbrüster, Yu. Grin, *Chem. Mater.* 2007, *19*, 1147.

MS11-T02

Structural studies of chalcocite Cu_2S at low temperatures

R. Gainov[1,2], M. Russina[1], R. Khassanov[2], A. Nikolaev[2], A. Dooglav[2], I. Pen'kov[2], V. Klekovkina[2]
[1]*Helmholtz-Zentrum Berlin fuer materialien und Energie GmbH, EM-ISFM, Berlin, Germany*
[2]*Kazan Federal University, Kazan, Russian Federation*

Nuclear resonance spectroscopic techniques, including nuclear quadrupole resonance (NQR) [1, 2], are considered as promising tools to study the typomorphic properties of minerals and real crystal structures, to identify and characterize the multiphase samples without destroying them. Chalcocite Cu_2S, as one of the most distributed specimen of copper ore, have been studied by ^{63,65}Cu NQR spectroscopy. The goals of measurements are the comparison of data (variability depending on the nature of samples), the clarification of crystal structural features of Cu_2S at different temperatures and the consideration of the NQR parameters as the possible database for perspective remote NQR spectroscopy for the purposes of geological exploration. Experimental data for Cu_2S at different temperatures and their analysis are presented [3], the impact of neutron instrumentation for further studies of Cu_2S is discussed. Calculations based on the available crystal structure model predict the NQR frequencies, which are close to experimental values. The investigation is a contribution to the database of copper minerals, which can be detected with NQR technique.

References:

[1] J.A. Lehmann-Horn, D.G. Miljak, T.J. Bastow, Solid State NMR, v.54 (2013) 8
[2] R.R. Gainov, A.V. Dooglav, F.G. Vagizov, I.N. Pen'kov, V.A. Golovanevskiy, A.Yu. Orlova, I.A. Evlampiev, V.V. Klekovkina, G. Klingelhöfer, V. Ksenofontov, N.N. Mozgova, Eur. J. Mineral., v.25 (2013) 569
[3] R.R. Gainov et al., submitted.

MS11-T03

Crystal structures and Li ion dynamics of Li_3PO_4-Li_4SiO_4 revisited by NMR

O. Pecher[1], S. Emge[1], Y. Deng[2,3], S. Islam[2], C. Masquelier[3], C. P. Grey[1]
[1]*University of Cambridge, Department of Chemistry, Cambridge, United Kingdom*
[2]*University of Bath, Department of Chemistry, Bath, United Kingdom*
[3]*Université de Picardie Jules Verne , LRCS, Amiens, France*

Solid solutions of Li_3PO_4 and Li_4SiO_4 - denoted as $Li_{3+x}(P_{1-x}Si_x)O_4$ in the following - are known to be good Li-ion conductors [1,2]. The end members come with similar crystal structures in terms of isolated anionic PO_4^{3-} and SiO_4^{4-} clusters with Li-ions in tetrahedral coordination (γ_{II}-Li_3PO_4, *Pnma*) and a highly ordered Li-ion distribution over available cationic sites (Li_4SiO_4, $P2_1/m$), respectively [3,4]. Significant improvements of the electrical conductivity are possible by replacing the anionic clusters and the respective tuning of the Li-ion (vacancies) mobility [4,5]. Due to a high chemical and physical stability associated with the Li mobility, Li_3PO_4 based systems can be reliable even as thin films for solid electrolyte materials within Li-ion batteries (LIBs) [6]. Hence, applications in all-solid-state thin film LIBs are envisioned.

We successfully synthesised $Li_{3+x}(P_{1-x}Si_x)O_4$ with $x = 0$, 0.25, 0.50, 0.75, and 1 by means of sol-gel methods and thermal annealing. Diffraction data proves single-phase materials with lattice parameters and trends similar to literature [3]. While $x = 0.25$ and $x = 0.50$ crystallise in the phosphate, the $x = 0.75$ phase possesses the orthosilicate structure. ^{7}Li magic angle spinning (MAS) nuclear magnetic resonance (NMR) experiments at ambient temperature show a slight shift of the ^{7}Li peak around 0 ppm (Figure). A broadening of the line shape with increasing *x*-values indicates the changes of the local Li environments with respect to the composition. Significant changes of the ^{7}Li spin-lattice relaxation times (T_1) - especially while comparing the end members vs. substituted phases - reflect the sensitivity of this value to the activation energy and Li-ion dynamics.

Ongoing variable temperature T_1 as well as ^{6}Li and ^{7}Li NMR line shape studies will shed light into bulk vs. grain boundary contributions to the conductivity on a local atomic scale, being only accessible by NMR. The discussion will benefit from alternating current (AC) impedance spectroscopy and molecular dynamics (MD) simulations on the Li-ion migration processes.

[1] Y.-W. Hu, I. D. Raistrick, R. A. Huggins, Mater. Res. Bull. 1976, 11, 1227; *J. Electrochem. Soc.* **1977**, *124*, 1240.
[2] R. D. Shannon, B. E. Taylor, A. D. English, T. Berzins, *Electrochem. Acta* **1977**, *22*, 783.
[3] W. H. Baur, T. Ohta, *J. Solid State Chem.* **1982**, *44*, 50.
[4] M. Smaihi, D. Petit, F. Gourbilleau, F. Chaput, J. P. Boilot, *Solid State Ionics* **1991**, *48*, 213.
[5] T. Asai, S. Kawai, Solid State Commun. 1980, 36, 891; *Solid State Ionics* **1982**, *7*, 43.
[6] C. Frayret, C. Masquelier, A. Villesuzanne, M. Morcette, J.-M. Tarascon, *Chem. Mater.* **2009**, *21*, 1861.

Figure 1

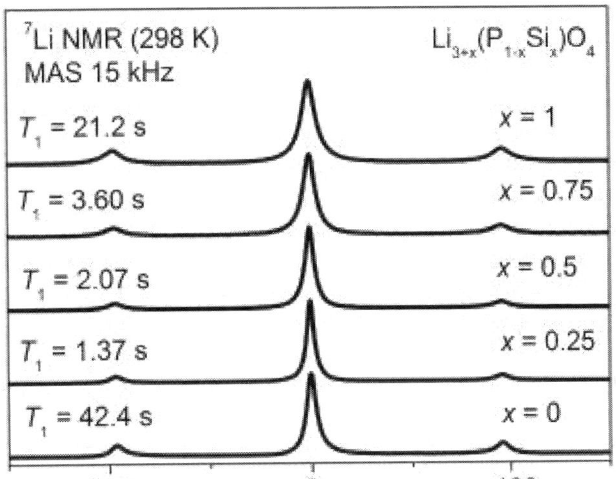

MS11-T04

Thermally induced increase of long range order in radiation damaged pyrochlore and titanite: Raman, NMR spectroscopy and X-ray diffraction

P. Zietlow[1], T. Beirau[1], L. A. Groat[2], C. Paulmann[1], B. Mihailova[1], U. Bismayer[1]
[1]*Universität Hamburg, Hamburg, Germany*
[2]*University of British Columbia, Vancouver, Canada*

Minerals with radiation damage are called metamict and show a structural pattern of crystalline, quasi amorphous and fully disordered clusters. Such clusters are the result of alpha decay processes of incorporated uranium and thorium ions in the minerals. Structural damage and the behaviour as a function of thermal annealing is of great interest for material sciences, for instance for the better understanding of long term embedding materials of actinides from radioactive waste. Previously discussed embedding materials include titanite and pyrochlore ceramics. In nature, such minerals can occur in the metamict state. [1], [2]

The analysis of pyrochlores and titanites exposed to different radiation doses on annealing using X-ray diffraction (XRD) and Raman spectroscopy provides insight into their ordering behaviour at different length scales. ^{29}Si magic angle spinning nuclear magnetic resonance (NMR) spectroscopy of damaged titanite is used to characterise local structural changes. Since XRD is sensitive to long range order while Raman and ^{29}Si NMR respond to the local structural vicinity, the methods are well complementary to better understand the transformation from amorphous to a better ordered state. [3]

[1] Ewing, R. C., Meldrum, A., Wang, L.M., Wang, S.X. (2000): Radiation-Induced Amorphization, Reviews in Mineralogy and Geochemistry, Ed. P.H. Ribbe, Min. Soc. America **39**, 319-361.
[2] Hawthorne, F.C., Groat, L.A., Raudsepp, M., Ball, N.A., Kimata, M., Spike, F., Gaba, R., Halden, N.M., Lumpkin, G.R., Ewing, R.C., and others. (1991): Alpha-decay damage in titanite. American Mineralogist, **76**, 370-396.
[3] Zietlow, P., Beirau, T., Groat, L. A., Paulmann, C., Bismayer, U. (2014): ^{29}Si MAS NMR and synchrotron XRD study of metamict Cardiff titanite. Z. Kristallogr. **229**, 551-554

MS11-T05

Pressure-induced spin collapse of octahedrally coordinated Fe^{3+} in andradite from experiment and theory

A. Friedrich[1], B. Winkler[1], W. Morgenroth[1], M. Koch-Müller[2], D. Rhede[3], J. Ruiz-Fuertes[1], V. Milman[4]
[1]*Goethe-University Frankfurt , Institute of Geosciences, Frankfurt am Main, Germany*
[2]*GFZ Potsdam, Sektion 3.3, Potsdam, Germany*
[3]*GFZ Potsdam, Sektion 4.2, Potsdam, Germany*
[4]*Dassault Systemes BIOVIA, Cambridge, United Kingdom*

Andradite, $Ca_3Fe_2[SiO_4]_3$, is a garnet endmember with Fe^{3+} in octahedral coordination. The high symmetry of the garnet structure (space group *Ia*-3*d*) and of the cation coordination polyhedra as well as the importance of garnet within the Earth's mantle make andradite an ideal model system to study pressure-induced spin-pairing transitions of Fe^{3+}. Specifically, the spin-pairing behavior in andradite may serve as a benchmark for that of Fe^{3+} located on the octahedral B-site of iron-bearing magnesium silicate perovskite [1]. As no structural phase transition has been reported in andradite on compression up to 36 GPa so far [2-6], we have studied andradite at considerably higher pressure.

Andradite single crystals were synthesized at 3 GPa and 1273 K in a multi-anvil press. The chemical composition was analyzed using the electron microprobe, while the water content was determined from synchrotron FTIR spectroscopy at BESSY (IRIS beamline, Berlin, Germany). Small single crystals were pressurized in diamond anvil cells up to 78 GPa using neon as pressure-transmitting medium. High-pressure intensity data for single-crystal structure analyses were collected at the PETRA III synchrotron facility (beamline P02.2, DESY, Hamburg, Germany). On decompression, Raman spectra were collected with a Renishaw micro-Raman spectrometer. Complementary quantum mechanical calculations based on density functional theory were performed using the program CASTEP [7].

An isosymmetric phase transition was observed in the pressure range between 60 and 70 GPa, which is associated with a gradual high spin-to-low spin electronic transition in Fe^{3+} [8]. Experimental structural data before and after the phase transition are in excellent agreement with the theoretically predicted structural compression of the high-spin and low-spin phases, respectively. Results will be discussed with respect to the structural and volume compression, overall and polyhedral bulk moduli, changes in vibrational properties and the computed spin-pairing energy.

Financial support from the DFG, Germany (Fr2491/2-1 within SPP1236), the BMBF, Germany (05K10RFA, 05KS7RF1, 05K13RF1), and the A. v. Humboldt Foundation, Germany (J. R.-F.) is gratefully acknowledged. Portions of this research were carried out at PETRA III (DESY) and at BESSY II, members of the Helmholtz Association. We thank H.-P. Liermann and U. Schade for support.

[1] J.-F. Lin et al., Rev. Geophys. **51**, 244 (2013).
[2] R.M. Hazen and L.W. Finger, Am. Mineral. **74**, 352 (1989).
[3] L. Zhang et al., Phys. Chem. Minerals **27**, 52 (1999).
[4] Z. Wang, T. Yage, and T. Kondo, J. Phys. Chem. Solids **60**, 441 (1999).
[5] A. Pavese, D. Levy, and V. Pischedda, Eur. J. Mineral. **13**, 929 (2001).
[6] A. Pavese et al., Phys. Chem. Minerals **28**, 242 (2001).
[7] S.J. Clark et al., Z. Kristallogr. **220**, 567 (2005).
[8] A. Friedrich et al., Phys. Rev. B **90**, 094105 (2014).

MS11-T06

The hydrogen defect in lithium niobate and lithium tantalate - a kinetics study

E. Mehner[1], T. Köhler[1], J. Hanzig[1], G. Gärtner[1], H. Stöcker[1], D. C. Meyer[1]
[1]TU Bergakademie Freiberg, Experimentelle Physik, Freiberg, Germany

$LiNbO_3$ and $LiTaO_3$ crystals are used in many optical devices and exhibit an interesting defect chemistry impacting crystal structure. Therefore, understanding defects is helpful to control optical and electrical properties.

A common and mobile defect in both materials is hydrogen. Its incorporation is investigated by FT-IR and UV/VIS spectroscopy with respect to crystallographic orientations and excitation polarisation. The aim of the study is the development of a structural model for the kinetics and the diffusion of the OH defect. The examined congruent crystals are cube shaped, cut and polished along the [2 -1 -1 0], [0 1 -1 0] and [0 0 0 1] directions. Whereas used stoichiometric crystals are plates, polished and cut in [0 0 0 1] direction.

Depending on the lithium-niobium stoichiometry the hydrogen defect causes an OH band with several sub-bands, at different spectral positions. In congruent material two sub-bands are detectable, whereas near stoichiometric material exhibits four. The composition of the OH band regarding formation or decline of sub-bands under oxidizing and reducing conditions shows the transport mechanisms within the crystals. The protonation in hot water shows a formation of new sub-bands at higher binding energies. We have found three and six sub-bands in congruent and near stoichiometric crystals, respectively and determined the according activation energies.

The results show that the hydrogen is disordered and migrate through the crystal via energetically different binding sites.

MS12 - Protein function and regulation

MS12-T01

You CAN teach an old dog new tricks: trypsin variants that catalyse peptide ligation

M. Schöpfel[1], C. Parthier[1], S. Liebscher[2], A. Shakhuukhen[2], F. Bordusa[2], M. T. Stubbs[1]
[1]*Institut für Biochemie und Biotechnologie, Martin Luther-Universität Halle Wittenberg, Physikalische Biotechnologie, Halle, Germany*
[2]*Institut für Biochemie und Biotechnologie, Martin Luther-Universität Halle Wittenberg, Naturstoffchemie, Halle, Germany*

The thermodynamic reversibility of enzymatic reactions means that proteases are in principle able to catalyse two opposing reactions: in addition to peptide hydrolysis, serine proteases should be able to carry out reverse proteolysis, the so-called peptide ligation reaction.1 Whereas the potential advantages (and applications) of enzyme-catalysed peptide ligation are numerous, the hydrolytic activity of proteases generally greatly outweighs that of ligation, so that enzyme-, substrate- or medium engineering is unavoidable. As an example, the subtilisin BPN' double mutant Ser221Cys/Pro225Ala, termed "subtiligase", is able to ligate esterified peptides site specifically onto the N termini of proteins or peptides in aqueous solution.2
We have developed a variant of trypsin, which we term trypsiligase, with two specific and opposing activities:3 (i) a site specific hydrolytic cleavage activity against peptide and protein substrates, and (ii) an N-terminal ligation activity under mild reaction conditions with quantitative product yields. X-ray crystallographic analyses of trypsiligase suggest that both substrate specificity and ligation activity are a result of a substrate-mediated order-disorder transition of the activation domain, an extension of the well-established trypsin zymogen activation mechanism.4

Literature

1 Wehofsky N, Koglin N, Thust S, Bordusa F: Reverse proteolysis promoted by in situ generated peptide ester fragments. J Am Chem Soc 2003, 125: 6126-6133.
2 Chang TK, Jackson DY, Burnier JP, Wells JA: Subtiligase: a tool for semisynthesis of proteins. Proc Natl Acad Sci U S A 1994, 91: 12544-12548.
3 Liebscher S, Schöpfel M, Aumüller T, Sharkhuukhen A, Pech A, Höss E, Parthier C, Jahreis G, Stubbs MT, Bordusa F: N-Terminal Protein Modification by Substrate-Activated Reverse Proteolysis. Ang. Chem. 2014, 53: 3024-3028.
4 Huber R, Bode W: Structural Basis of Activation and Action of Trypsin. Accounts of Chemical Research 1978, 11: 114-122.

MS12-T02

The Crystal Structure of FtsH from Aquifex aeolicus: of disordered and C2-symmetric AAA rings

M. Vostrukhina[1], E. Brunstein[1], M. Schacherl[1], U. Baumann[1]
[1]*University of Cologne, Institute of Biochemistry, Cologne, Germany*

The crystal structure of a truncated, soluble quadruple-mutant of FtsH from Aquifex aeolicus comprising the AAA and the protease domains has been determined at 2.96 Å resolution in space group I222. Large crystals were obtained from the beginning but reasonable diffraction (< 3.5 A) could only be achieved by extensive additive and cryoprotectant screening. The protein crystallizes as a hexamer with the protease domain forming layers in the ab plane. Contacts between these layers are mediated by the AAA domains, which are disordered in one crystal form, thus preventing a complete structure determination for a long time. The fortuitous finding of a new crystal form with a shorter c axis eventually allowed structure determination of the whole construct spanning AAA and protease domains. Here, adenosine diphosphate (ADP) is bound to each subunit and the AAA ring exhibits two-fold symmetry. The arrangement is different from the ADP-bound state of an analogously truncated, soluble FtsH construct from Thermotoga maritima. The pore is completely closed and the phenylalanine residues in the pore line a contiguous path. The protease hexamer is very similar to those described for other FtsH structures. To resolve certain open issues regarding a conserved glycine in the linker between AAA and the protease domain, as well as on the active-site switch ß-strand, mutations in the full-length membrane-bound protein have been introduced. Activity analysis of these point mutants reveals the crucial importance of those residues and is in accord with our previous interpretation of the active-site switch and the importance of the linker glycine residue.

MS12-T03

Oligomeric assembly states of the AAA+ ATPases Reptin/Pontin for diverse macromolecular machines

K. Lakomek[1], G. Stoehr[1], K.- P. Hopfner[1]
[1]*Ludwig-Maximilians University Munich, Gene Center and Department of Biochemistry, Munich, Germany*

The highly conserved orthologs Reptin (Rvb1) and Pontin (Rvb2) are components of diverse essential macromolecular machines and crucial for a whole plethora of cellular activities including cancer-related processes. They belong to the family of AAA+ proteins (ATPases associated with various cellular activities) and as characteristic of this family can form hexameric rings. Yet, the oligomeric assembly state and architecture of Rvb1-Rvb2 complexes are controversially debated, and their functions remain unclear.

Two new crystal structures of full-length Reptin-Pontin complexes in distinct nucleotide-binding states recently determined in our lab provide a structural framework for the functional analyses of Reptin-Pontin hexamers or dodecamers in different cellular contexts. Reptin and Pontin assemble into stable dodecamers that are composed of two hexameric rings of alternating molecules as validated based on the anomalous scattering of inherent sulphur atoms. The structures uncover novel inter-ring interfaces with significant rearrangements between nucleotide-free and -bound states highlighting the conformational plasticity of the complex. Strikingly, Rvb1's and Rvb2's characteristic oligonucleotide-binding OB-fold domains (DIIs) occupy unequal places relative to the compact AAA+ core ring. The conformational and positional flexibility of DIIs and the other domains will have major impact on assembly and function.
Complementary ATPase data provide interesting mechanistic insights with respect to the interdependency of ATP binding and hydrolysis in the active sites of neighbouring subunits in the ring.

Figure 1

MS12-T04

The crystal structure of apo ArnA features an unexpected central binding pocket and provides an explanation for enzymatic cooperativity

C. Grimm[1], S. Hertlein[1], U. Fischer[1]
[1]*University of Würzburg, Würzburg, Germany*

Abstract The bacterial protein ArnA is an essential enzyme in the pathway leading to the modification of lipid A with the pentose sugar 4-amino-4-deoxy-L-arabinose. This modification confers resistance to polymyxins, antibiotics used as a last resort to treat infections with multiple drug resistant gram-negative bacteria. ArnA contains two domains with distinct catalytic functions, a dehydrogenase and a transformylase domain. The protein forms homohexamers organized in a dimer of trimer fashion. Here, we present the crystal structure of apo ArnA and compare it to its ATP and UDP-glucuronic acid-bound counterpart. The comparison reveals major structural rearrangements in the dehydrogenase domain that lead to the formation of a previously unobserved binding pocket at the centre of each ArnA trimer in its apo state. In our crystal structure, this pocket is occupied by a DTT molecule. We show that formation of the pocket is linked to a cascade of structural rearrangements that emerges from the NAD^+ binding site. Based on these findings we postulate a small effector molecule that binds to the central pocket and modulates the catalytic properties of ArnA. Furthermore, the discovered conformational changes provide a mechanistic explanation for the strong cooperative effect recently reported for the ArnA dehydrogenase function.

Figure 1

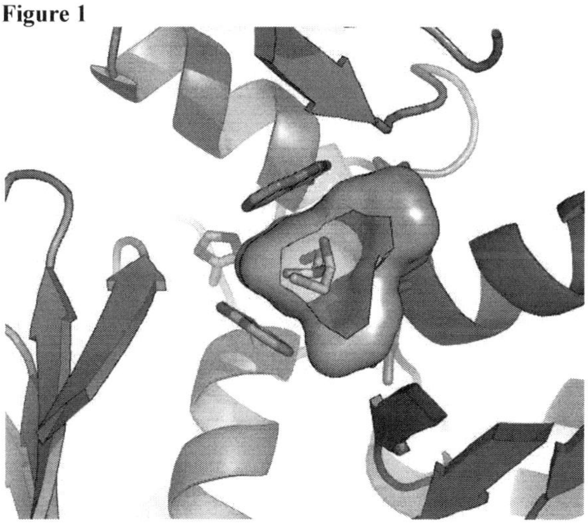

Figure 2

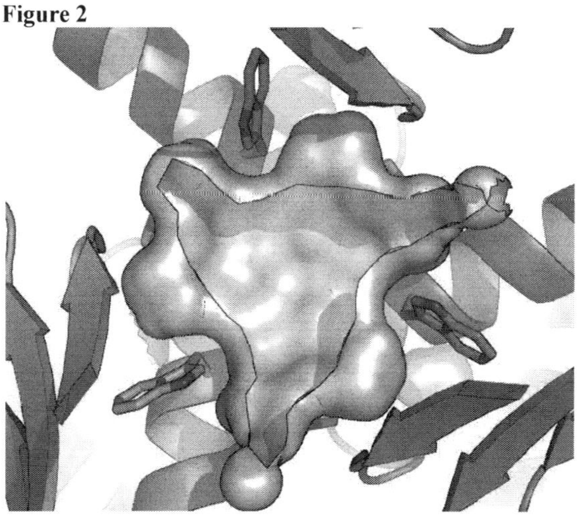

MS12-T05

Structural Insights into Terpenoid Biosynthesis

P. Baer[1], P. Rabe[2], J. S. Dickschat[2], M. Groll[1]
[1]*Technische Universität München, Department of Chemistry, Garching, Germany*
[2]*Rheinische Friedrich Wilhelms-Universität Bonn, Bonn, Germany*

With the molecular life frontier being pushed by the era of the transcriptome and the proteome, we reached the era of the metabolome nowadays. Whereas DNA and proteins mainly target the intramolecular operation of an organism, metabolites partly serve the interaction with the organism's environment, thus acting on an intermolecular level. To achieve a complex interaction it is necessary to access a huge arsenal of chemical distinct molecules. This prerequisite is perfectly matched by the enzyme class of terpene cyclases, which provide thousands of different molecules (>50,000 identified). This astonishing number derives from a few linear polyprenyl diphosphate precursors whose conversion is controlled by the enzyme's active site contour. Simple variations of the active site may result in considerably distinct products featuring new biological activities. Based on detailed structural data, we provide fundamental new insights into the enzymatic mechanism of class I terpene cyclases[1,2]. The terpenoids' wide distribution in nature and its' chemical elegance both mark them as perfect molecules to start the development of novel and biological active, target- orientated pharmaceuticals. To apply the class of terpene synthases to this approach a comprehensive understanding of the sophisticated enzymatic mechanism is necessary.

[1] P. Baer, P. Rabe, K. Fischer, C. A. Citron, T. A. Klapschinski, M. Groll, J. S. Dickschat, Induced Fit Mechanism in Class I Terpene Cyclases, Angew. Chem. Int. Ed. 2014, 53, 7652-7656.
[2] P. Baer, P. Rabe, C. A. Citron, C. C. de Oliveira Mann, N. Kaufmann, M. Groll, J. S. Dickschat, Hedycaryol Synthase in Complex with Nerolidol Reveals Terpene Cyclase Mechanism, ChemBioChem 2014, 15, 213-216.

MS12-T06

Insights into enzymatic CO_x-transformations at true-atomic resolution

J. Fesseler[1], J.- H. Jeoung[1], H. Dobbek[1]
[1]*HU Berlin, Strukturbiologie / Biochemie, Berlin, Germany*

Carbon dioxide (CO_2) is considered as major cause of global warming and directly reflecting anthropogenic usage of carbon-based fuels since the beginning of the industrial era. Despite its high abundance in atmosphere, chemical usage of carbon dioxide is restricted due to its low reactivity. [1] A major target for the shift towards a more sustainable future would be the usage of CO_2 as carbon feedstock for chemical industry. Still reasonable catalysts are lacking that could facilitate energy efficient transformations under mild reaction conditions.

Nature's prime catalyst for CO_2 activation and reduction is cluster C of Ni,Fe-CO dehydrogenase (CODH). CODHs employ a Ni-substituted cubane Fe_3S_4 -cluster with a Fe-exo ligand as active site. Turnover numbers of up to 40 s^{-1} have been reported for CO_2 reduction and 31,000 s^{-1} for CO-oxidation for CODH from *Carboxydothermus hydrogenoformans* (CODHII$_{Ch}$), making it a world class catalyst for CO_x-transformations (Equation 1). [2]

$$CO_2 + 2e^- + 2H^+ \quad CO + H_2O$$

Crystal structure analysis identified CO_2 bridged between Ni and Fe_1 atoms as a key intermediate in the catalytic cycle.[3] However, the resolution of 1.5 Å allowed only a rough estimation of bond distances and angles.

To achieve a more detailed description of the CO_2-bound intermediate, we extended the crystallographic studies to true-atomic resolution (d_{min} < 1.1 Å) and beyond. At this resolution, atomic positions can be accurately refined and bond distances and angles determined with their individual standard deviations. Here we report high-resolution crystal structures of CO_2-bound cluster C, unraveling unprecedented insights into the catalytic strategy of nature's gold standard catalyst for CO_2-reduction.

References:
[1] A.M. Appel, et al., *Chemical Reviews* 2013, *113*,6621--6658.
[2] V. Svetlitchnyi, et al., *J Bacteriol* 2001, *183*, 5134-5144.
[3] J.H. Jeoung, H. Dobbek, *Science* 2007, *318*, 1461-1464.

MS13 - Inorganic Structural Chemistry II

MS13-T01
More on Common Structural Features in bcc and Ni_2In Type Superstructures

C. J. Müller[1], S. Lidin[1]
[1]*Lund University, CAS, Lund, Sweden*

There is no common higher symmetric space group for any cubic and any hexagonal space group amongst the 230 space groups. However, both crystal systems contain a threefold axis as one of their constituting symmetry operators. In the cubic case, threefold rotation axes are present along all of the room diagonals. In fact, this symmetry operation is irremissible for metrical cubic unit cells to be described in any cubic space group. In the hexagonal case, the threefold rotation axis is equal to the *c* axis of the unit cell, as all sixfold axes are threefold axes as well. Moreover, cubic and hexagonal crystal structures can consist of similar building units such as polyhedra or layers. But, the conjunction of these building units is usually different.

One of these cases with similar building units is the cubic γ-brass type structures and pseudohexagonal Cu_7In_3 type structures. It is well known that γ-brass resembles a bcc defect packing as well as that Cu_7In_3 is a Ni_2In superstructure where all octahedral and all trigonal-bipyramidal voids are occupied, and thus, vacancies in the distorted hcp network structure are present. On the other hand, Lidin and coworkers showed that a bcc defect packing is a valuable alternative description for Cu_7In_3.[1] Surprisingly, it has been the only Ni_2In superstructure that showed these distinct structural features.

We present a new bcc superstructure that contains building units from Cu_7In_3 along its threefold axis. Further, we introduce a new Ni_2In superstructure that can be described either by using the classical Ni_2In building units or a bcc defect packing.

[1] S. Lidin, L. Stenberg, M. Elding-Pontén, *J. Alloys Comp.*, **1997**, *255*, 221-226.

MS13-T02
Crystal chemistry of the synthetic alluaudite-like arsenates

T. Đorđević[1], A. Wittwer[1]
[1]*Institut für Mineralogue und Kristallographie, Universität Wien, Wien, Austria*

Alluaudite-group compounds is a large group of minerals and synthetic materials adopting the general formula *A2A2'A1A1'A1"*$_2$*M1M2*$_2$[*X*O_4]$_3$, where *A* are large cations and *M* are distorted octahedral centres and *X* are mostly small pentavalent cations [1] or *A1A2M1M2*$_2$(*X*1O_4)(*X*2O_3OH)$_2$ for protonated members. In minerals *A*-positions are mostly occupied with Na, Ca, Mn, *M*-positions with Mn, Fe, Al, Mg, and *X* with P and As. Synthetic members show much broader chemical variability. Crystal structures of alluaudite materials (most common space group is *C*2/*c*) consist of chains of edge-sharing *M*1$O_4(OH)_2$ and *M*2O6 octahedra that are stacked parallel to (1 0 1) and are connected by the AsO_4 tetrahedra. These chains produce two types of channels parallel to the *c*-axis. *A*2 atoms are located in channels 2, while *A*1 are situated in channels 1. The understanding of the crystal chemistry of alluaudite-type materials is necessary because they represent potential materials for practical applications, such as corrosion inhibition, passivation of metal surfaces, catalysis, and energy storage [2 and references therein].

During present crystal-chemical investigations on synthetic alluaudite arsenates, $NaZn_3(AsO_4)(AsO_3OH)_2$ was synthesised [space group *C*2/*c*, $a = 12.023(2)$, $b = 12.385(3)$, $c = 6.750(2)$ Å, $\beta = 112.79(3)°$, $V = 926.7(4)$ Å^3, $Z = 4$] and its structure was refined using single-crystal X-ray diffraction data (293 K, Mo*Ka*, $2q_{max} = 80°$) to $R_1(F) = 0.025$.

Alluaudite group compounds presently count ten arsenic minerals and over 23 synthetic arsenates. *A*1 and *A*2 sites in synthetic arsenate alluaudites are mostly occupied by Na^+, Ag^+ and H^+. In protonated members, *A*1, *A*1' and *A*2 sites are vacant, and the protons always occupy *A*1'' site. The *A*2' site is mostly occupied by alkali ions, Na^+, Ag^+, K^+. The further increase of the cation size on the *A* positions in the alluaudite type structure by using Tl^+ or Cs^+ leads to the break-up of the three-dimensional framework and to the formation of layered structures [3]. On the contrary, the non-protonated synthetic arsenate alluaudites, have *A*1'' vacant and Na^+, Ag^+, K^+ and Ca^{2+} atoms are located on both *A*1 and *A*2' positions. The insertion of Cd^{2+} in the alluaudite-type materials mainly takes place on the *M*1 site, while in extremely rare cases it can be found on *M*2 site in minor amounts. In $Cd_{1.16}Zn_{2.34}(AsO4)_{1.5}(HAsO_4)(H_2AsO_4)_{0.5}$ and $Cd_{0.74}Mg_{2.76}(AsO_4)_{1.5}(HAsO_4)(H_2AsO_4)_{0.5}$ [5] besides *M*1 site, Cd also occupies the *A*2' position.

Financial support of the Austrian Science Foundation (FWF) (Grant V203-N19) is gratefully acknowledged.

[1] Hatert, F., Keller, P., Lissner, F., Antenucci, D., Fransolet, A.-M. (2000) Eur. J. Mineral., 12, 847-857.
[2] Rondeux, M., Hatert, F. (2010) Am. Mineral., 95, 844-852.
[3] Stock, N., Stucky, G.D., Cheetham, A.K. (2001) J. Phys. Chem. Solids., 62, 1457-1467.
[4] Stojanović, J., Đorđević, T., Karanović, Lj. (2012) J. Alloys Compd., 520, 180-189.

MS13-T03
Formation and crystal-chemical studies of mullite-type aluminum borate compounds

K. Hoffmann[1,2], M. M. Murshed[2], M. Fischer[1], T. J. Hooper[3], J. V. Hanna[3], O. Dolotko[4], H. Schneider[1,5], T. M. Gesing[2], R. X. Fischer[1]
[1]*Universität Bremen, Kristallographie, FB05, Bremen, Germany*
[2]*Universität Bremen, Chemische Kristallographie fester Stoffe, Institut für Anorganische Chemie, Bremen, Germany*
[3]*University of Warwick, Department of Physics, Coventry, United Kingdom*
[4]*Technische Universität München, Heinz Maier-Leibnitz Zentrum (MLZ), Garching, Germany*
[5]*Universität Köln, Institut für Kristallographie, Köln, Germany*

Aluminum borates studied here belong to the family of mullite-type materials [1]. In the binary phase diagram of Al_2O_3-B_2O_3 [2] two thermodynamically stable phases are described to exist: $Al_4B_2O_9$ (A_2B) and $Al_{18}B_4O_{33}$ (A_9B_2). In addition, Mazza et al. [3] published a series of compounds, with a composition given as $Al_{6-x}B_xO_9$ ($1 \leq x \leq 3$) having average structures described to crystallize in the orthorhombic space group *Pbam* with a pseudo-tetragonal metric. The polymorph of A_2B thermodynamically stable up to 1308 K is reported to crystallize in the monoclinic space group *C*2/*m* [4] in an enlarged unit cell with all lattice parameters doubled with respect to the *Pbam* structure; those of the aluminum-rich phase A_9B_2 crystallize in the orthorhombic space group *A*2$_1$*am* [5].

We found $Al_{6-x}B_xO_9$, with the structure described by Mazza et al. [3], to crystallize at 1173 K in the compositional range between the two end members A_9B_2 and A_2B. X-ray diffraction patterns show broad reflections, for instance around 20° 2*θ* or 18° 2*θ*, indicating poorly crystalline A_9B_2 or A_2B, respectively, for samples with high and low B contents. Rietveld refinements show the lattice parameters to be clearly orthorhombic with a continuous change between the end members of $Al_{6-x}B_xO_9$ at x = 1 and x = 2. Therefore, we assume that there is a solid solution in which aluminum is partially replaced by boron, accompanied by slight structural changes. Fourier transform infrared (IR) and ^{11}B MAS NMR spectra confirm an increase of BO_4 tetrahedra with increasing boron content in the precursor phase. Distance least squares (DLS) modeling and neutron diffraction were performed to determine probable split positions for oxygen, as no position for boron could be modeled leading to acceptable B-O distances.

Series of $A_{5-B}B_{1+B}$ (≥ 1373 K) and $A_{2-B}B_{1+B}$ (1223 K) powder samples were synthesized as a function of boron content in the precursor phases. The ^{11}B MAS NMR signal indicates a decrease of the relative ratios of the integrated area of the BO_3 site compared to that of the BO_4 site with increasing boron content in the precursors of both phases. From the MAS NMR data a maximum of ~2 % BO_4 and 25 % BO_4 are estimated for A_9B_2 and A_2B, respectively. Details of the structural features, in particular, the oxygen disordering could not be solved yet. However, DLS modeling and density functional theory calculations indicate a mixture of slightly different structures which differ in the environment of the positions of the channel oxygen atoms O5 and O10.

[1] R.X. Fischer, H. Schneider, Eur. J. Mineral. 20 (2008) 917. [2] P.J.M. Gielisse, W.R. Foster, Nature 195 (1962) 69. [3] D. Mazza, M. Vallino, G. Busca, J. Am. Ceram. Soc. 75 (1992) 1929. [4] R.X. Fischer, V. Kahlenberg, D. Voll, K.J.D. MacKenzie, M.E. Smith, B. Schnetger, H.-J. Brumsack, H. Schneider, Am. Mineral. 93 (2008) 918. [5] M. Garsche, E. Tillmanns, H. Almen, H. Schneider, V. Kupcik, Euro. J. Mineral. 3 (1991) 793.

MS13-T04

Hydrogenation Properties of $BiPd_3$ and Related Selenium Substituted Compounds

A. Götze[1], H. Kohlmann[1]
[1]*Universität Leipzig, Insititut für Anorganische Chemie, Leipzig, Germany*

Intermetallic palladium rich compounds like MPd_3 (M: Mg, Mn, In; $ZrAl_3$ type) form hydrides under hydrogen pressure.[1-3] These compounds are interesting, because a mechanism of the hydrogenation could be established. $BiPd_3$[4] as another candidate for hydrogenation was synthesized from the elements. Iodine was added to enhance the reactivity by means of chemical vapor transport. The compound was studied during the hydrogenation by in situ thermal analysis (DSC) at starting hydrogen pressures of 50 MPa and temperatures up to 430°C. X-ray powder diffraction before and after the hydrogenation showed a rearrangement of the intermetallic structure to a tetragonal $ZrAl_3$ type (a = 4.0445(1) Å, c = 16.6195(5) Å, *I4/mmm*) with an increase in unit cell volume per formula unit by 1.8 %. Both structure types, before and after hydrogenation, are a fourfold superstructure of the Cu type structure. In autoclaves at a higher starting hydrogen pressures of 200 MPa and temperatures up to 500°C the unit cell volume per formula unit increase up to 2.3 %. This product is stable at air. $BiPd_3$ is the first binary palladium rich intermetallic with an element of the group 15th or 16th that reacts with hydrogen. To ascertain the influence of the electronic phenomena and to understand the effect of the size of an atom and the number of electrons, the composition of the system were substituted with selenium. Hydrogenation properties near the composition $Pd_3Bi_{0.6}Se_{0.4}$[5] were performed.

[1] H. Kohlmann, N. Kurtzemann, R. Weihrich, T. Hansen, *Z. Anorg. Allg. Chem.* **2009**, *635*, 2399-2405.
[2] P. Oennerud, Y. Andersson, R. Tellgren, P. Nordblad, F. Bourée, G. André, *Solid State Comm.* **1997**, *101*, 433-437.
[3] H. Kohlmann, *J. Solid State Chem.* **2010**, *183*, 367-372.
[4] K. Schubert, S. Bhan, T. K. Biswas, K. Frank, P. K. Panday, *Naturwissenschaften* **1968**, *55*, 542-543.

[5] M. El-Boragy, M. Ellner, K. Schubert, *Z. Metallkde.* **1989**, *80*, 197-200.

MS13-T05

Vacancy ordering in Fe^{3+} containing perovskite compounds and structure stabilizing factors

O. Clemens[1,2]
[1]*Technische Universität Darmstadt, Joint Research Laboratory Nanomaterials, Darmstadt, Germany*
[2]*Karlsruher Institut für Technologie, Institut für Nanotechnologie, Eggenstein Leopoldshafen, Germany*

The perovskite structure (ABX_3) is well known to be capable of accommodating a large amount of vacancies [1,2]. Such vacancy formation results in the lowering of coordination number for the B-site cation, resulting in a change from octahedral to square pyramidal, trigonal bipyramidal, square and tetrahedral coordination.

The ferrite compounds $SrFeO_{2.5}$ and $CaFeO_{2.5}$ have been well studied and are known to adopt the brownmillerite type structure [3, 4]. In contrast, the detailed structure of $BaFeO_{2.5}$ has been unknown until recently, and we managed to solve this structure by means of a combined Rietveld analysis of X-ray and neutron powder diffraction data [5]. In comparison to $SrFeO_{2.5}$ and $CaFeO_{2.5}$, the structure of $BaFeO_{2.5}$ is highly complex, containing 28 formula units per unit cell. In a recent study, we showed that this compound can be fluorinated to form $BaFeO_{2.333}F_{0.333}$ ($Ba_3Fe_3O_7F$) [6], again showing a complex crystal structure with differences to perovskite type $La_{0.333}Sr_{0.667}FeO_{2.667}$. Although all those compounds contain Fe^{3+} as the only B-site cations, the structural flexibility is tremendous comparing compounds with the same amounts of vacancies.

To understand such differences in vacancy ordering we compared a large variety of ferrite (Fe^{3+}) perovskite compounds, among them also some compounds which crystallize in the hexagonal perovskite structure (6H, 15R). Summarizing such influences, we here report on the structure chemical factors for local and overall stabilization of vacancies, trying to create a deeper understanding why one structure type can be favoured over the other for different A site cations.

References

[1] M.T. Anderson, J.T. Vaughey, K.R. Poeppelmeier, *Chem. Mater.* 5 (1993) 151-165.
[2] S. Stolen, E. Bakken, C.E. Mohn, *Phys. Chem. Chem. Phys.* 8 (2006) 429-447.
[3] P. Berastegui, S.G. Eriksson, S. Hull, *Mater. Res. Bull.* 34 (1999) 303-314.
[4] C. Greaves, A.J. Jacobson, B.C. Tofield, B.E.F. Fender, *Acta Crystallogr.* B31 (1975) 641-646.
[5] O. Clemens, M. Gröting, R. Witte, J.M. Perez-Mato, C. Loho, F.J. Berry, R. Kruk, K.S. Knight, A.J. Wright, H. Hahn, P.R. Slater, *Inorg. Chem.* 53 (2014) 5911-5921.
[6] O. Clemens, *J. Solid State Chem.* http://dx.doi.org/10.1016/j.jssc.2014.12.027 (2015).

MS14 - Structural chemistry at extreme conditions

MS14-T01

A Simple EoS for Structural Phase Transitions

R. Angel[1]
[1]*University of Padova, Geosciences, Padova, Italy*

Framework structures, built of relatively rigid atomic clusters, are of central importance to mineralogy, materials science, chemistry and physics. They include perovskites, zeolites and metal-organic frameworks. Critical to the use of all of these materials is the fact that the physical, chemical and thermodynamic properties of frameworks are controlled in large part by the changes in tilting of the relatively rigid polyhedral units that comprise the frameworks. The intrinsic flexibility of frameworks means that displacive phase transitions involving discontinuous changes in the tilting of the polyhedral units is very common. These transitions give rise to many of the excess properties of industrial importance and modify the catalytic activity of materials such as zeolites. The appropriate description of phase transitions is the Landau model which has been applied very successfully to isobaric studies of phase transitions. Its extension in a fully self-consistent form to characterize the effects of pressure on phase transitions is challenging [1] because of non-linear elasticity and associated finite strains, and the expected variation of coupling terms with pressure. Further difficulties arise because of the need to integrate the resulting elastic terms over pressure in order to achieve a description of the P-V-T equation of state.

This is problematic as the primary data on materials at high pressures is usually volume data. I will show that a fully self-consistent description of the volume and unit-cell parameter variation of frameworks in the neighbourhood of structural phase transitions can be achieved by a Landau-type purely phenomenological approach to describe the spontaneous strains and that, for simple materials such as quartz, the P-V-T EoS is completely consistent with the Landau coupling terms being invariant with pressure. Fitting of P-V, T-V, and P-V-T data of materials undergoing phase transitions has been incorporated in to the EosFit-7c program [2].

[1] Troester et al. (2014) Physical Review X, 4, 031010
[2] Angel et al. (2014) Zeitschrift fuer Kristallographie 229, 405-419.

MS14-T02

Improving the quality of high-pressure structure refinements by implementing NPD anvils

R. Miletich[1], D. R. Allan[2], M. Warren[2], T. Pippinger[1,3], T. Yagi[4], T. Irifune[4]
[1]*Universität Wien, Institut für Mineralogie und Kristallographie, Wien, Germany*
[2]*Diamond Light Source, Didcot, United Kingdom*
[3]*Stoe & Cie GmbH, Darmstadt, Germany*
[4]*Ehime University, GRC, Matsuyama, Japan*

In-situ single-crystal XRD measurements using DACs in general are characterized by relatively lower quality of the data due to spatial restrictions giving lower resolution, high absorption effects due to DAC components, and interferences owing to parasitic diffraction effects from individual DAC components. In particular effects of parasitic diffraction (i.e. through diamonds, any anvil or backing-plate (BP) material, and primary-beam intensity drops related to single-crystal anvil materials used) have a strong influence on the quality of structure refinements. The development of nano-polycrystalline diamond (NPD) material with excellent mechanical and optical properties suggests an alternative experimental approach to reduce the amount of parasitic diffraction phenomena. A pair of anvils in Böhler-Almax geometry was used for comparative single-crystal experiments. First tests to collect XRD data sets were performed using inhouse (*Stoe Stadivari*, Mo-micro-source) and synchrotron (*Diamond Light Source*, beamline I19) X-ray sources and a 300K Pilatus pixel detector for all data collections. Results of several refinements based on data collections with this new type of nanopolycrystalline anvils will be presented with a focus on the obtained improvments on the quality of refinements.

MS14-T03

Pressure-induced phase transitions in the Wadsley-type bronze b-$Li_{0.33}V_2O_5$

A. Grzechnik[1], Y. Ueda[2], M. Hanfland[3], P. Hering[4], V. Potapkin[4], K. Friese[4]
[1]*RWTH Aachen University, Institute of Crystallography, Aachen, Germany*
[2]*University of Tokyo, Institute for Solid State Physics, Tokyo, Japan*
[3]*ESRF, Grenoble, France*
[4]*Forschungszentrum Jülich, Jülich, Germany*

b-$A_{0.33}V_2O_5$ bronzes (A = Li, Na, Ag) have a crystal structure (C2/m, Z = 6) [1] built of zigzag double strings of distorted VO_6 octahedra forming layers by joining corners. The adjacent layers are linked by chains of edge-sharing VO_5 tetragonal pyramids resulting in tunnels along the *b* axis. Each tunnel contains one symmetrically independent site that is partially occupied by the A^+ cations.

The low-dimensional b-$A_{0.33}V_2O_5$ bronzes exhibit metal-insulator phase transitions with successive charge-spin ordering at atmospheric conditions [2]. They are superconducting under high pressure, possibly due to a phase transition from the charge ordered to the superconducting phase [3]. The superconductivity occurs at about 7 GPa in b-$Na_{0.33}V_2O_5$ and b-$Ag_{0.33}V_2O_5$ and at about 9 GPa in b-$Li_{0.33}V_2O_5$. The exact mechanism of the superconductivity in these bronzes has not been presented so far. One of the most fundamental issues to be resolved is the determination of the underlying crystal structures.

Here we report on the high-pressure behaviour of b-$Li_{0.33}V_2O_5$ studied with synchrotron single-crystal diffraction in a diamond anvil cell at room temperature. This material undergoes a series of first-order phase transitions at about 9 and 11 GPa due to relative displacements of the adjacent octahedral layers. We were able to determine the low- (<9 GPa) and high-pressure polymorphs (>11 GPa). In the intermediate phase between 9 and 11 GPa very strong one-dimensional diffuse scattering is observed indicating the presence of stacking faults. The structural refinements of the data above 11 GPa reveal that the layers of VO_6 octahedra remain essentially intact. The relative position of the chains of edge-sharing VO_5 tetragonal pyramids with respect to the octahedral layers is changed. As a result, the tunnels populated by the Li^+ cations collapse on compression. The structure of the high-pressure polymorph of b-$Li_{0.33}V_2O_5$ is distinct from the charge-ordered polymorphs of b-$Na_{0.33}V_2O_5$ at low temperatures [4].

References

[1] A.D. Wadsley, Acta Cryst. 8, 695 (1955).

[2] K. Ohwada et al., Phys. Rev. B 85, 134102 (2012).

[3] T. Yamauchi, Y. Ueda, Phys. Rev. B **77**, 104529 (2008).

[4] J.-I. Yamaura et al., J. Phys. Chem. Solids 63, 957 (2002).

MS14-T04

Urea at high pressures and high temperatures

N. Rademacher[1], L. Bayarjargal[1], W. Morgenroth[1], B. Winkler[1]
[1]*Goethe University, Institute of Geosciences, Frankfurt am Main, Germany*

Chemical reactions at high pressures and temperatures are of interest because extreme conditions can lead to reaction pathways which result in novel compounds with unusual structural features and interesting properties [1]. Reactions of molecules from the chemical system C-H-N-O are of particular interest for example in astrophysics, condensed-matter physics or chemistry [2]. This study presents high pressure and high temperature experiments of urea (CH_4N_2O). Urea has previously been studied at high pressures and elevated temperatures by means of diffraction and spectroscopy [e.g. 3,4]. Olejniczak et al. [3] reported on three phase transformations of urea at pressures up to 12 GPa and ambient temperature. The ambient pressure phase I transforms to phase III at around 0.5 GPa, phase IV at around 3 GPa and to phase V at around 7 GPa [3]. Bridgman [4] found an additional phase transformation of phase I to phase II while heating urea above 373 K at around 0.6 GPa. However, a chemical reaction, e.g. polymerisation, has not been reported to occur at high pressures and high temperatures yet.
We have performed diamond anvil cell (DAC) experiments at pressures up to 47 GPa in order to study the stability field of urea. The experiments were performed with neon as pressure transmitting medium. We have followed the structural changes during pressure increase by means of micro-Raman spectroscopy and found at least one additional pressure-induced phase transformation. Up to 47 GPa we have not found any indication for a pressure- or irradiation-induced (photochemical) reaction between the urea molecules. The experiments at high pressures and temperatures up to 600 K are still ongoing.

The authors gratefully acknowledge financial support from the DFG (project RA 2585/1-1) and the BMBF (project 05K13RF1).

Figure 1: Section of micro-Raman spectra of urea up to 47 GPa (the spectra are not background corrected). The known phase transformations are marked. Significant changes can be seen at >30 GPa.

[1] R. Bini, M. Ceppatelli, M. Citroni and V. Schettino, Chemical Physics 398, 262 (2012).
[2] F. Datchi and G. Weck, Z. Kristallogr. 229, 135 (2014).
[3] A. Olejniczak, K. Ostrowska and A. Katrusiak, J. Phys. Chem. C 113, 15761 (2009).
[4] P. W. Bridgman, Proc. Am. Acad. Arts Sci. 52, 91 (1916).

Figure 1

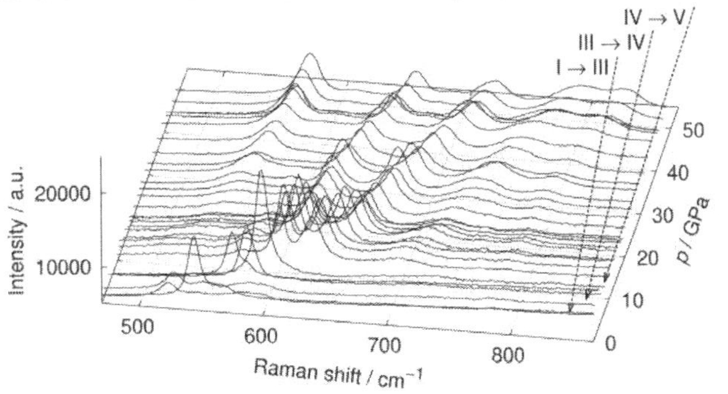

MS14-T05

Unveiling the transformation of α-cyclodextrin at high pressure: a combined X-ray diffraction and molecular dynamics study

R. Granero-García[1], B. Corry[2], F. P. A. Fabbiani[1]
[1]*GZG, Georg-August-Universität, Kristallographie, Göttingen, Germany*
[2]*Australian National University, Research School of Biology, Canberra, ACT, Australia*

In-situ high-pressure crystallisation has proved to be an effective technique to obtain polymorphs and hydrates of organic molecules [1]. While knowledge of the crystal structures of these new forms is interesting by itself, understanding how the transformations take place is even more compelling, for instance to improve crystal engineering and crystal structure prediction techniques [2,3].

We have recently explored the high-pressure behaviour in water of α-cyclodextrin, a cyclic derivative of starch composed of six glucose units. By means of single-crystal X-ray diffraction we have determined that α-cyclodextrin undergoes two major structural changes under compression up to 0.65 GPa: as pressure is applied, water is gradually incorporated in the cavity; and one of the terminal hydroxyl groups becomes increasingly disordered, expanding the available volume inside the cavity [4]. Application of pressure usually promotes opposite changes in crystal structures: enhancement of order and decrease of void volume [5].

These observations on α-cyclodextrin have motivated us to study the system by molecular dynamics using two complementary approaches. First, we have simulated an infinite crystal of α-cyclodextrin in vacuum, at ambient pressure and 0.65 GPa, to see how pressure influences the disorder of the primary hydroxyl groups. We have also used umbrella sampling to obtain the free energy curves for the rotation of several hydroxyl groups at both pressures. Second, we have simulated an arrangement of up to 20 unit cells immersed in water at variable pressure to see how water molecules are incorporated in the crystal and how they move into the cavities; a metadynamics simulation reproduces the energy landscape inside the unit cell to show the paths of minimal energy. The analysis of the trajectories and derived energies has let us draw a mechanism for the structural changes that is in agreement with the experimental observations.

[1] F. P. A. Fabbiani and C. R. Pulham, Chem. Soc. Rev. 2006, 35, 932-942.
[2] C. C. Fischer, K. J. Tibbetts, D. Morgan and G. Ceder, Nature Mater., 2006, 5, 641-646.
[3] J. Fayos, Cryst. Growth Des., 2009, 9, 3142-3153.
[4] R. Granero-García, F. J. Lahoz, C. Paulmann, S. Saouane and F. P. A. Fabbiani, CrystEngComm, 2012, 14, 8664-8670.
[5] A. Olejniczak, A. Katrusiak and A. Vij, CrystEngComm, 2009, 11, 1073-1080

MS15 - Structures of small biologically active molecules and protein-inhibitor complexes

MS15-T01
Invariom based Electron Density Studies of Sildenafil and Iso-Sildenafil, a comparison.

P. Luger[1], B. Dittrich[2]
[1] *Institute for Chemistry and Biochemistry, FU Berlin, Berlin, Germany*
[2] *University, Inorganic Chemistry, Hamburg, Germany*

It raised exceptional attention in the broad public, when in the 1990ies sildenafil (Viagra ©), an agent against male erectile disfunction came into use. The first X-ray analysis on a sildenafil compound was published in 1999 on iso-sildenafil [1], which is not in clinical use and it took surprisingly until 2005 when a first structure of sildenafil itself as a citrate salt appeared [2], followed one year later by a detailed study on 12 solvates of sildenafil saccharinate [3]. The structure of the free base of sildenafil was reported at first in 2012 [4]. From an application of the invariom formalism, the electron densities of sildenafil and the iso analogue were derived from their known low resolution X-ray data sets [1], [4], in order to provide in addition to steric properties electronic evidences for the different pharmacological behaviour of the two compounds. Inspection of the electrostatic potentials (ESP's) on the iso-electron density surfaces at 0.0067 eÅ^{-3} shows already clear ESP differences in the region of the pyrazolopyrimidine ring system, where both compounds are differently methylated, Fig.1. Further topological properties will be compared.

[1] M. M. El-Abadelah, S. S. Sabri, M. A. Khanfar, W. Völter, C. Maichle-Mössmer, *Z. Naturforsch.*, 1999, *54b*, 1323 - 1326.
[2] H. S. Yathirajan, B. Nagaraj, P. Nagaraja, M. Bolte, *Acta Cryst*, 2005, *E61*, o489-o491.
[3] R. Banerjee, P. M. Bhat, G. R. Desiraju, *Cryst. Growth Des.*, 2006, *6*, 1468 - 1478.
[4] D. Stepanovs, A. Mishnev, *Z. Naturforsch.*, 2012, *67b*, 491 - 494.
[5] C. B. Hübschle, P. Luger, *J. Appl. Crystallogr.*, 2006, 39, 901-904.

Fig. 1. Electrostatic potentials of sildenafil (left), Fig.2. iso-sildenafil (right), MOLISO representations [5].

Figure 1

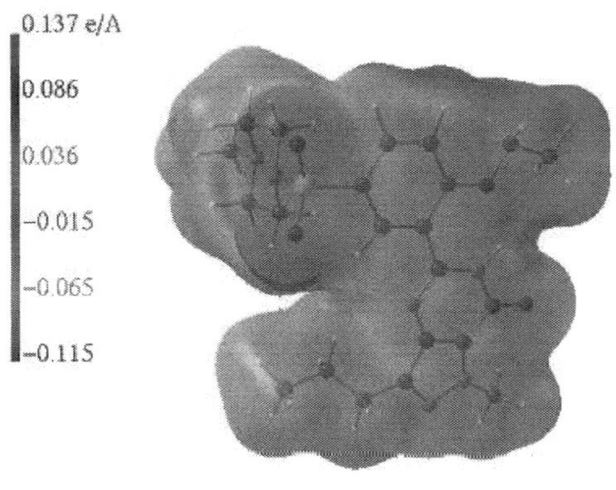

Figure 2

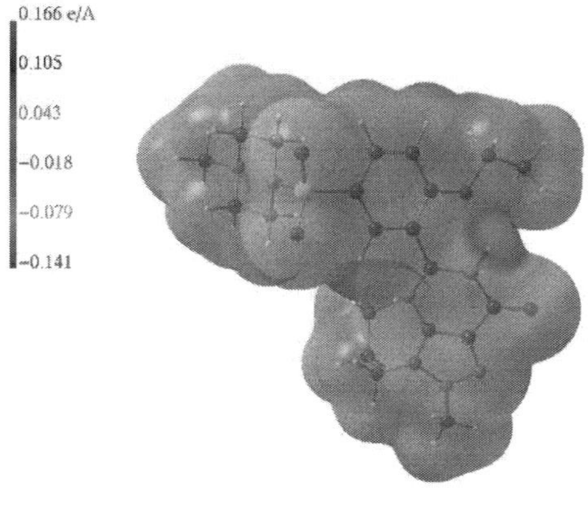

MS15-T02
Driving forces of co-crystallization: "in silico" *vs* experimental screening

L. Kuleshova[1], D. W. M. Hofmann[1]
[1] *CRS4, Biomedizine, Pula(CA), Italy*

Pharmaceutical cocrystallization represents a promising approach to generate novel crystal forms of known API with improved (or alternating) physicochemical and pharmacokinetic properties. Together with experimental methods, the various theoretical approaches have been suggested to minimize the costs of experimental efforts and introduce a "virtual filter" to the selection of an appropriate coformer (**CF**) already on a very early stage of research. The theoretical approaches mainly are based on the comparison of thermodynamic characteristics (free energy, enthalpy, electrochemical potentials and so on) of pure compounds *vs* cocrystals (**CC**). The supramolecular synthons also are considered as a driving force of cocrystallization, which can play important role while planing of cocrystallization strategy.

To get insight into driving forces, guiding cocrystallization, we performed full volume analysis of experimental and *in silico* screening results for several purine derivatives, which play an important role in chemistry of natural compounds (purine bases of DNA and RNA; coenzyme NAD; alkaloids, caffeine, theophylline and theobromine, toxins, saxitoxin and related compounds, uric acid), and thereby, in pharmaceutics. As the coformers pharmaceutically acceptable carboxylic acids have been tried. The free energy of experimental and hypothetical crystal structures have been calculated with *FlexCryst* program, according to the methodology, developed in[1]. The experimental[2] and *in silico* results were found matching very well with each other. The correlations of physical properties (melting points, solubility) with lattice energies are discussed. The peculiarities of intermolecular interactions in cocrystals are analysed regarding "seeding" role of synthons.

[1] Kuleshova, L. N.; Hofmann, D. W. M.; Boese, R. *Chem. Phys. Lett.* 2013, *564*, 26-32.

[2] Stepanovs, D., Kuleshova, L. N., Hofmann, D. W. M., Mishnev A. Crystal Growth and Design. submitted

Fig.1. Solubility and Lattice Free Energy Correlation for CF and CC of pentoxifylline

Fig.2. The strongest interactions in cocrystals of pentoxyfylline with aspirin

Figure 1

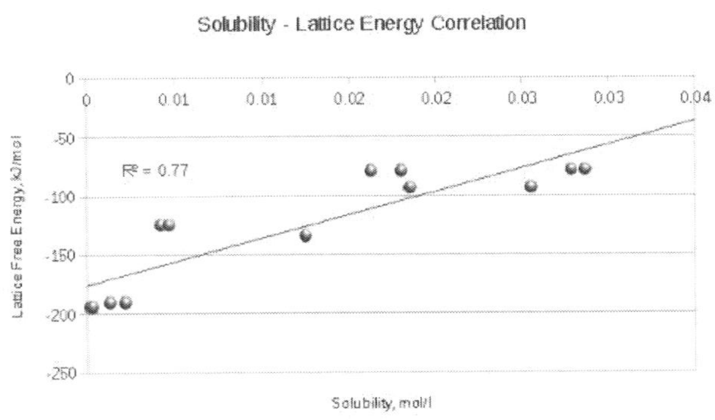

Figure 2

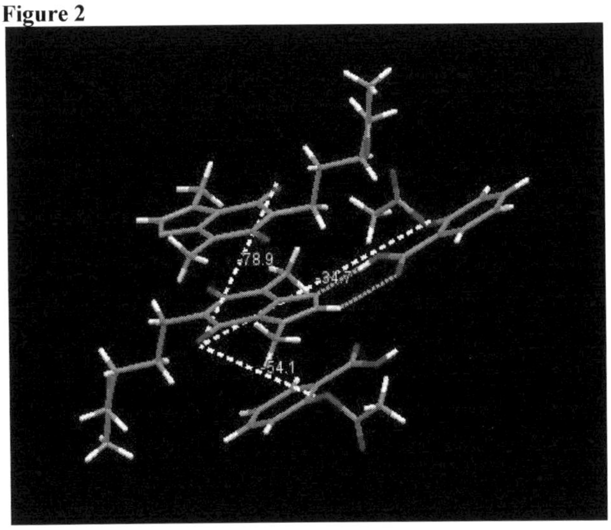

MS15-T03

Structural characterisation of novel potent inhibitors for treatment of TB infections

S. Eltschkner[1], W. Yu[2], A. Pschibul[1], P. Tonge[2], C. Kisker[1]
[1]Rudolf-Virchow-Zentrum, Universität Würzburg, Strukturbiologie, Würzburg, Germany
[2]Stony Brook University, Department of Chemistry, New York, United States

Due to the rapid development of resistances against antimicrobial agents for treatment of tuberculosis (TB) infections and the emergence of multidrug-resistant (MDR-TB) and extensively drug-resistant (XDR-TB) strains of *Mycobacterium tuberculosis*, TB infections have become a life-threatening issue during the last years. Therefore new and effective antibiotics are urgently needed.

The bacterial fatty acid synthesis pathway II (FAS II) is an important metabolic pathway since it is essential for the biosynthesis of mycolic acids as part of the outer layer of the mycobacterial cell wall. This layer prevents many antibiotics from penetrating into the cytosol. Compounds which impair the synthesis of mycolic acids are thus promising leads for the development of anti-mycobacterial agents.

The enoyl-ACP-reductase of *Mycobacterium tuberculosis*, InhA, which catalyses the last step in the elongation cycle of fatty acids in FAS II, serves as a target for the current first line drug isoniazid (INH) in TB infection treatment. Due to its rate-limiting character in the FAS II cycle this enzyme is an attractive candidate for drug development.

A compound's residence time plays an important role in its *in-vivo* efficacy. Based on the analysis of several diphenyl ethers it was shown that slow-onset inhibitors, in contrast to rapid-reversible compounds, lead to an ordering and closure of the substrate-binding loop (SBL) of InhA, resulting in a longer residence time.

Novel diaryl ether compounds with different substitution patterns were designed, aiming to prolong the residence time in the enzyme's active site. Here, we structurally characterise the state of the SBL of InhA in complex with these inhibitors using X-ray crystallography to understand the molecular basis underlying their longer residence times.Our crystallographic data reveal insights into the SBL-ordering by displaying different snapshots along the reaction coordinate of this process. Additionally, the structural basis for synergistic effects of certain substituents at different positions of the diphenyl ether scaffold, leading to an improvement of the inhibitors' residence times, can be visualised.

References:

Luckner S, Liu N, am Ende CW, Tonge PJ, Kisker C; A Slow, Tight Binding Inhibitor of InhA, the Enoyl-Acyl Carrier Protein Reductase from *Mycobacterium tuberculosis*; JBC 2010; 285-19: 14330-14337

Pan P, Tonge PJ;Targeting InhA, the FASII Enoyl-ACP Reductase: SAR Studies on Novel Inhibitor Scaffolds;CTMC 2012; 12: 672-693

Pan P et al.;Time-Dependent Diaryl Ether Inhibitors of InhA: Structure-Activity Relationship Studies of Enzyme Inhibition, Antibacterial Activity, and in vivo Efficacy;ChemMedChem 2014; 0000-00: 1-17

Li HJ, Lai CT, Pan P, Yu W, Liu N, Bommineni GR, Garcia-Diaz M, Simmerling C, Tonge PJ;A Structural and Energetic Model for the Slow-Onset Inhibition of the Mycobacterium tuberculosis Enoyl-ACP Reductase InhA;ACS Chem.Biol. 2014; 9 (4): 986-993

MS15-T04

Structure-based Design of Selective Ecto-5'-nucleotidase Inhibitors

J. Pippel[1], K. Knapp[1], M. Zebisch[1,2], A. El-Tayeb[3], C. E. Müller[3], N. Sträter[1]
[1]Institute of Bioanalytical Chemistry, Center for Biotechnology and Biomedicine, Leipzig, Germany
[2]University of Oxford, Division of Structural Biology, Oxford, United Kingdom
[3]University of Bonn, Pharmaceutical Chemistry I, Bonn, Germany

The eukaryotic ecto-5'-nucleotidase (e5NT) is an extracellular enzyme attached to the cell membrane and constitutes part of the purinergic signaling pathways. It catalyzes the hydrolysis of AMP and thereby generates the signaling molecule adenosine. The effects of adenosine are often opposite to those of ATP giving e5NT an important regulatory role in processes such as inflammation, hypoxia and certain types of cancer. Especially in the context of cancer, e5NT has been intensively studied. Recent results demonstrate that the enzyme suppresses the antitumor immune response via activation of A2A adenosine receptors in mice [1]. As a consequence, anti-e5NT antibodies administration inhibited tumor growth and metastasis in these animals [1,2]. Such results underline the high pharmaceutical potential of e5NT as drug target.

Structures of e5NT revealed an open and closed conformation for the enzyme [3,4]. The structural characterization showed that each subunit of e5NT is composed of a C- and N-terminal domain and that conformational change is achieved by a large (~100°) rotation of the N-terminal domain [3].

Based on these structures of e5NT we aim to support rational inhibitor design by determining complex structures. We therefore detected several potential e5NT inhibitors via high-throughput screening as well as in-silico screening and docking experiments resulting in some compounds with inhibitory effects up to the nM range. For a set of these compounds structural details of the inhibition mode could be unveiled by co-crystal structures. Those results were then used for rational modifications. Until now, a derivative of the substrate-mimicking compound AMPCP showed the most promising results. The π-stacking interactions appear to be important for the improved K_i value. Further modifications could include extension into a large water-filled pocket next to the base binding site [3].

References

1. Loi, S., Pommey, S., Haibe-Kains, B., Beavis, P. A., Darcy, P. K., Smyth, M. J. and Stagg, J. (2013). CD73 promotes anthracycline resistance and poor prognosis in triple negative breast cancer. PNAS *vol. 110 no. 27*, 11091-11096.
2. Stagg, J., Divisekera, U., McLaughlin, N., Sharkey, J., Pommey, S., Denoyer, D., Dwyer, K. M. and Smyth, M. J. (2010). Anti-CD73 antibody therapy inhibits breast tumor growth and metastasis. PNAS *vol. 107 no. 4*, 1547-1552.
3. Knapp, K., Zebisch, M., Pippel, J., El-Tayeb, A., Müller, C. E. and Sträter, N. (2012). Crystal Structure of the Human ecto-5'-Nucleotidase (CD73): Insights into the Regulation of Purinergic Signaling. Structure *20*, 2161-2173.
4. Heuts, D. P. H. M., Weissenborn, M. J., Olkov, R. V., Shaw, A. M., Levy, C. and Scrutton, N. S. (2012).Crystal Structure of a Soluble Form of Human CD73 with Ecto-5'-Nucleotidase Activity. ChemBioChem *13*, 2384-2391.

Figure 1

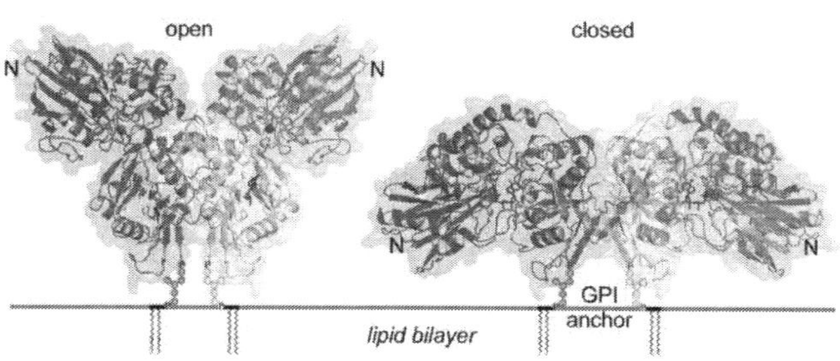

MS15-T05

Crystallographic fragment screening: challenges, opportunities and lessons learned

A. Heine[1], J. Schiebel[1], N. Radeva[1], A. Metz[1], F. Huschmann[1,2], M. Uhlein[2], K. Sparta[2], M. Weiss[2], U. Mueller[2], G. Klebe[1]
[1]*Philipps-University Marburg, Institute of Pharmaceutical Chemistry, Marburg, Germany*
[2]*Helmholtz-Zentrum Berlin für Materialien und Energie, Macromolecular Crystallography, Berlin, Germany*

Nowadays, fragment screening is an established method for lead development in pharmaceutical drug research. Library design and applied screening methods are critical for successful hit identification. We generated a small library, consisting of 364 compounds, which does not strictly adhere to the Astex rule of three often applied for library design.[1]

The pepsin-like aspartyl protease endothiapepsin (EP) was selected for library validation. EP is a model system for proteins involved in serious diseases such as malaria (plasmepsins), hypertension (renin) and Alzheimer's disease (ß-secretase) and, therefore, can aid in further drug development.

Fragments tend to exhibit only low affinity toward their target protein due to their limited size. This hampers hit identification and, therefore, often multiple screening methods are applied. We used a biochemical as well as various biophysical assays as pre-screening methods. Surprisingly, the results of the different screening methods showed only limited overlap. Thus, we decided that crystallographic screening of the entire library is ultimately required to identify all fragments that bind to EP. In any case, the resulting structural information is the basis for further synthesis by medicinal chemists.

Crystallographic screening of an entire library requires a steady supply of crystals, reproducible soaking conditions and a sufficient amount of beamtime. The latter was provided by the HZB BESSY II MX beamlines,[2] in combination with automation in initial data processing and refinement.

More than 60 fragment complex structures were obtained which will be compared to results from other screening methods. The binding modes of the resulting fragments will be discussed and provide an ideal basis for further development of endothiapepsin inhibitors.

1. H. Köster, T. Craan, S. Brass, C. Herhaus, M. Zentgraf, L. Neumann, A. Heine, G. Klebe; A Small Nonrule of 3 Compatible Fragment Library Provides High Hit Rate of Endothiapepsin Crystal Structures with Various Fragment Chemotypes, J. Med. Chem.54, 2011, 7784-7796.

2. U. Mueller, N. Darowski, M. R. Fuchs, R. Forster, M. Hellmig, K. S. Paithankar, S. Puhringer, M. Steffien, G. Zocher, M. S. Weiss; Facilities for macromolecular crystallography at the Helmholtz-Zentrum Berlin, J. Synchrotron Radiat. 19, 2012, 442-449.

MS16 - Material Science IV

MS16-T01
The polysulfates $[S_nO_{3n+1}]^{2-}$: Where is the *n*d? The first hexasulfate $Rb_2[S_6O_{19}]$

L. V. Schindler[1], M. S. Wickleder[1]
[1]*Universität Oldenburg, Institut für Chemie, Oldenburg, Germany*

Although sulfuric acid and its derivatives are some of the most important chemical compounds in research and industry their chemistry is still not fully elucidated.[1] One of the topics of interest in our group is the investigation of polysulfuric acids $H_2S_nO_{3n+1}$ and their salts. These acids are believed to form in condensation reactions in SO_3 rich H_2SO_4, however, structural knowledge is scarce. Meanwhile, our group presented various polysulfates with chain length up to n = 4.[2] By now the highest polysulfate and the only alkaline metal polysulfate with a chain length longer than n = 2 is potassium pentasulfate $K_2[S_5O_{16}]$.[3] Recently we successfully reproduced the mentioned compound and analyzed it with state of the art methods (ICSD 428773). Moreover, we were able to synthesize further alkaline metal polysulfates depending on the reaction conditions, namely $Na_2[S_3O_{10}]$ (ICSD 428772) and $K_2[S_3O_{10}]$ (ICSD 428771). Aside from these we present here another pentasulfate, $Cs_2[S_5O_{16}]$ (ICSD 428769) and the first hexasulfate, $Rb_2[S_6O_{19}]$ (ICSD 428770). The compounds were obtained from the reaction of the respective alkaline metal carbonate with neat SO_3 in torch-sealed glass ampoules at 80°C. $Rb_2[S_6O_{19}]$ crystallizes in the monoclinic space group $P2_1/c$ (a = 1581.97(5) pm, b = 722.00(2) pm, c = 14.5257(5) pm, β = 102.614(2)° V = 1619.06(9) 10^6 pm^3), $Cs_2[S_5O_{16}]$ in the monoclinic space group $P2_1$ (a = 892.96(3) pm, b = 859.72(3) pm, c = 978.30(4) pm, β = 101.443(2)° V = 736.11(5) 10^6 pm^3). Figure 1 shows the respective anions of these structures. Previously we discussed the correlation of the S-O bond lengths within the terminal S-O-S bridge and the number of $[SO_4]$ units in the polysulfate anions $[S_nO_{3n+1}]^{2-}$ based on theoretical investigations and demonstrated that the asymmetry of the terminal S-O-S bridge is assumed to increase with growing chain length.[4] In the crystal structure of $Cs_2[S_5O_{16}]$ we find this asymmetry in the S1-O121-S2 bridge (190.1(1) pm / 150.2(1) pm) in accordance with the calculations, whereas in the S4-O451-S5 bridge (148.5(1) pm / 195.9(1) pm) it is a little bit more distinctive than expected. In the crystal structure of the hexasulfate the asymmetry of the S1-O121-S2 bridge (196.18(6) pm / 149.04(6) pm) corresponds to the expectations, within the S5-O561-S6 bridge (145.22(6) pm / 220.49(6) pm) it exceeds the prediction by far.

Figure 1: Structure and labeling scheme of the anions in the crystal structures of $Cs_2[S_5O_{16}]$ and $Rb_2[S_6O_{19}]$. The displacement ellipsoids are drawn at an 85% probability level.

[1] M. S. Wickleder, C. Logemann, in *Handbook on Chalcogen Chemistry, Vol. 1* (Eds.: F. Devillanova, W.-W. du Mont), Royal Society of Chemistry, London, 2013, pp. 307-345.

[2] a) J. Bruns, T. Klüner, M. S. Wickleder, *Angew. Chem.* 2013, *125*, 2650-2652; b) C. Logemann, T. Klüner, M. S. Wickleder, *Z. Anorg. Allg. Chem.* 2012, *638*, 758-762.

[3] R. de Vries, F. C. Mijlhoff, *Acta Crystallogr.* 1969, *B25*, 1696-1699.

[4] C. Logemann, T. Klüner, M. S. Wickleder, *Angew. Chem.* 2012, *124*, 5082-5085.

Figure 1

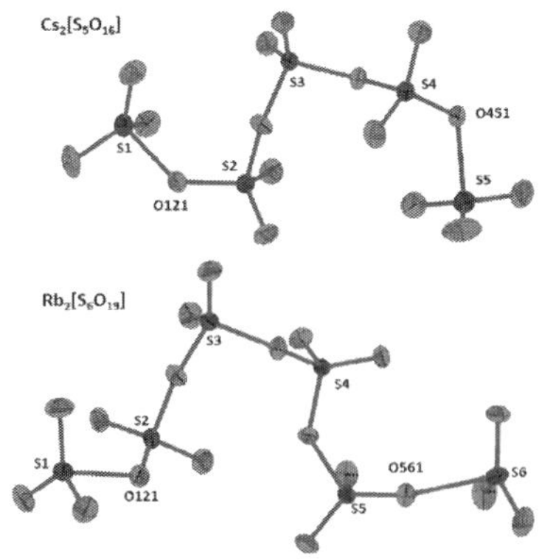

MS16-T02
Oxidation behaviour of Zintl phases MTt, M=Sr, Ba; Tt=Si, Ge; under hydrogen pressure

H. Auer[1], P. Wenderoth[2], H. Kohlmann[1]
[1]*Universität Leipzig, Chemistry, Leipzig, Germany*
[2]*Universität der Saarlandes, Chemistry, Saarbrücken, Germany*

Zintl phases may take up hydrogen by at least two reaction path ways that lead either to an oxidized Zinlt anion or a polyanionic hydride. In the former case hydrogen is coordinated by the electropositive element while in the latter case hydrogen binds covalently to the polyanion [1]. A prominent example for the oxidation of a Zintl anion is $CaSiD_x$ x=1.0-1.2 where deuterium occupies tetrahedral voids [2] and the analogous compounds $MSiD_x$, M=Sr, Ba [3]. We reinvestigated the structure of $MSiD_x$ and determined hydrogen position from neutron powder diffraction. Deuterium occupies tetrahedral M_4 and in addition M_3Si voids [4]. In advance we investigated the hydrogenation behaviour of the heavier germanides MGe, M=Sr, Ba. For these systems we present thermal data for the formation and decomposition of hydrogenated phases under 50bar H_2 pressure which were collected on a high pressure differential scanning calorimetry instrument (HP-DSC, Q1000 TA Instruments). For the Ba-Ge-H system we can distinguish 4 steps of the reaction of BaGe with hydrogen until the segragation into $BaGe_2$ and BaH_2. All indermediate steps are stable at room temperature without H_2 pressure, thus, *ex situ* characterisation is possible. The second phase, $BaGeH_x$, could be indexed and is isopointal to $CaSiD_x$ with lattice parameters a = 15.852(2), b = 4.2375(6), c = 12.988(2). Reaction step one and three lead to powders of poor crystallinity. From pressure dependent measurements on the HP-DSC we assume that the $BaGeH_x$ phase has the highest hydrogen content. In the Sr-Ge-H system a corresponding phase could be isolated as well with lattice parameters a = 15.3741(9), b = 4.0602(2), c = 11.8347(6). Thermal data suggest a further reaction at higher temperatures. This step is reversible with a temperature hysteresis of 80°C.

[1] U. Häußermann, V. F. Kranak, K. Puhakainen, *Struct. Bond.* 2011, *139*, 143-162.
[2] H. Wu, W. Zhou, T. J. Udovic, J. J. Rush, T. Yildirim, *Phys. Rev. B: Condens. Matter Mater. Phys.* 2006, *74*, 224101.
[3] M. Armbruster, M. Wörle, F. Krumeich, R. Nesper, *Z. Anorg. Allg. Chem.* 2009, *635*, 1758-1766 .
[4] P. Wenderoth, PhD thesis, *Untersuchungen zur Hydridbildung von Zintl-Phasen der Erdalkalimetalle mit Aluminium, Gallium und Silicium* , Universität des Saarlandes, 2014.

MS16-T03

Microstructural evolution of ferroelectric $BiScO_3$-$PbTiO_3$ throughout its morphotropic phase boundary

K. Datta[1], A. Richter[2], M. Göbbels[2], R. Neder[1]
[1]*Friedrich-Alexander Universität Erlangen/Nürnberg , Kristallographie und Strukturphysik, Erlangen, Germany*
[2]*Friedrich-Alexander Universität Erlangen/Nürnberg, Lehrstuhl für Mineralogie, Erlangen, Germany*

Morphotropic phase boundary (MPB) is an important theme in the research of ferroelectric materials since materials often exhibit enhanced physical properties at MPB. Although an established understanding of creating MPB behaviour relies on invoking instability of average crystal structure or a low-symmetry monoclinic phase in the system driven by composition [1], the actual mechanism is far more complex and heavily dependent on the local ordering of the cations as evidences found in recent diffuse scattering studies [2] on either lead-free or lead-based based systems.

We have studied an important and popular ferroelectric system $x$$BiScO_3$-(1-$x$) $PbTiO_3$, which was first reported in 2001 with MPB characteristics [3], through total scattering technique to understand its structure evolution as a function of composition in the vicinity of the MPB. Total scattering technique which is essentially the analysis of the pair distribution function (PDF) of a system provides crucial structural information at the microscopic level which are not easily available from conventional structural analysis like Rietveld refinement. Both x-ray and neutron powder diffraction experiments were carried out on six different compositions of BS-PT in the range $0.30<x<0.40$ and neutron PDFs were analysed through RMC simulations to extract the behaviour of individual cations. It was observed that locally there was no abrupt change in any of the cation-cation correlations to assign the MPB unambiguously, however specific features were noticed in the shifts of the cations with respect to their oxygen framework.

Reference :

[1] R. Guo, L. E. Cross, S-E. Park *et al* Phys. Rev. Lett. 84, 5423-5426 (2000).
[2] R. E. Eitel, C. A. Randall, T. R. Shrout *et al*, Jpn. J. Appl. Phys. 40, 5999 (2001).
[6] D. J. Goossens , ISRN Materials Science, 2013, 107178/1-17 (2013).

MS16-T04

The high pressure structural phase transformation of coffinite, $USiO_4$ studied by experimental and computational methods

J. Bauer[1], S. Labs[2], W. Morgenroth[1], V. Milman[3], H. Curtius[2], B. Winkler[1]
[1]*Goethe Universität Frankfurt am Main, Geowissenschaften - Kristallographie, Frankfurt am Main, Germany*
[2]*Forschungszentrum Jülich GmbH, Inst. f. Energie- und Klimaforschung – Nukleare Entsorgung (IEK-6), Jülich, Germany*
[3]*Dassault Systèmes BIOVIA, Cambridge, United Kingdom*

Coffinite, $USiO_4$, can form under reducing conditions from UO_2 in contact with silica-rich waters (Langmuir's criterion) [1]. Spent nuclear fuel (SNF) consists to > 90% of UO_2, therefore the safety assessment for a final repository in deep geological formation will benefit greatly if coffinite is taken into account as a potential secondary phase. While high pressures are not of specific relevance for a final repository for SNF, its structural behavior at high pressures is of general interest to understand the phase stabilities and to benchmark model calculations.

The high pressure behavior of coffinite has been studied on natural and synthetic samples. A pressure-induced irreversible phase transformation from the zircon- to the scheelite-type structure was found at about 15 GPa [2]. We have performed pressure dependent synchrotron powder diffraction (PETRA III, DESY) and Raman measurements on synthetic coffinite powder in the diamond anvil cell. In addition, DFT+U calculations were performed [3]. The bulk moduli of the low and high pressure phases were determined from the diffraction experiments. The bulk modulus of the high pressure scheelite-type phase is significantly lower than the value proposed earlier [2]. We explain that by non-hydrostatic conditions using an alcohol-water mixture by Zhang [2] in contrast to quasihydrostatic conditions in the new experiments using neon as a pressure medium [3]. The experimental results are in very good agreement with the data from our DFT+U calculations.

Raman spectra could be measured up to a pressure of about 19 GPa. At pressures above approx. 3 GPa, we could observe the "activation" of a high-frequency infrared-active mode in the Raman spectra. The mode Grüneisen parameters derived from the experimental data are in good agreement with the theoretical results. The DFT+U calculations further reveal that one of the silent (neither Raman- nor IR-active) vibrational modes is a soft mode having an imaginary frequency at high pressures and triggering the structural phase transition.

Financial support from BMBF, Germany, grants 02NUK019C, 02NUK019E and 05K13RF1 is gratefully acknowledged.

References

[1] D. Langmuir: *Geochim. Cosmochim. Acta* 1978, **42**, 547-569
[2] F.X. Zhang et al.: *Am. Mineral.* 2009, **94**, 916-920
[3] J.D. Bauer et al.: *J. Phys. Chem. C* 2014, **118**, 25141-25149

MS16-T05

Multiferroic switching kinetics of $MnWO_4$ and $TbMnO_3$

J. Leist[1], M. Baum[2], J. Stein[2], T. Cronert[2], T. Finger[2], K. Schmalzl[3], L.- P. Regnault[3], A. Hiess[3,4], P. Becker[5], L. Bohaty[5], G. Eckold[1], M. Braden[2]
[1]*Georg-August-Universität Göttingen, Institut für Physikalische Chemie, Göttingen, Germany*
[2]*Universität zu Köln, II. Physikalisches Institut, Köln, Germany*
[3]*Institut Laue-Langevin, Grenoble, Sweden*
[4]*present adress: European Spallation Source, Lund, Sweden*
[5]*Universität zu Köln, Institut für Kristallographie, Köln, Germany*

Multiferroic crystals are characterized by a coexistence of both ferroelectric and magnetic order in the same phase. This is of particular interest for the construction of novel memory devices. In materials like $MnWO_4$ or $TbMnO_3$ the magnetic moments of the manganese ions form an elliptical spiral. This induces a spontaneous electric polarization due to the inverse Dzyaloshinskii-Moriya interaction [1]. Moreover, the chirality of the spiral can be changed by the application of an electric field [2].

Polarized neutron scattering was used to determine the chiral ratio of the magnetic structure. The experiments were performed at the cold three-axes-spectrometers IN12 and IN14 at the ILL neutron source in Grenoble. The main focus of our investigation was to study the time dependence of the transition with stroboscopic scattering techniques [3]. The direction of an electric field of up to 15 kV/cm was reversed periodically and the intensity of the scattered neutrons recorded as a function of time in order to investigate how fast the magnetic structure adapts to the pulsed field. In addition static magnetoelectric hysteresis loops were recorded to characterize the samples.

In $MnWO_4$ the transition is quite slow within the multiferroic phase between 7.0 and 12.3 K. Approaching both transition temperatures a faster response to the electric field is observed. The relaxation times are found to be between 2 and 30 ms. The kinetics also depend on the direction of the electric field applied during cooling [4]. Fig. 1 shows an example of a stretched exponential function, fitted to the time evolution of the chiral ratio.

In $TbMnO_3$ the transitions slows down considerably below the Curie-temperature at 28.0 K. The relaxation times show a logarithmic temperature dependence over 5 decades.

[1] H. Sagayama et al., Phys. Rev. B 77, 220407 (2008).
[2] T. Finger et al., Phys. Rev. B 81, 054430 (2010).
[3] G. Eckold, H. Schober, S. E. Nagler, (Ed.), Studying Kinetics with Neutrons, Springer, 2010.
[4] Baum et al., Phys. Rev. B 89, 144406 (2014).

Fig. 1: Time evolution of the chiral ratio in $MnWO_4$.

Figure 1

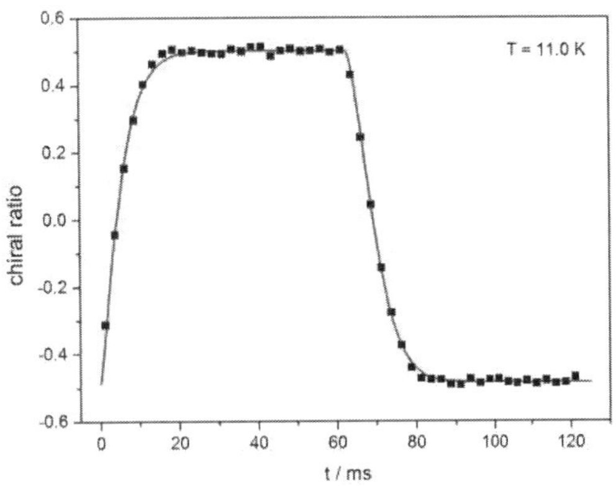

MS16-T06

X-ray diffraction study of inhomogeneous deformations in α-quartz single crystal, impacted by a nanosecond electric pulse

H. Choe[1], S. Gorfman[1], M. Zilokowski[1], M. Vogt[1], S. Heidbrink[1], P. Gaal[2], W. Leitenberger[3], U. Pietsch[1]
[1]*University of Siegen, Department of Physics, Siegen, Germany*
[2]*Helmholtz-Zentrum Berlin für Materialien und Energie, Wilhelm-Conrad-Röntgen Campus, BESSY II, Berlin, Germany*
[3]*Universität Potsdam, Institut für Physik und Astronomie, Potsdam, Germany*

Piezoelectric effect is the physical property describing the ability of some materials to convert electrical energy to mechanical energy or vice versa. Despite the great technological importance of piezoelectric materials, many fundamental aspects of piezoelectricity are still quite poorly understood. For example, very little are known about reaction of a piezoelectric crystal to ultra-short (e.g. nanosecond) electric pulses on the macroscopic and mesoscopic length scales. The aim of this work is to investigate the reaction of α-quartz - most comprehensively studied materials - to a nanosecond electric pulse by time-resolved X-ray diffraction.

We report on the measurement of time-dependent rocking curves of selected Bragg reflections from the 0.05mm thick α-quartz (α-SiO_2) cut perpendicular to [100] direction. The measurements were performed at the XPP/KMC3 @ BESSY beamline. We applied 2ns electric pulse reaching the maximum amplitude of 30 kV/mm and collected diffraction intensity as a function of time delay after the applied pulse. The measurement was performed stroboscopically at the repetition rate of 2 kHz [1-3]. Figure 1 shows the intensity map of 6 -3 0 Bragg rocking curve as a function of time (passed after the pulse) and rocking angle. We observed the range of phenomena, corresponding to the changes on different time and length scales, including: (1) the change of integrated intensity, (2) oscillation of peak position, caused by the piezoelectric strain relaxation [4-5], and (3) change of full width half maxima. The change of intensity can be caused by both, the rearrangement within the unit cell (change of the structure factors) or alternatively by the change induced of the mosaicity (change of the extinction parameter). The time-dependence of above phenomena will be discussed in the presentation.

[1] S. Gorfman, O. Schmidt, M. Ziolkowski, M. Kozierowski, and U. Pietsch, J. Appl. Phys. 108, 064911 (2010)
[2] S. Gorfman, O. Schmidt, V. Tsirelson, M. Ziolkowski, and U. Pietsch, Z. Anorg. Allgem. Chem. 639(11), 1953-1962 (2013)
[3] S. Gorfman, Crystallogr. Rev. 20(3), 210-232 (2014)
[4] S. J. van Reeuwijk, A. Puig-Molina, O. Mathon, R. Tucoulou, and H. Graafsma, J. Appl. Phys.94, 6708 (2003)
[5] H. Navirian, H. Enquist, R. Nueske, A. Jurgilaitis, C. v. KorffSchmising, P. Sondhauss, and J. Larsson, Phys. Rev. B81, 024113 (2010)

Figure 1 The dynamics of α-quartz 6 -3 0 Bragg rocking curves as function of delay time and the time-resolved Bragg intensity map by induced nanosecond electric pulse

Figure 1

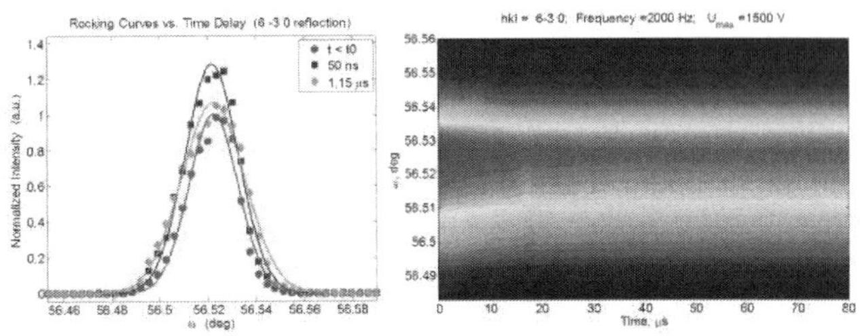

MS17 - Spectroscopy II

MS17-T01

High-pressure induced phase transition of $MnWO_4$ studied by single crystal XRD and Raman spectroscopy

J. Ruiz-Fuertes[1], A. Friedrich[1], O. Gomis[2], D. Errandonea[3], W. Morgenroth[1], J. A. Sans[2], D. Santamaría-Pérez[3,4]
[1]*Goethe-Universitaet Frankfurt, Frankfurt am Main, Germany, Germany*
[2]*Universitat Politècnica de València, Valencia, Spain, Spain*
[3]*Universitat de València, Valencia, Spain, Spain*
[4]*University College London, London, UK, Spain*

Manganese tungstate ($MnWO_4$) is a wolframite-type (*P2/c*) multiferroic compound that shows three antiferromagnetic phases below 13.7 K [1]. The intermediate AF2 phase, found between 7.6 and 12.7 K, is in addition ferroelectric as a consequence of the helicoidal spin arrangement that lifts the center of inversion. According to Chaudhury et al. [2] the AF2 phase is frustrated beyond a critical pressure of 4 GPa, but the Néel temperature of the first antiferromagnetic AF1 phase increases with pressure at dT_N/dP = 2.1(1) K/GPa. Extrapolating this result one could speculate that upon sufficient compression the AF1 phase could be stabilized at temperatures close to 293 K. Thus, high pressure could open new avenues to the study of $MnWO_4$. Even though there have been some efforts to study the optical [3] and structural [4] properties of $MnWO_4$, these have been limited to 10 GPa. Regarding the vibrational properties, Raman spectroscopic studies have revealed the presence of a structural phase transition at around 25 GPa that remains controversial [5,6].

In this work we study the structural behavior of $MnWO_4$ up to 31 GPa with single-crystal x-ray diffraction performed at the P02.2 Extreme Conditions Beamline at PETRA III (DESY, Hamburg). We have found that $MnWO_4$ transforms from a monoclinic (*P2/c*) structure to a triclinic (*P*-1) phase at 20.1 GPa that consists of at least two twin domains of identical triclinic unit-cells but rotated with respect to the parent monoclinic structure. In addition, the symmetry difference between the two types of Raman active modes observed on monoclinic $MnWO_4$ has been used to select only half of the Raman-active modes in oriented single crystals to solve the controversy about the onset of the phase transition.

J.R.-F. thanks the Alexander von Humboldt Foundation for a postdoctoral fellowship and Prof. M. M. Gospodinov for providing the samples. A.F. acknowledges financial support from the DFG, Germany (Fr2491/2-1). W.M. thanks the BMBF (05K13RF1). This research was partially supported by the Spanish government MINECO (MAT2013-46649-C4-1/2-P). DESY Photon Science is gratefully acknowledged. PETRA III at DESY is a member of the Helmholtz Association (HGF).

[1] T. Finger, D. Senff, K. Schmalzl, W. Schmidt, L. P. Regnault, P. Becker, L. Bohaty, and M. Braden, Phys. Rev. B 81, 054430 (2010).
[2] R. P. Chaudhury, F. Yen, C. R. De la Cruz, B. Lorenz, Y. Q. Wang, Y. Y. Sun, and C. W. Chu, Physica B 403, 1428 (2008).
[3] J. Ruiz-Fuertes,S. Lopez-Moreno, J. Lopez-Solano, D. Errandonea, A. Segura, R. Lacomba-Perales, A. Muñoz, S. Radescu, P. Rodriguez-Hernandez, M. Gospodinov, L. L. Nagornaya, C. Y. Tu, Phys. Rev. B 86, 125202 (2012).
[4] J. Macavei and H. Schulz, Z. Kristallogr. 207, 193 (1993).
[5] R. C. Dai, X. Ding, Z. P. Wang, and Z. M. Zhang, Chem. Phys. Lett. 586, 76 (2013).
[6] J. Ruiz-Fuertes, D. Errandonea, O. Gomis, A. Friedrich, F. J. Manjon, J. Appl. Phys. 115, 043510 (2014).

MS17-T02

Pressure-induced transformation processes in ferroelastic $Pb_3(P_xAs_{1-x})O_4)_2$, x = 0 and 0.8

B. Mihailova[1], R. J. Angel[2], U. Bismayer[1]
[1]*University of Hamburg, Earth Sciences, Hamburg, Germany*
[2]*University of Padua, Geosciences, Padua, Italy*

The quest for smart ferroic materials has been motivating studies of solid solutions at non-ambient conditions. Palmierite-type $Pb_3(P_{1-x}As_xO_4)_2$ is a textbook example of ferroelastic material which exhibits two-mode behaviour, allowing for studying structural transformations in ferroics with relatively weak correlation between the chemically distinguishable building units. Here we report on the pressure evolution of the Raman-active phonon modes of $Pb_3(P_xAs_{1-x})O_4)_2$, x = 0 and 0.8. The derived characteristic pressures were compared with the monoclinic-to-rhombohedral critical pressures determined by x-ray diffraction: 1.8 GPa and 1.0 GPa for x = 0 and 0.8, respectively. The critical pressure in $Pb_3(PO_4)_2$ is clearly mirrored by the softening and damping of the in-plane Pb-localized phonon modes, broadening of the in-plane antisymmetric stretching and bending TO_4 modes, vanishing of the splitting of the in-plane TO_4 bending, and a kink in the pressure dependence of the wavenumber w of the symmetric and in-plane antisymmetric TO_4 modes. The pressure evolution of the splitting of the TO_4 bending mode, which is indicative of the local tetrahedral monoclinicity, suggests a second-order structural transition in terms of Landau theory. By strong contrast to the case of pure $Pb_3(PO_4)_2$, the change in the long-range order at p_c = 1 GPa is barely seen by the lattice dynamics of $Pb_3(P_{0.2}As_{0.8})O_4)_2$. The only similarity with $Pb_3(PO_4)_2$ is that at 1 GPa there is a kink of $w(p)$ of the in-plane antisymmetric stretching of the dominant AsO_4 tetrahedral unit. Additionally, the $w(p)$-dependence for the symmetric stretching of the tetrahedral PO_4 unit of a minority also shows a kink at 1 GPa. The absence of rearrangement in the Pb sublattice of $Pb_3(P_{0.2}As_{0.8})O_4)_2$ near p_c may be the reason for the zero non-symmetry-breaking strain revealed by the pressure evolution of the unit cell parameters. The splitting of the AsO_4 bending persists well above p_c, indicating substantial order-disorder phenomena. The in-plane Pb-localized modes exhibit significant softening within the pressure range 4-7 GPa, suggesting the occurrence of a second phase transition at high pressures.

MS17-T03

Evaluation of Structure Proposals of modulated Ho_2PdSi_3 using Diffraction Anomalous Fine Structure

M. Nentwich[1], M. Zschornak[1], C. Richter[1,2], D. Novikov[2], D. C. Meyer[1]
[1]*TU Bergakademie Freiberg, Experimentelle Physik, Freiberg, Germany*
[2]*DESY, Photon Science, Hamburg, Germany*

Amongst the family of intermetallic compounds R_2PdSi_3 (with the rare earth R) only few do not form a superstructure of the hexagonal aristotype RSi_2 with space group *P6/mmm* (191). There are different models for the particular superstructures involving diverse combinations of occupational and displacive modulation. These modulations manifest themselves by changes in the electronic structure compared to the aristotype. Here, the site-, element- and orbital-selective method Diffraction Anomalous Fine Structure (DAFS) was used to study the local electronic structure of the representative Ho_2PdSi_3. The experimental part was conducted at beamlines E2 and BW1 of former synchrotron DORIS III in Hamburg. Simulation of DAFS, XAFS and LDOS of different structure proposals has been performed using FDMNES.
The experimental spectra of XAFS and DAFS, *i.a.* at a satellite reflection, show a better agreement with the 2 × 2 × 8 super structure models compared to the 2 × 2 × 1 model. Therefore, Ho_2PdSi_3 exhibits the biggest unit cell of the R_2PdSi_3 family known so far.

MS17-T04

The site preferences of Al and Ga in (Al, Ga)-doped garnet-type $Li_7La_3Zr_2O_{12}$ as deduced from ^{27}Al and ^{71}Ga MAS NMR at ultrahigh magnetic fields

D. Rettenwander[1], J. Langer[2], W. Schmidt[2], K. Harris[3], V. Terskikh[4], G. Goward[3], M. Wilkening[2], G. Amthauer[1]
[1]*Universität Salzburg, Materials Research & Physics, Salzburg, Austria*
[2]*Graz University of Technology, Christian Doppler Laboratory for Lithium Batteries, Institute for Chemistry and Technology of Materials, Graz, Austria*
[3]*McMaster University, Department of Chemistry & Chemical Biology, Hamilton, Ontario, Canada*
[4]*National Ultrahigh-Field NMR Facility for Solids, Ottawa, Ontario, Canada*

$Li_7La_3Zr_2O_{12}$ (LLZO) garnet is a promising Li ion conducting electrolyte for all-solid state Li ion and Li-air batteries.[1] LLZO has two polymorphic structures with largely different Li ion conductivities. The conductivity of the cubic phase (*Ia-3d*) ranges from 10^{-3} to 10^{-4} S/cm, which is two orders of magnitude higher than that of the tetragonal phase ($I4_1/acd$).[2-4] Unfortunately, the poorly conducting tetragonal polymorph is the thermodynamic stable phase at room temperature (RT). The cubic modification, however, can be stabilized at RT by supervalent doping (*e.g.*, 3 $Li^+ \leftrightarrow Al^{3+}/Ga^{3+}$).[5,6] The associated change in defect chemistry is expected to strongly influence both Li self-diffusivity and, therefore, Li ion conductivity. Hence, considerable efforts have been undertaken to understand the role of dopant cations that are responsible for the stabilization of highly conducting cubic LLZO. Contrary to the 96*h* sites, the 24*d* voids connect the loops of the Li-ion pathways formed by the 96*h* (48*g*) and 24*d* sites in the garnet structure. Thus, one may anticipate that the occupation of these sites by dopant cations act as blockades for the mobile Li ions. In literature, the variation in ionic conductivity is frequently attributed to the different site occupation of Al and Ga based on crystal chemical consideration as well as NMR studies at lower field.[6] Our latest high-field NMR studies on Al and Ga doped LLZO carried out at 21 Tesla, however, reveal two distinct ^{71}Ga (spin-3/2) NMR central lines differing in line widths because of distinct second order quadrupolar broadening. This shows that similarly to Al^{3+}, Ga^{3+} occupies both the 24*d* and 96*h* sites. The result is in contrast to previous NMR studies performed at lower magnetic fields.[6] At lower fields the rather broad ^{71}Ga line, which can be attributed to Ga ions on 24*d* sites, turns out to be almost invisible. This inevitably raises the question about the significance of the above mentioned blocking effect on Li-ion diffusivity. It should be noted that preliminary ^7Li NMR spectra reveal a decrease of the line width with increasing Ga content even though Ga ions occupy both the 24*d* and 96*h* sites.

[1] R. Murugan, V. Thangadurai and W. Weppner *Angew. Chem. Int. Ed.*, 2007, 46, 7778.
[2] J. Awaka, N. Kijima, H. Hayakawa and J. Akimoto, *J. Solid State Chem.*, 2009, 182, 2046.
[3] A. Kuhn, S. Narayanan, L. Spencer, G. Goward, V. Thangadurai and M. Wilkening, *Phys. Rev. B*, 2011, 83, 094302.
[4] J. Awaka, A. Takashima, H. Hayakawa, N. Kijima, Y. Idemoto and J. Akimoto, *Key Eng. Mater.*, 2011, 485, 99.
[5] H. Buschmann, J. Dölle, S. Berendts, A. Kuhn, P. Bottke, M. Wilkening, P. Heitjans, A. Senyshyn, H. Ehrenberg, A. Lotnyk, V. Duppel, L. Kienle and J. Janek, *Phys. Chem. Chem. Phys.*, 2011, 13, 19378.
[6] D. Rettenwander, C. A. Geiger, M. Tribus, P. Tropper and G. Amthauer, *Inorg. Chem.*, 2014, **53**, 6264.

MS17-T05

Halogen Bonds easily revealed with IR-Spectroscopy

V. Vasylyeva[1,2], L. Catalano[2], S. Nayak[2], C. Nervi[3], G. Terraneo[2], P. Metrangolo[2], G. Resnati[2]
[1]*RWTH Aachen, Aachen, Germany*
[2]*Politecnico di Milano, Milan, Italy*
[3]*University of Turin, Turin, Italy*

In recent years halogen bonds (XBs) of the type R-X···Y (X=I, Br, Cl; Y=O, N) were analyzed by various techniques and identified as robust enough to be employed in the design of supramolecular architectures.[1] The infrared (IR) spectroscopy, however, is not commonly used to prove the formation of XBs due to the low vibration frequencies and weak intensities of the C-X bond. Our goal is to demonstrate that IR spectroscopy, particularly in the far-IR region, can be a simple, sensitive, and easily accessible tool to detect the occurrence of the XB.

Our first study was devoted to the case where hydrogen bonds (HBs) and XBs arise separately, sharing the same electron donor site, *i.e.* in an orthogonal manner as shown for the protein-ligand complexes.[2] Geometrically perpendicular XB/HB synthon was used to rationally design complexes of a peptide bond model N-methylacetamide (NMA) with halogen bond donors dihalootetrafluorobenzenes, that mimics an orthogonal relationship of XBs to the classical HBs in b-sheet structured proteins and peptides.

IR-spectroscopy gives a first hint for the formation of the desired complexes, showing remarkable concurrent shifts of the C=O (red shift) and N-H (blue shift) vibrational modes of the NMA molecule. The molecular co-crystal structures contain discrete supramolecular assemblies based on the same primary perpendicular XB/HB synthon that confirms its robustness. Moreover XB donors can be introduced in this structural arrangement in an bioorthogonal manner without violating of the pre-existing HBs and herewith serve as a powerful tool in biological molecular recognition.[3] (Fig.1)

Further, a series of prototypical halogen-bonded 1:1 complexes of dihaloperfluoro-arenes with various pyridine derivatives was studied.[4] Assignments of the normal vibrational modes related to the functional groups involved in the formation of the XB were performed by quantum chemical calculations at density functional theory level.

The far-IR analysis reveals that compared to the starting materials the assigned bands undergo noticeable shifts and in some cases an increase of the intensity up to 25%, providing a clear indication of the XB formation. Thermal analysis and crystallographic study approve the formation of self-assembled co-crystals[4, 5, 6] driven by short N···Br and N···I halogen bonds.

References:

1. P. Metrangolo, G. Resnati, *Halogen Bonding. Fundamentals and Applications*, Springer Berlin Heidelberg, Berlin, 2008.
2. A. R. Voth, P. Khuu, K. Oishi, P. S. Ho, *Nat. Chem.* 2009, 1, 74.
3. V. Vasylyeva, S. K. Nayak, G. Terraneo, P. Metrangolo, G. Resnati, *CrystEngComm*, 2014, 16, 8102.
4. V. Vasylyeva, L. Catalano, C. Nervi, R. Gobetto, P. Metrangolo, G. Resnati, *in prep.*
5. A. Wasilewska, M. Gdaniec, T. Polonsky, *CrystEngComm* **9** (2007) 203.
6. R. B. Walsh, C. W. Padgett. P. Metrangolo, G. Resnati, T. W. Hanks, W. T. Pennington, *Cryst. Growth Des.* **1** (2007) 165.

Fig. 1: Molecular assembly of NMA (left) and its co-crystals with 1,4-dibromotetrafluorobenzene (middle) and 1,4-dibromotetrafluorobenzene (right) in the solid state.

Figure 1

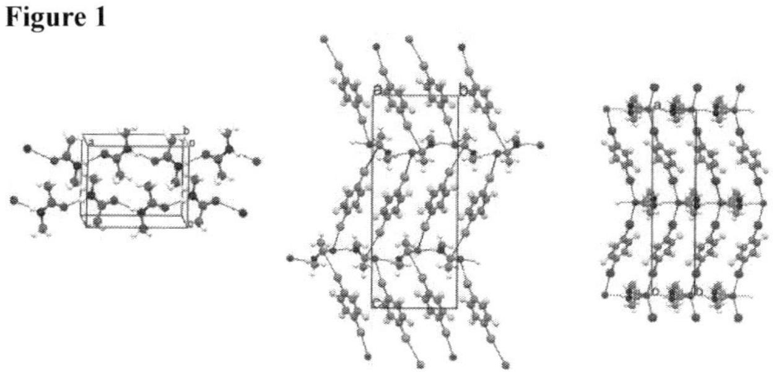

MS17-T06
Investigation of the ultra-fast structural changes in metal-organic complexes: comparison spectroscopy and time-resolved XRD

D. Raiser[1], S. Thekku Veedu[2], D. Storozhuk[2], S. Techert[1,2]
[1]*Max Planck Institute, Biophysical Chemistry, Goettingen, Germany*
[2]*Deutsches Elektronen Synchroton, FS-SCS, Hamburg, Germany*

Our aim is a detailed understanding of the ultrafast structural dynamics and the underlying basic mechanisms of molecular switches under optical excitation, which are still not fully understood in its variety. In order to monitor the occurring structural changes in molecules and alloys we are taking advantage of time-resolved x-ray diffraction measurements. But since the important electronic changes due to optical excitation take place on very short ps- and sub-ps timescales it gives rise to the necessity of faster time-resolved measurements than those which are currently possible at third generation synchrotrons. Besides new attempts of directly monitoring the structural changes with shorter x-ray pulses by femtosecond FEL- and XPS-pump-probe experiments, we are currently gaining more insight in the structural dynamics by indirect measurements using our fs-time-resolved transient absorption spectroscopy. Therefore we will present our recent spectroscopic data of a metal-organic complex which will be related to certain structural changes, while the near future aim is to prove these by further x-ray experiments.

MS18 - Protein-nucleic acid complexes

MS18-T01

Symportin 1 chaperones 5S RNP assembly during ribosome biogenesis

F. Rodriguez-Calvino[1], S. Kharde[1], A. Ori[2], K. Wild[1], M. Beck[2], I. Sinning[1]
[1]*Universität Heidelberg, Biochemie-Zentrum, Heidelberg, Germany*
[2]*EMBL, Heidelberg, Germany*

Eukaryotic ribosome biogenesis requires the ordered assembly of about 80 ribosomal proteins (r-proteins) and four ribosomal RNAs (rRNAs). Many r-proteins form functional clusters within the ribosome or assemble at distinct entry points during ribosome assembly. An important example is the 5S RNP which is formed by the r-proteins RpL5 and RpL11 together with the 5S RNA. In 60S maturation, the 5S RNP assembles into a recently characterized pre-60S particle. 5S RNP Docking involves the interaction of RpL11 with helix 84 (H84) of the 25S RNA. However, how 5S RNP is assembled prior to its recruitment into the pre-60S is not known. We determined the crystal structure of a ternary symportin Syo1/RpL5-N/RpL11 complex and provide biochemical and structural insights into 5S RNP assembly. Syo1 chaperones the 25S RNA binding surface on RpL11 and competes with H84 for binding. 5S RNA binding has been analyzed by EMSAs, pull-down and cross linking-MS experiments experiments. Together, our data indicate the formation and dynamics of a pre-5S RNP and underline the dual role of Syo1 in ribosomal protein transport and as an assembly platform for 5S RNP.

MS18-T02

Structures of two bacterial resistance factors mediating tRNA-dependent aminoacylation of phosphatidylglycerol with lysine or alanine

J. Krausze[1], S. Hebecker[2], J. Reichelt[1], J. Moser[2], D. Heinz[3]
[1]*Helmholtz Center for Infection Research, Structure and Function of Proteins, Braunschweig, Germany*
[2]*Technische Universität Braunschweig, Institute of Microbiology, Braunschweig, Germany*
[3]*Helmholtz Center for Infection Research, Braunschweig, Germany*

In microbes, the cytoplasmic membrane is probably the most important physical barrier to the environment.. Aminoacylation of the polar head group of the phospholipid phosphatidylglycerol (PG) catalyzed by Ala-tRNA^{Ala}-dependent alanyl-phosphatidylglycerol synthase (A-PGS) or by Lys-tRNA^{Lys}-dependent lysyl-phosphatidylglycerol synthase (L-PGS) enables bacteria to cope with susbtances that are harmful for the integrity of the cell membrane such as cationic host defense peptides. Accordingly, these synthases are also referred to as multiple peptide resistance factors (MprF). They consist of a separable C-terminal catalytic domain and an N-terminal transmembrane flippase domain. We present the X-ray crystallographic structure of the catalytic domain of A-PGS from the opportunistic human pathogen *Pseudomonas aeruginosa* and of L-PGS from *Bacillus licheniformis*, the latter in complex with the substrate analogue L-lysine amide. Both proteins feature a continuous tunnel that allows the hydrophobic lipid substrate PG and the polar aminoacyl-tRNA substrate to access the catalytic site in its center from opposite directions. The presented structural work in combination with experiments using artificial tRNA or artificial lipid substrates reveal the tRNA acceptor stem, the aminoacyl-moiety and the polar head group of PG as the main determinants for substrate recognition. These findings have broad implications for the design of L-PGS and A-PGS inhibitors that would render microbial pathogens more susceptible to antimicrobial compounds.

Figure 1

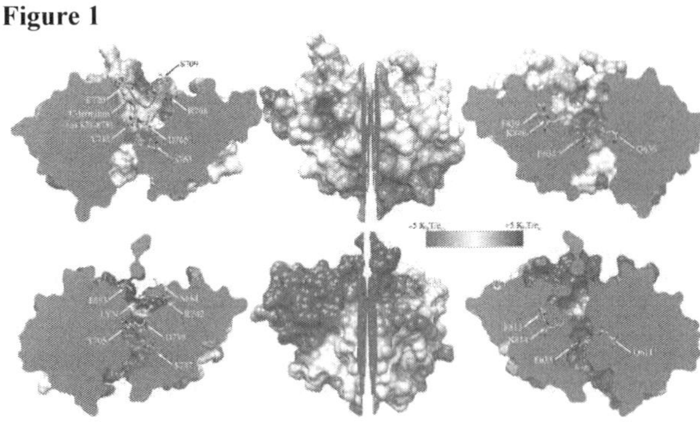

Figure 2

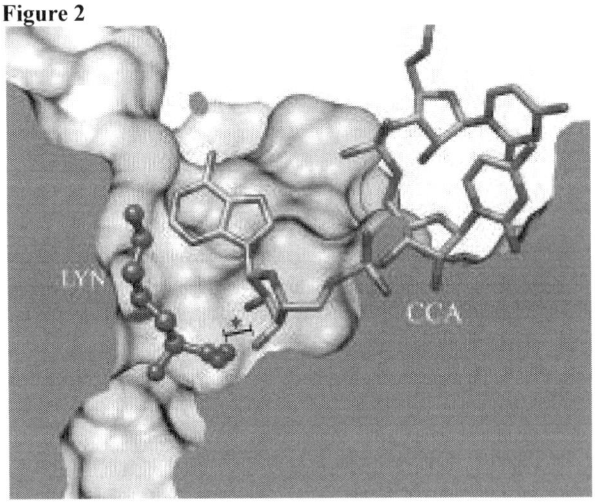

MS18-T03

Crystal structures of GTP-bound eIF5B provide novel insight into the mechanisms of GTP-dependent conformational switching and GTP hydrolysis in translational GTPases

B. Kuhle[1], R. Ficner[1]
[1]*Goerg-August-Universität Göttingen, Abteilung Molekulare Strukturbiologie, Göttingen, Germany*

Translational GTPases from bacteria to eukarya share the G domain as common evolutionarily conserved structural core, which couples GTP binding and hydrolysis to the specific biological function of the translation factor on the ribosome. Despite its importance for the understanding of the translation process, the mechanism of GDP/GTP-dependent conformational switching has remained obscure for most trGTPases at the molecular level. This includes the eukaryal translation initiation factor 5B (eIF5B), which catalyzes the joining of the two ribosomal subunits in the final step of the translation initiation pathway. We used a combination of X-ray crystallographic and biochemical methods to study the GTP-induced conformational switch in eIF5B. High-resolution structures of eIF5B in its apo, GDP- and GTP-bound forms reveal how eIF5B is activated for ribosomal subunit joining by transformation of local GTP-induced conformational rearrangements in the G domain into an increase of internal flexibility of the overall factor. Based on this mode of activation, it is possible to envisage how eIF5B discriminates against erroneous pre-initiation complexes that do not contain the Met-tRNAi and thus performs its role as final checkpoint in the formation of elongation competent 80S ribosomes [1, 2]. Moreover, we show that a monovalent cation, bound in the catalytic GTPase center of eIF5B acts as a critical, but previously unrecognized structural and catalytic cofactor required for the conformational switch in eIF5B. Further structural and biochemical studies on other trGTPases demonstrate that monovalent cation dependency is most likely common among members of this G protein family. These observations provide important new insight into the universal mechanisms of nucleotide-dependent conformational switching and GTP hydrolysis in trGTPases, and have methodologically relevant implications for the interpretation of biochemical and structural data [3].

[1] Kuhle, B. and R. Ficner (2014), eIF5B Employs a Novel Domain Release Mechanism to Catalyze Ribosomal Subunit Joining. The EMBO Journal 33: 1177-1191.
[2] Kuhle, B. and Ficner, R. (2014), Structural insight into the recognition of amino-acylated initiator tRNA by eIF5B in the 80S initiation complex. BMC Structural Biology 33: 1177-1191.
[3] Kuhle, B. and Ficner, R. (2014), A monovalent cation acts as structural and catalytic cofactor in translational GTPases. The EMBO Journal 33(21): 2547-2563.

MS18-T04

Structural studies of the GPN-loop GTPase Npa3

J. Niesser[1], A. Heuer[1], D. Kostrewa[1], R. Beckmann[1], P. Cramer[2]
[1]*Center for Integrated Protein Science CIPSM, Ludwig-Maximilians-Universität München, Gene Center and Department of Biochemistry, Munich, Germany*
[2]*Max-Planck-Institute for Biophysical Chemistry, Am Faßberg 11, Göttingen, Germany*

Although gene transcription by RNA polymerase II (Pol II) has been studied extensively during the last decades, surprisingly only very little is known about a crucial process that takes place upstream of this events, namely biogenesis of Pol II. Recently, several studies identified a novel subfamily of GTPases called GPN-loop GTPases to be involved in biogenesis of Pol II since silencing of these essential proteins leads to a cytoplasmic accumulation of Pol II both in yeast [1] and in human [2]. Here we report the first structure of a eukaryotic GPN loop GTPase in different nucleotide states at 1.85-2.25 Å resolution. The crystal structures of truncated Npa3 from Saccharomyces cerevisiae revealed an open and a closed conformation. Whereas the open conformation was found to bind hydrophobic substrates via a highly conserved, hydrophobic binding pocket, the substrate gets released upon closing of the enzyme. Our results demonstrate detailed molecular insights into the complete GTP hydrolysis cycle of eukaryotic GPN-loop GTPases.

References

[1] GTP-dependent binding and nuclear transport of RNA polymerase II by Npa3 protein. Staresincic L et al., 2011, JBC, 286 (2011), 35553-35561
[2] The protein interaction network of the human transcription machinery reveals a role for the conserved GTPase RPAP4/GPN1 and microtubule assembly in nuclear import and biogenesis of RNA polymerase II. Forget D et al., 2010, Mol. Cell Proteomics, 9 (2010), 2827-2839

MS18-T05

Short helical interaction modules in a novel spliceosomal subcomplex

A. Ulrich[1], T. Schütze[1], M. Seeger[1], M. C. Wahl[1]
[1]*FU Berlin, Biochemistry, Berlin, Germany*

Pre-mRNA splicing is catalyzed by a large and highly dynamic RNA-protein (RNP) molecular machine, the spliceosome, which is composed of five small nuclear (sn) RNPs and numerous non-snRNP proteins. For each round of splicing, a spliceosome is assembled, catalytically activated and, after splicing catalysis, disassembled in a stepwise fashion [1]. Transitions between functional stages in this reaction pathway are characterized by the ordered recruitment and release of specific snRNPs and non-snRNP proteins [2]. Initially, U1 and U2 snRNPs recognize the 5' splice site and branch point region of an intron, respectively, forming the A complex. Subsequently, the U4, U5 and U6 snRNPs join as a pre-formed tri-snRNP, giving rise to the pre-catalytic B complex. The B complex is then catalytically activated, yielding the B* complex that can carry out the first of two transesterification reactions of a splicing event. After step one of splicing the C complex catalyzes the second transesterification reaction, subsequent to which the spliceosome is disassembled in an ordered fashion. In higher eukaryotes a group of B complex-specific non-snRNP proteins joins the spliceosome at the stage of B complex formation [3]. Their restriction to higher eukaryotes and the implication of several members in alternative splicing [4] suggest that B complex proteins are key splicing regulators. However, their precise functions and the underlying molecular mechanisms are presently unknown. We have characterized the interactome of the B complex-specific protein Prp38 and shown that Prp38 acts as a major interaction hub of the spliceosome. Based on these results, we have identified and structurally analyzed a novel spliceosomal subcomplex of B-specific proteins centered on Prp38. The interactions are mediated by short helical modules that bind Prp38 with very high affinity. Our results suggest that several B complex-specific proteins are recruited as a preformed subcomplex and thus might act as a functional unit in the spliceosome.

References

1. Wahl, M.C., Will, C.L., and Lührmann, R. (2009). The spliceosome: design principles of a dynamic RNP machine. Cell *136*, 701-718.

2. Brow, D.A. (2002). Allosteric cascade of spliceosome activation. Annu. Rev. Genet. *36*, 333-360.

3. Agafonov, D.E., Deckert, J., Wolf, E., Odenwalder, P., Bessonov, S., Will, C.L., Urlaub, H., and Lührmann, R. (2011). Semiquantitative proteomic analysis of the human spliceosome via a novel two-dimensional gel electrophoresis method. Mol Cell Biol *31*, 2667-2682.

4. Park, J.W., Parisky, K., Celotto, A.M., Reenan, R.A., and Graveley, B.R. (2004). Identification of alternative splicing regulators by RNA interference in Drosophila. Proc Natl Acad Sci U S A *101*, 15974-15979.

MS18-T06

Quaternary structure of Prp19 provides mechanistic insight into the regulation of its ubiquitin ligation activity

T. Rocha de Moura[1], J. Schmitzova[1], C. Kibedi[1], S. Mozaffari Jovin[2], M. Kachala[3], D. Svergun[3], R. Lührmann[2], V. Pena[1]
[1]*MPI-BPC, Macromolecular Crystallography, Gottingen, Germany*
[2]*MPI-BPC, Cellular Biochemistry, Göttingen, Germany*
[3]*EMBL , Hamburg, Germany*

Pre-mRNA splicing is an essential step of gene expression, in which non-coding sequences called introns are removed from the pre-mRNA and the coding sequences named exons are ligated to form a mature mRNA. Splicing occurs on the spliceosome, a multi-Megadalton machinery that assembles step-wise on the pre-mRNA substrate, first as an inactive particle. Subsequently, the spliceosome undergoes dramatic compositional and conformational changes that lead to the formation of an active center able to catalyze the chemistry of splicing.

The highly conserved splicing factor Prp19 is a scaffold of the so-called Nineteen complex (NTC) - a large building block of the spliceosome that is essential for the formation of the spliceosomal catalytic center, both in yeast and human. Prp19 forms a tetramer of about 220kDa *in vivo* and *in vitro* and its quaternary structure is essential for Prp19 function as a protein scaffold. In addition, owing to the U-box domains of the Prp19 component, NTC regulates splicing and DNA repair by its ubiquitin ligation activity [1; 2]. We demonstrate that the Prp19 homotetramer becomes inactive in the absence of the other NTC components. Crystal structure of Prp19 tetrameric core reveals how a long quadruple coiled-coil sequesters the four U-box domains, providing a snapshot of Prp19 in the inactive state. Further analysis of the full-length Prp19 as well as of its tetrameric core by small-angle X-ray scattering are fully consistent with the crystallographic analysis and the previously published EM analysis [3]. This work elucidates the quaternary structure of a multifunctional protein and provides insight into a novel mode of regulating the ubiquitination signaling pathway.

[1] Song, E. J., *et al.*, (2010) Genes Dev. 1;24(13): 1434-47
[2] Maréchal, A., *et al.*, (2014) Mol. Cell. 23;53(2):235-46.
[3] Ohi, M. D., *et al.*, (2005) Mol. Cell. Biol. 25, 451-460.

Poster

BIO - Biostructures

BIO-P01
Facilities for Macromolecular Crystallography at the HZB

M. Weiss[1], R. Förster[1], M. Hellmig[1], F. Huschmann[1], A. Kastner[1], E. Kolibacz[1], P. Malecki[1], M. Röwer[1], K. Sparta[1], M. Steffien[1], M. Ühlein[1], P. Wilk[1], U. Mueller[1]
[1]*HZB, Macromolecular Crystallography, Berlin, Germany*

The Macromolecular Crystallography (MX) group at the Helmholtz-Zentrum Berlin (HZB) has been in operation since 2003. Since then, three state-of-the-art synchrotron beam lines (BL14.1-3) for MX have been built up on a 7T-wavelength shifter source [1,2]. Currently, the three beam lines represent the most productive MX-stations in Germany, with more than 1300 PDB depositions (Status 12/2014). BLs14.1 and 14.2 are energy tuneable in the range 5.5-15.5 keV, while beam line 14.3 is a fixed-energy side station operated at 13.8 keV. All three beam lines are equipped with state-of-the-art detectors: BL14.1 with a PILATUS 6M detector and BLs14.2 and 14.3 with large CCD-detectors. BL14.1 and BL14.2 are in regular user operation providing close to 200 beam days per year and about 600 user shifts to approximately 70 research groups across Europe. BL14.3 has been equipped with a HC1 crystal dehydration device in 2011. In addition to serving the user community mainly as a screening and test beam line, it is currently the only MX beamline in Europe with a HC1 device permanently installed. Additional user facilities include office space adjacent to the beam lines, a sample preparation laboratory, a biology laboratory (safety level 1) and high-end computing resources. On the poster, a summary on the experimental possibilities of the beam lines and the ancillary equipment provided to the user community will be given.

[1] Heinemann U., Büssow K., Mueller, U. & Umbach, P. (2003). Acc. Chem. Res. 36, 157-163.

[2] U. Mueller, N. Darowski, M. R. Fuchs, R. Förster, M. Hellmig, K. S. Paithankar, S. Pühringer, M. Steffien, G. Zocher & M. S. Weiss (2012). J. Synchr. Rad. 19, 442-449.

BIO-P02
Crystal structure of the catalytic domain of the ethylene receptor ETR1 from Arabidopsis thaliana.

S. Panneerselvam[1], J. Mueller Dieckmann[2]
[1]*DESY, Hamburg, Germany*
[2]*University of Hamburg, Biozentrum, hamburg, Germany*

Plants employ ethylene to regulate many developmental processes such as seed germination, root growth, fruit ripening, and senescence (1). Arabidopsis thaliana perceives ethylene by a group of five partly redundant, membrane bound receptors (ETR1, ETR2, ERS1, ERS2, EIN4). Ethylene binds at the N terminal hyrophobic domain of the receptors and initiate the signalling events via its cytosolic domain. The receptor cytosolic domain contains several different domains which are triggering the signallig events by interacting with different proteins such CTR-1 kinase (2). In case of ETR1, the cytosolic domain is made of a GAF (cGMP-specific phosphodiesterase, adenylyl cyclases and FhlA) domain followed by a histidine kinase dimerization (dHp) and catalytic domain (CA) and a receiver domain (RD).

Sequence analysis of ethylene receptors (ER) indicated similarity to bacterial two-component systems (TCS) by their C-terminal cytosolic histidine kinase (HK) and receiver domains (3). Based on the integrity of signature motifs in the catalytic domain of the HK, the receptors are further assigned to two subfamilies. Members of subfamily 1 (ETR1 and ERS1) possess all sequence motifs of canonical HK domains and also histidine kinase activity. Subfamily 2 receptors (ETR2, ERS2 and EIN4) have incomplete motifs and lack histidine kinase activity(4) .While all receptors participate in ethylene signal transduction , members of subfamily 1 seem to play a predominant role.

To understand the kinetic and regulatory mechanism of ethylene receptors, We have successfully crystallized and solved the structure of catalytic domain of ETR1. The protein was crystallized with various nucleotides. Crystals obtained in the presence of ADP belong to the space group I212121 with one molecule per asymmetric unit and diffract x-ray radiation to 1.85 Å resolution. The overall structure assumes an α/β sandwich fold and closely resembles the CA domain of HK853.The crystal structure reveals, the flexibility of ATP lid and the role of metal specificity of subfmaily 1 receptors through the unusual binding of an metal ion with adenine moiety of ADP and residue C573 from G3 motif (5).

References:

1. Bleeker, A.B. & Kende, H. (2000). Annu. Rev. Cell Dev. Biol. 16, 1-18
2. Mayerhofer, H., Panneerselvam, S., and Mueller-Dieckmann, J. (2012). J. Mol. Biol. 415, 768-779.
3.Chang, C., Kwok, S., Bleecker, A., and Meyerowitz, E. (1993). Science 262, 539 -544.
4.Moussatche, P. & Klee, H. J. (2004). Journal of Biological Chemistry. 279, 48734 -48741.
5. Mayerhofer, H., Panneerselvam, S., Kaljunen, H., Tuukkanen, A., Mertens, HD., and Mueller-Dieckmann, J. (2014). Journal of Biological Chemistry (in press)

BIO-P03
Exploring the role of CooC2 in the maturation of Acetyl-CoA Synthase

C. Wörmann[1], J.- H. Jeoung[1], S. Goetzl[1], H. Dobbek[1]
[1]*Humboldt Universität zu Berlin, Strukturbiologie/Biochemie, Berlin, Germany*

BIO-P04
Solving structures from the anaerobic degradation pathways

S. Weidenweber[1], U. Demmer[1], E. Warkentin[1], T. Weinert[2], K. Schühle[3], J. Heider[3], S. Huwiler[4], C. Loeffler[4], M. Boll[4], U. Ermler[1]
[1]*Max Planck Institute of Biophysics, Molecular Membrane Biology, Frankfurt am Main, Germany*
[2]*Paul Scherrer Institut , Villigen, Germany*
[3]*University of Marburg, Marburg, Germany*
[4]*University of Freiburg, Freiburg, Germany*

Aromatic compounds represent the second most abundant class of organic molecules generated in biological processes. Moreover, environmental pollutants, in particular, from the BTEX group (benzene,toluene,ethylbenzene andxylene) accumulateduring human oil production activities [1]. Specialized microorganisms are able to metabolize large varieties of them by funneling them into a few central key intermediates like benzoyl-CoA. Challenging enzymatic processes are required to overcome - especially in anaerobic habitats - the inertness of the C-H and C-C bonds and the resonance stability of the aromatic ring [2]. Our aim is to understand the mechanism of the catalytic key degradation reaction under anoxic conditions by structural characterization of the involved enzyme complexes.

In the anaerobic ethylbenzene degradation pathway we are working on the structure determination of the acetophenone carboxylase (Apc). This huge multi-subunit complex with a molecular weight over 500 kDa carboxylates the methyl group of acetophenone to benzoylacetate in a biotin independent way with 2 ATP [3]. In anaerobic toluene degradation we analyze the structures of the benzoylsuccinyl-CoA thiolase (BbsAB) and the succinyl-CoA:benzylsuccinate CoA-transferase (BbsEF). BbsEF activates benzylsuccinate to the respective CoA ester. BbsAB catalyzes the cleavage of benzoylsuccinyl-CoA to succinyl-CoA and benzoyl-CoA. [4].

The central intermediate benzoyl-CoA is dearomatized by benzoyl-CoA reductases (BCRs) in an ATP-dependent or -independent manner. The $Bam(BC)_2$ subcomplex of the ATP-independent BCR containing a tungsten-cofactor could be structurally established in a Zn^{2+} stabilizing but inactive state and substrate bound active state [5]. A radical Birch-like enzymatic mechanism was proposed and analyzed with respect of its catalytic mechanism. Crystallization attempts under anaerobic conditions were performed with different multi-subunit BCR complexes. The structure of the catalytic BamBC subunits with a tungsten-cofactor could be solved with substrate, inhibitor and product.

References
[1] Fuchs, G., Boll, M. & Heider, J. (2011). *Nature reviews.*; 9:803 - 816.
[2] Boll, M., Fuchs, G. & Heider, J. (2002). *Curr Opin Chem Biol.*; 6:604-11.
[3] Jobst, B., Schühle, K., Linne U. and Heider, J. (2010). *J Bacteriol.*; 192:1387-94
[4] Leuthner, B., Heider, J. (2000). *J Bacteriol.*; 182:272-7.
[5] Weinert, T. et al. in preparation

BIO-P05
Optimizing Data Collection in the Home Lab

S. Freisz[1], M. Benning[2]
[1]*Bruker AXS, Karlsruhe, Germany*
[2]*Bruker AXS, Madison, United States*

Many of the projects currently under investigation in biological research labs focus on macromolecules that are difficult to crystallize such as: complexes, multi-domain and membrane proteins. Typically, crystallization trials can produce small, weakly diffracting crystals that may also have other challenging attributes. Recent hardware and software developments have improved in-house data quality on a wide range of samples. Small and highly focused x-ray beams allow one to select the best diffracting portion of a larger crystal and reduce background scatter for much smaller samples. Shutterless data collection helps to reduce instrument error resulting from shutter jitter and allows fine slicing of data runs without frame to frame dead time penalties while the practice of dealing with multiple, cracked or twinned crystals has improved greatly due to software enhancements. Results from in-house data collection including: shutterless operation, optimization of crystal orientation and collection parameters will be discussed.

BIO-P06
Structural Biology of the *Pseudomonas aeruginosa* Quinolone Signaling System

F. Witzgall[1], W. Blankenfeldt[1]
[1]*Helmholtz Centre for Infection Research, Structure and Function of Proteins, Braunschweig, Germany*

The Gram-negative bacterium *Pseudomonas aeruginosa* is a human pathogen that causes roughly 10% of all nosocomial infections. It mainly infects persons with a weakened immune system such as cancer or cystic fibrosis patients leading to severe infections of the eyes, the ears, the skin, the lungs or the urinary tract. A serious problem in the treatment and the control of *P. aeruginosa* infections is the rising occurrence of bacterial strains that are virtually resistant towards all commonly used antibiotics. Therefore it is very urgent to find alternatives to conventional drugs. A promising strategy to combat multidrug-resistant strains is the disruption of the quorum sensing (QS) network in *P. aeruginosa*, because the processes that are controlled by QS include biofilm formation and virulence factor production among others. The bacterium harbours at least four different QS systems that regulate about 10% of the whole *P. aeruginosa* genome. One of these cell-to-cell communication networks is the *pqs* system that is composed of the transcriptional regulator PqsR and the associated signaling molecule PQS. HHQ, which is the precursor of PQS, is synthesized by PqsA-D that are part of a common operon. While the roles of PqsA and PqsD in HHQ biosynthesis are mostly resolved, the precise functions of PqsB and PqsC remain unclear. In 2013 Dulcey *et al.* brought some light into the darkness as they demonstrated that PqsB and PqsC form an active protein complex and disproved the long-standing hypothesis that β-ketododecanoic acid is the biologically relevant precursor of HHQ. Based on the results of Dulcey *et al.* (2013) we co-purified PqsBC by affinity chromatography and gel filtration. Afterwards the complex was crystallized and the structure of the heterodimer was determined by a combination of molecular replacement and isomorphous replacement techniques. The crystal structure of PqsBC represents another jigsaw piece to reveal the secrets of PQS signaling in *P. aeruginosa.*

References
1. Driscoll, J. A. et al.. The epidemiology, pathogenesis and treatment of *Pseudomonas aeruginosa*
infections. *Drugs* 67, 351-368 (2007)
2. Rutherford, S. T. and Bassler, B. L. Bacterial quorum sensing: its role in virulence and
possibilities for its control. *Cold Spring Harb. Perspect. Med.* **2,** (2012)
3. Lee, J. and Zhang, L. The hierarchy quorum sensing network in *Pseudomonas aeruginosa. Protein*
Cell (2014)
4. Dulcey, C. E. *et al.* The end of an old hypothesis: the *Pseudomonas* signaling molecules 4-
hydroxy-2-alkylquinolines derive from fatty acids, not 3-ketofatty acids. *Chem. Biol.* 20, 1481- 1491 (2013)

BIO-P07
Thermal stabilization of Polyhydroxybutyrate

B. M. Liebeck[1], P. K. Manvi[2], T. Vad[2], G. Roth[1]
[1]*RWTH Aachen, Institut für Kristallographie, Aachen, Germany*
[2]*RWTH Aachen, Institut für Textiltechnik, Aachen, Germany*

The utilization of bio-based polymers is an efficient approach to reduce the negative impacts of petro-based products, e.g. CO_2 emission and global pollution. Polyhydroxybutyrate (PHB) is a biopolymer produced by bacterial fermentation from biomass. This thermoplastic polymer shows potential for diverse industrial applications due to its high degree of crystallinity and mechanical properties similar to polypropylene. However, the applications of

PHB are limited due to its thermal instability above the melting point and the low nucleation density during crystallization [1].

Due to these drawbacks it is necessary to either stabilize the PHB melt or to shift the melting point below the decomposition temperature and thus prevent thermal degradation. It has been shown that it is possible to improve the thermal stability via chemical modification with specific reactants, e.g. anhydrides and to affect the transition temperatures as well as the flow behaviour of PHB through plasticization [2, 3].

In this contribution we investigate the influence of maleic anhydride (MA) as stabilizer on the thermal behaviour and structure of PHB via x-ray diffraction, thermal analysis and optical microscopy. Our results reveal that maleic anhydride can prevent the thermal decomposition of PHB and shows a remarkable influence on its melting behaviour. The stabilizer also influences the crystallization behaviour as can be seen in the image below.

In addition the influence of dioctyl phthalate (DEHP), a common representative of polymer plasticizers, is investigated in terms of a modification of the thermal and crystalline properties of PHB. We are able to show that DEHP significantly decreases both its melting and crystallization temperatures and thus influences the spherulite structure as shown below.

[1] Erkske, D.; Viskere, I.; Dzene, A.; Tupureina, V.; Savenkova, L.:
Biobased polymer composites for films and coating
Proceeding of the Estonian Academy of Sciences, Chemistry, **55** (2006) 2, P. 70-77
[2] Hong, S.G.; Lin, Y.C.; Lin, C.H.:

Improvement of the thermal stability of polyhydroxybutyrates by grafting with maleic anhydride by different methods: Differential scanning calorimetry, thermogravimetric analysis, and gel permeation chromatography

Journal of applied polymer science, 110 (2008), 5, P. 2718-2726
[3] Wang, L.; Zhu, W.; Wang, X.; Chen, X.; Chen, G.Q.; Xu, K.: Processability Modifications of Poly(3-hydroxybutyrate) by Plasticizing, Blending, and Stabilizing
Journal of applied polymer science 107, (2008), P. 166-173

Figure 1

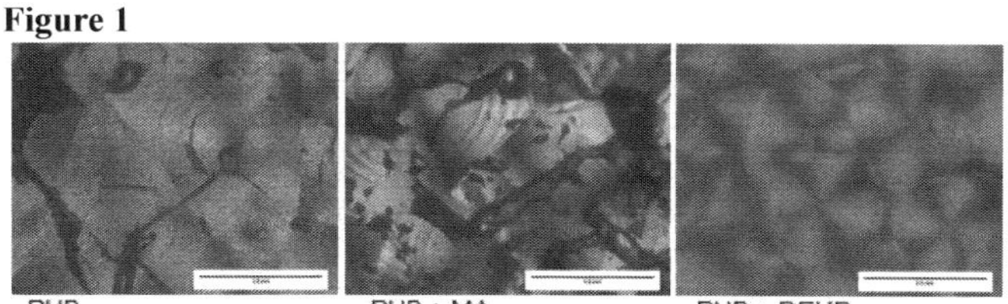

BIO-P08

Structural and mechanistic studies of RNA-ligating deoxyribozymes

A. Ponce-Salvatierra[1], V. Pena[1], C. Höbartner[1]
[1]*Max Planck Institute for Biophysical Chemistry, Göttingen, Germany*

Deoxyribozymes are synthetic single stranded DNA molecules that can catalyze chemical reactions with high selectivity. Since the discovery of the first deoxyribozyme (Breaker et al, 1994) a wide variety of them have been isolated by in vitro selection. The applications of deoxyribozymes run from analytical tools to therapeutic agents and reagents for synthesis. However the information about their catalytic mechanism is limited and there is no three dimensional structure available of any deoxyribozyme in an active conformation.

Therefore, the key question that we wish to address is: how is DNA catalysis possible? As a mean to understand the structural basis underlying their fold and the mechanistic principles that govern DNA catalysis, we decided to determine the crystal structure of a deoxyribozyme in complex with its substrate by X-ray crystallography.

As a first step we selected a biochemically well-characterized deoxyribozyme that acts on RNA substrates (Wachowius et al, 2011). This deoxyribozyme, denoted 9DB1 employs single stranded extensions to recognize two RNA strands by specific base-pairing and then it catalyzes their ligation, resulting in a single RNA strand (Purtha et al, 2005).

We succeeded to crystallize 9DB1 in complex with an RNA strand that mimics the post-catalytic state of the reaction. Concomitant presence of the deoxyribozyme strand and of the RNA in the crystal were confirmed.

The initial crystals were reproducible and diffracted up to 9 angstroms. Having learned the sequence of the complex that is able to form crystal contacts, we designed a new series of constructs by inserting small variations that may induce a higher internal order of the crystal lattice and therefore a better diffraction. So far, out of 30 different complexes subjected to crystallization trials, 15 of them yielded crystals. We are now working on the optimization of the crystals in order to make them suitable for derivatization.

BIO-P09

CT441 from *Chlamydia trachomatis*: Structural basis of proteolytic and chaperone activity

F. Kohlmann[1], K. Shima[2], R. Hilgenfeld[1], W. Solbach[2], J. Rupp[2], G. Hansen[1]
[1]*University of Lübeck, Institute of Biochemistry, Lübeck, Germany*
[2]*University of Lübeck, Institute of Medical Microbiology and Hygiene, Lübeck, Germany*

The prokaryotic obligate intracellular pathogen *Chlamydia trachomatis* is the most prevalent cause of preventable blindness affecting approximately 6 million people worldwide. In addition, *C. trachomatis* is the most commonly reported sexually transmitted infection in Europe and the US causing pelvic inflammation, ectopic pregnancy, and infertility. As for other intracellular pathogens, proteases play crucial roles during most stages of the complex life cycle of *Chlamydia*. CT441 is a chlamydial protease reported to interfere with immune response and estrogen signaling of the host cell. Our structural and mutational analysis shows that proteolytic activity of CT441 depends on a conserved Ser/Lys/Gln catalytic triad and a functional substrate-binding site within a flexible, accessory PDZ domain. It has been suggested that CT441 is involved in the disruption of estrogen signaling pathways of the host cell by interacting with SRAP1, a co-activator of the estrogen receptor α. However, our results show that although *in vitro* CT441 exhibits proteolytic activity against SRAP1, CT441-mediated SRAP1 degradation is not observed during the intracellular developmental cycle. Interestingly, we found that CT441 has a chaperone activity indicating a role in prokaryotic protein-quality control processes.

Fig. 1. Structural architecture and proteolytic site of CT441. (A) Predicted domain organization of CT441 with signal peptide (SP;

grey), N-terminal domain (NTD; red), PDZ domain (green), and C-terminal domain (CTD; blue). Residues of the proteolytic site and the previously annotated DUF3340 subdomain are indicated. (B) Overall structure of CT441 in ribbon representation with residues of the proteolytic site shown as sticks (yellow). The PDZ domain and several loop regions were too flexible to be modelled into electron density and their approximate position is indicated by a green box and dashed lines, respectively. (C) Topology diagram of CT441 with residues of the proteolytic site and domain boundaries indicated.

Fig. 2. CT441 co-localizes with SRAP1 only after disruption of the chlamydial inclusion. Infected HEK293 cells overexpressing SRAP1 were stained with antibodies against SRAP1 (green) and CT441 (red), DNA was stained with DAPI (blue). Representative confocal images 32 h or 48 h p.i. are shown (white scale bar = 5 µm). The right column displays fluorescence distribution profiles along broken lines in the merged confocal images. Most cells imaged 32 h p.i. do not show overlapping profiles of SRAP1 and CT441 (upper panel). However, in some cells, overlapping fluorescence profiles were observed as early as 32 h p.i. (middle panel) and in most cells, this was the case at 48 h p.i. (lower panel). This indicates a progressing co-localization of SRAP1 and CT441 at late stages of the infection, most likely due to the disruption of the inclusion membrane before egress of C. trachomatis from the host cell.

Figure 1

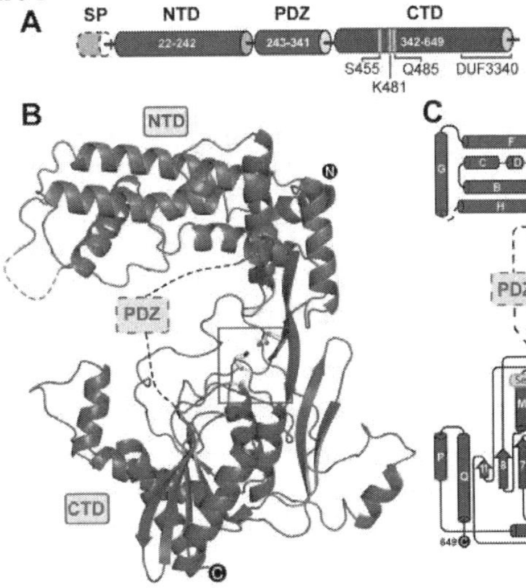

Figure 2

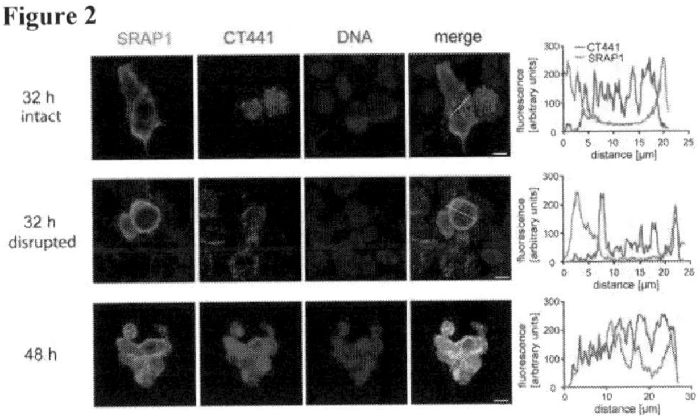

BIO-P10

Introduction to a novel genome browser based on structure prediction by Phyre^2

J. Poppe[1], J. Reichelt[1], C. Feiler[1], W. Blankenfeldt[1]
[1]*Helmholtz Centre for Infection Research, SFPR, Braunschweig, Germany*

Understanding the mechanisms of life on a molecular basis requires a profound look at its roots, namely the genes and their translations into proteins. In order to identify new proteins to investigate (targets), biochemists have to rely in most cases on correct genome annotation as the only information available. Unfortunately, the *in-silico* annotation based on sequence homology is often hard to interpret as 30-40% of all annotated genes have only putative functions. Others cannot be predicted, while those with wrong annotations bring up misleading interpretations. In addition, the growing availability of fully sequenced genomes leads to an ever-increasing excess of unexplored genes [1]. This can be very confusing for scientists looking for new protein challenges for further biochemical and structural characterization. Therefore, in order to complement the genetic information of a target protein, biochemists use protein structure prediction tools like Phyre^2 [2,3]. These programs try to construct a 3D model based on sophisticated sequence analysis with respect to known protein structures.

Searching a whole genome for interesting targets and feeding the results into a structure prediction tool is a tedious, time consuming task. To ease this target identification we constructed a genome browser that connects genome information and structure prediction. It allows whole genome searches for genes matching criteria such as belonging to an operon, protein size as well as structural predictability based on Phyre^2 predictions. Proteins that match a query are assembled in a hit list showing the gene and its predictability, a 3D model and background data like number of amino acids and presence/absence of transmembrane helices. All information used to create the hit list is extracted from public databases and gene bank entries. If necessary, they can be directly traced back to their source by a simple click.

To illustrate the usefulness of our browser we crystallized a protein,whose gene came up during a low similarity sequence homology search of the *P. aeruginosa* genome. The prediction showed that only a C-terminal α-helix of the protein could be modeled with high confidence. The x-ray structure later on revealed that the unpredicted sequence contains a novel fold. So far our database has been set up for the genomes of *P. aeruginosa*, *C. difficile* and *S. aureus*, but more strains can easily be added upon request.

[1] Galperin, M.Y. and Koonin, E.V. (2004) 'Conserved hypothetical' proteins: prioritization of targets for experimental study. *Nucleic Acids Res*, 32, 5452-5463
[2] Moult, J., Fidelis K., Kryshtafovych A., Schwede T. and Tramontano A. (2012) Critical assessment of methods of protein structure prediction(CASP)-round X. *Proteins*, 82, 1-6
[3] Kelley, L.A. and Sternberg, M.J.E. (2009) Protein structure prediction on the web: a case study using the Phyre server. *Nature Protocols*, 4, 363-371

BIO-P11

Terminal Uridylyl Transferases (TUTases): Steps towards understanding pre-*let-7* miRNA degradation.

A. Garg[1], U. Heinemann[1]
[1]*Max-Delbrück-Centrum für Molekulare Medizin (MDC), crystallography, berlin, Germany*

In last few decades, microRNAs (miRNAs) have emerged as important regulators of gene expression in eukaryotes that affect multiple cellular processes such as tissue development, differentiation, proliferation and innate and adaptive immune response [1].

These miRNAs are ~22-nucleotide long non-coding RNAs which bind with target mRNA via complementary base pairing following translational silencing, often coupled to degradation of the mRNA [1,2]. *Let-7* miRNAs are highly conserved regulators of developmental timing and stem cell differentiation and by down-regulating the expression of several oncogenes (including *RAS, HMGA2, cMYC, CDC25A* and *CDK6*) *let-7* miRNA acts as a tumor suppressor [3]. Biogenesis of the *let-7* miRNAs is regulated by different enzymes depending on the developmental state of the cell. The cytoplasmic terminal uridylyl transferase (TUTase/TUT) bind to the pre-*let-7* molecule either in coordination with Lin28 or by itself, followed by the polyuridylation or monouridylation of the miRNA in undifferentiated or differentiated cells, respectively. In contrast to monouridylation of pre-*let-7*, which is necessary for the stabilization and further processing by dicer enzyme, polyuridylation of pre-*let-7* leads to its destabilization, making it resistant to dicer cleavage, and signals for the exonuclease DIS3L2 to degrade the pre-*let-7* molecule [4]. This addition of non-templated uridines to the 3'-ends of miRNAs is an emerging form of RNA control that influences RNA stability and processing. Despite the presence of various TUTases in eukaryotes, only TUT4 and TUT7 are capable of uridylating pre-*let-7*, as these share several unique features over other TUTases. Apart from the catalytic motif, which is common for all TUTases, TUT4 and TUT7 have three CCHC Zn-knuckles, a second, functionally inactive nucleotidyl transferase (NTP*) domain linked to a PAP-associated domain (PAP-asso), a putative C_2H_2 Zn-finger at the N terminus. It has been demonstrated that the NTP*/PAP-associated domain and N-terminal Zn-finger are essential for the Lin28-mediated pre-*let-7* polyuridylation, explaining why only TUT4 and TUT7 can regulate *let-7* biogenesis. We are particularly interested to understand the molecular mechanism of uridylation of pre-*let-7* miRNA by TUTase, alone and in coordination with Lin28, and how the relatively small protein Lin28 accommodates a comparatively very large TUTase molecule in order to regulate the biogenesis of *let-7* miRNAs.

References:

1) Thornton, J.E., Chang, H.M., Piskounova, E., Gregory, R.I. Lin28 mediated control of let-7 microRNA expression by alternative TUTase Zcchc11(TUT4) and Zcchc6(TUT7). RNA, 18, 1875-1885 (2012).
2) Schmidt, M.J., West, S., Norbury, C.J. The human cytoplasmic RNA terminal U-transferase Zcchc11 targets histone mRNAs for degradation. RNA, 17, 39-44 (2011).
3) Mayr, F., Schuetz, A., Doege, N., Heinemann, U. Lin28 cold shock domain remodels pre-let7 miRNA . Nucleic Acids Res, 40, 7492-7506 (2012).
4) Mayr, F., Heinemann, U. Mechanism of Lin28 mediated miRNA & mRNA regulation-A structural and functional perspective. Int J Mol Sci, 14, 16532-16553 (2013).

BIO-P12

βαβββ-module-containing proteins in resistance and metabolism of *Pseudomonas aeruginosa*

M. Popp[1], A. Vit[1], W. Blankenfeldt [1]
[1]*Helmholtz-Zentrum für Infektionsforschung, Braunschweig, Germany*

Pseudomonas aeruginosa is an opportunistic pathogen that is responsible for a variety of fatal hospital acquired infections. Treatment of these infections becomes increasingly difficult because of rising levels of multi drug-resistant strains while lacking new effective therapeutics[1]. Gaining detailed insights into resistance mechanisms might deliver new approaches in the battle against *P. aeruginosa* infections. One protein family that is often involved in resistance mechanisms consists of βαβββ-module proteins that are found in all kingdoms of life showing a highly conserved structure, based on the characteristic, name giving βαβββ-motif[2]. Some of them catalyse metal-dependent degradation processes of antibiotic compounds while others have chaperone-like functions only mediating the binding of toxic compounds through π-stacking[3]. A BackPhyre[4] search with the βαβββ-module protein EhpR from *Enterobacter agglomerans*[3] against the *P. aeruginosa* genome discovered 22 genes coding for potential βαβββ-module proteins. They are classified as metal-dependent (Class I) and aromatic compound binding proteins (Class II). Besides, there are proteins that can't be assigned to any of these groups. These are summarized as Class III. So far only four candidates of the metal-dependent group are fully characterized as antibiotic degrading enzymes and it was shown that Class II proteins are able to bind small, toxic and aromatic compounds through π-stacking. However, no suggestions for the function of Class III proteins are available at present. Class III βαβββ-module proteins are PA1358 (PDB-ID: 1U71), PA1672, PA2721 (PDB-ID: 1U69), PA4518 and PA4183. Moreover PA1358, PA2721 and PA4183 are expressed in one transcriptional unit with the neighbouring genes PA1359, PA2722, PA2723 and PA4184. Next to the functional characterization of these proteins crystallographic experiments are in progress to determine the missing structures for Class III βαβββ-module proteins and for the genes in the same operon.

References

1. Driscoll, J. A., *et al.*, The epidemiology, pathogenesis and treatment of *Pseudomonas aeruginosa* infections. *Drugs* **67,** 351-368 (2007).
2. Bergdoll, M., *et al.*, All in the family: structural and evolutionary relationships among three modular proteins with diverse functions and variable assembly. *Protein Sci. Publ. Protein Soc.* **7,** 1661-1670 (1998).
3. Yu, S. *et al.*, Atomic resolution structure of EhpR: phenazine resistance in *Enterobacter agglomerans* Eh1087 follows principles of bleomycin/mitomycin C resistance in other bacteria. *BMC Struct. Biol.* 11, 33 (2011).
4. Kelley, L. A. & Sternberg, M. J. E. Protein structure prediction on the Web: a case study using the Phyre server. *Nat. Protoc.* 4, 363-371 (2009).

BIO-P13

EMBL Beamlines for Macromolecular Crystallography at PETRA III

T. Schneider[1], G. Bourenkov[1], M. Cianci[1], J. Kallio[1], G. Pompidor[1], S. Fiedler[1]
[1]*EMBL c/o DESY, Hamburg, Germany*

EMBL Hamburg operates two beamlines for macromolecular crystallography - P13 and P14 - at the high-brilliance synchrotron radiation source PETRA III at DESY (Hamburg, Germany).

P13 offers a photon flux of $>10^{13}$ph/sec from 16 keV to 6 keV (0.8 - 2.0 Å). Energies between 4 and 6 keV (3.1 - 2.0 Å) are accessible with a photon flux between $>10^{12}$ph/sec providing unique conditions for experimental phasing. Using adaptive focusing mirrors, the full beam (~1.5 x 1 mm^2) can be focused to sizes between 30 x 20 μm^2 and 150 x 100 μm^2. An MD2 diffractometer with a mini-kappa goniostat allows for rapid and accurate rotation of crystals in the X-ray beam. A MARVIN-robot is installed for rapid (< 30 s cycle time) loading and unloading of crystals.

P14 offers a photon flux of $>10^{13}$ph/sec between 7 and 20 keV (1.8 - 0.6 Å). Using adaptive focusing mirrors, a beamsize of 5 x 4 μm^2

with a total flux of >10^{12}ph/sec can be used for crystallographic experiments. The focusing mirrors can be moved into and out of the beam in less than half a minute allowing to rapidly toggle between a fully focused small beam and an unfocused large (~1.5 x 1 mm^2) beam to optimize data collection parameters for differently sized crystals. A high-precision vertically mounted MD3 diffractometer with a sphere of confusion of less than 100 nm is available for data collection.

With funding from the Röntgen-Ångstroem-Cluster, we are currently upgrading the optics of P14 with Compoud Refractive Lenses to have a higher flexibility in terms of beam size and to achieve an increased photon flux under micro-focussing conditions.

Both MX beamlines are equipped with PILATUS 6M-F detectors and dedicated data processing computers. The beamlines are embedded into the 'Integrated Facility for Structural Biology' offering facilities for sample preparation and characterization, a laboratory specifically equipped for the preparation of heavy atom derivatives, and downstream facilities for data evaluation. Applications for beamtime can be made at smis.embl-hamburg.de. Access support can be obtained via www.biostruct-x.eu. User operation will resume on 27-Apr-2015.

We will report about the status of the beamlines, describe typical experimental situations (small crystals, serial crystallography, low-energy phasing and others), and discuss the new opportunities becoming available once the Compound Refractive Lenses are in place.

BIO-P14

On the electron density distrubution of a double zwitterionic dipeptide

C. Ruhmlieb[1], B. Dittrich[1]
[1]*Universitaet Hamburg, Institut f. Anorganische und Angewandte Chemie, Hamburg, Germany*

The dipeptide L-Arginyl-L-Glutamic acid was initially crystallized to study its electron density distribution (EDD) by high-resolution single-crystal X-ray diffraction, and it was known from the literature to crystallize [1] with two water solvent molecules. Re-crystallization by vapor diffusion methods with acetone as anti-solvent from water instead afforded a new tetrahydrate structure, for which Bragg data to high-resolution were collected at the synchrotron. Data quality is high, but an oblique-incidence correction [2] was required to reach it. Due to occurrence of extinction a classical charge density study was not carried out. Instead the invariom model [3] was chosen to adequately model bonding and lone-pair EDD in the molecule. The system is interesting not only for its EDD but for the fact that it crystallizes as a double zwitterion, since the arginine side chain is protonated and the glutamic acid residue deprotonated in addition to the terminal amino and carboxylate functionalities. We have carried out a database search in the Cambridge Structural Database for other similar oligopeptides involving histidine, lysine or arginine on one hand and aspartic acid and glutamic acid on the other and found that structures of this kind reported so far all form double zwitterions as well. The database search extends to oligopeptides up to heptapeptides. We will report the results of this literature search with respect to occurrence, common folds / other structural aspects, solubility and crystallizability and discuss how these properties could be controlled by pairing the abovementioned amino acids in a suitable manner. Our findings could be used for crystal engineering and might be applied to improve the solubility of suitably modified oligopeptides of biological interest.

[1] J. Pandit, T. P. Seshadri, M. A. Viswamitra, Acta Cryst. C39, 1983, 1669
[2] G. Wu, B. L. Rodrigues and P. Coppens, J. Appl. Cryst. 35, 2002, 356
[3] B. Hübschle, K. Pröpper, F. Dietrich, T. Stolper, J. J. Holstein, Acta Crystallogr
B69, 2013, 91

BIO-P15

Human Prolidase Mutations - Structural basis of enzymatic activity loss

P. Wilk[1,2], M. Ühlein[1], H. Dobbek[2], M. Weiss[1], U. Mueller[1]
[1]*HZB, Macromolecular Crystallography (BESSY-MX), Berlin, Germany*
[2]*Humboldt-Universität zu Berlin, Lebenswissenschaftliche Fakultät, Institut für Biologie - Strukturbiologie/Biochemie, Berlin, Germany*

Prolidase is a multifunctional enzyme whose biological relevance, its structure and mechanism of catalysis are still only partially understood. In humans prolidase is the only metalloenzyme that cleaves the iminodipeptides containing a proline or hydroxyproline residue at the C-terminal end. Prolidase deficiency (PD) is a rare recessive disorder characterized by diminished prolidase activity and manifested by variety of clinical symptoms. Several mutations responsible for loss of prolidase activity were identified, but the structural basis of the enzyme inactivation mechanism remains unknown. The aim of this study is to determine the influence of single amino acids substitutions or deletions on prolidase structure. Those structures will help in understanding the mechanism of enzyme inactivation. In this study selected mutants will be studied by X-ray diffraction.

BIO-P16

Purification and Crystallization of the SiiAB-Complex

P. Kirchweger[1], S. Weiler[1], C. Egerer-Sieber[1], Y. Muller[1]
[1]*FAU Erlangen, Erlangen, Germany*

Recently a type 1 secretion system (T1SS) in *Salmonella* was identified to be necessary for invasion of polarized epithelial cells. This T1SS is encoded by the *Salmonella* pathogenic Island 4 (SPI4), which consists of the genetic information of the SiiAB-complex, an outer membrane pore (SiiC), a periplasmatic adaptor protein (SiiD), a giant nonfimbral adhesion (SiiE) and the inner membrane ATP-binding cassette (SiiF; Gerlach *et al.* 2007). The SiiAB complex is thought to form a proton channel through the inner cell membrane (Wille *et al.* 2014). Since the structure is unknown and the function remains unclear, it is our goal to solve the crystal structure of the SiiAB complex to gain further insights into the function. So far, we succeeded in solving the crystal structure of N-terminal domain of SiiA. To purify the SiiAB-complex fluorescence size exclusion chromatography (FSEC) is our method of choice for analyzing expression and purification. Constructs of SiiAB together with a cleavable GFP fusion domain have been designed for this purpose. Here, we will present the crystal structure of the N-terminal SiiA domain and preliminary results from the purification and crystallization of the SiiAB complex using FSEC.

References

Gerlach, R. G., D. Jäckel, B. Stecher, C. Wagner, A. Lupas, W.-D. Hardt and M. Hensel (2007). "Salmonella Pathogenicity Island 4 encodes a giant non-fimbrial adhesin and the cognate type 1 secretion system."Cellular Microbiology9(7): 1834-1850.

Wille, T., C. Wagner, W. Mittelstädt, K. Blank, E. Sommer, G. Malengo, . . . R. G. Gerlach (2014). "SiiA and SiiB are novel type I secretion system subunits controlling SPI4-mediated adhesion of Salmonella enterica." Cellular Microbiology 16(2): 161-178.

BIO-P17

Structural and functional characterization of Invasin-like proteins

P. Sadana[1], R. Geyer[2], P. Dersch[2], A. Scrima[1]
[1]*Helmholtz centre for Infection Research, Structural Biology of Autophagy, braunschweig, Germany*
[2]*Helmholtz centre for Infection Research, Department of Molecular Infection Biology., braunschweig, Germany*

QUESTION

Yersinia spp. and other enteric bacteria utilise a set of diverse adhesin molecules, which are essential virulence factors involved in the infection process. Invasin-like proteins represent a subgroup of the adhesin family that specifically interacts with receptors on the host cell plasma membrane. Invasin-like proteins share a common architecture: they consist of i) an N-terminal beta-barrel domain, ii) a central region of varying length consisting of highly homologous repeats of Ig-like domains and iii) the C-terminal part, which varies significantly among Invasin-like proteins.This C-terminal part is responsible for binding to the host-cell receptors, thereby promoting tight attachment to the host cell and inducing internalization of the bacterial pathogen.

The aim of this work is to get insight into the structure and function of the C-terminal domain of InvasinD (InvD). These studies will improve our understanding of the receptor-specificity of InvD and its role in the internalization and infection process.

METHODS-RESULTS

To achieve this, InvD was cloned and expressed in *E. coli* and was purified by a two-step purification protocol to obtain protein of high yield and purity. Protein was subsequently used for screening of crystallization conditions at 20ºC and 4ºC. We obtained initial hits with crystals diffracting to approx. 3.5Å resolution. To solve the phase problem, crystals were soaked in heavy atom solution. Phasing was successful using a partial molecular replacement model combined with single wavelength anomalous dispersion (MR-SAD). Upon further improvement of crystal quality, the structure was solved to a final resolution of 2.6Å.

CONCLUSION

The N-terminal domains of crystallized InvD reveal similarities to the Immunoglobulin Superfamily (IgSF) fold. In contrast, the C-terminal domain adopts a novel fold, suggesting a role in targeting InvD to a specific, yet unidentified receptor on the host cell plasma membrane.

BIO-P18

Native-like Photosystem II Superstructure at 2.44 Å Resolution through Detergent Extraction from the Protein Crystal

M. Bommer[1], J. Hellmich[1], A. Burkhardt[2], M. Ibrahim[1], J. Kern[3], A. Meents[2], F. Mueh[4], H. Dobbek[1], A. Zouni[1]
[1]*Humboldt Universitaet Berlin, Strukturbiologie, Berlin, Germany*
[2]*DESY, Hamburg, Germany*
[3]*Berkeley Lab, Berkeley, United States*
[4]*Universitaet Linz, Linz, Austria*

Photosystem II (PSII) is the site of sunlight-driven water oxidation in plants, algae and cyanobacteria, which generates atmospheric oxygen. It is embedded in the thylakoid membrane. A single PSII (dimer) is a 12 nm size membrane-embedded protein-cofactor complex (Figure: alternating blue or green background).

We present a crystallographic model of such PSII dimers aligned in a 100 nm scale superstructure as exists in the native photosynthetic membrane, obtained by extracting detergent from the protein crystal. The extended alignment of light-harvesting internal antennae (Figure: clusters of blue cylinders) plausibly allows sharing of excitation energy across the superstructure to direct energy from absorbed photons of sunlight to an 'open' reaction centre - one which is at the right stage of the catalytic cycle.

Figure 1

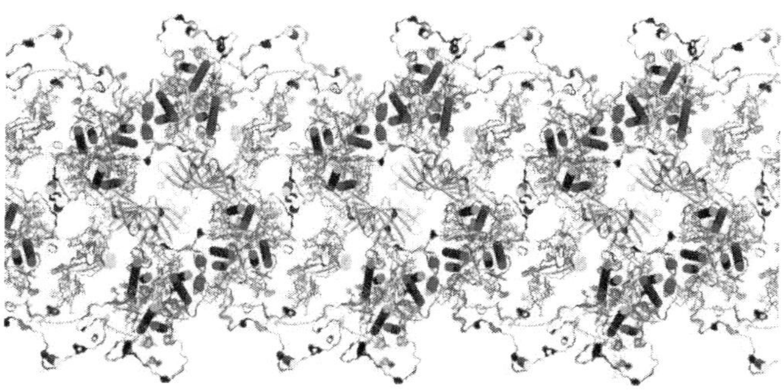

BIO-P19

mtsslSuite: A PyMOL based toolkit for the integration of macromolecular crystallography and EPR long-range distance measurements

G. Hagelueken[1], D. Abdullin[1], O. Schiemann[1]
[1]*University of Bonn, Bonn, Germany*

The focus of structural biology shifts to more and more complex systems. Often, important functional questions can only be answered by a combination of methods. Recently, the combination of X-ray crystallography with long-range EPR distance constraints has proven a successful combination, for example to unravel conformational states of membrane transporters during their transport cycle. The software package mtsslSuite provides crystallographers and EPR spectroscopists with a set of tools for the translation of experimental EPR distance distributions into structural information. The package is based on the PyMOL molecular graphics system and is therefore very intuitive to use. The mtsslSuite includes software for the *in silico* spin labelling of structural models with various spin labels, for distance constrained rigid body docking and for the trilateration of spin centers within macromolecules. The mtsslSuite is tested on a number of challenging test cases and its strengths and weaknesses are evaluated.

BIO-P20

Test Experiments for Future Laue Crystallography at PETRA III

P. Roedig[1], A. Burkhardt[1], P. Fischer[1], E. Weckert[1], A. Meents[1]
[1]*DESY, Photon Science, Hamburg, Germany*

At synchrotrons conventional single-crystal diffraction experiments using monochromatic X-rays are a standard technique for structure determination of protein crystals. Crystals are typically mounted on a goniometer and rotated during exposure to the X-rays. Due to the relatively long timescale for data acquisition of at least several seconds this approach becomes infeasible when it comes to the investigation of dynamical processes which proceed with timescales of milliseconds or less. This hindrance could be overcome by either using femtosecond pulsed X-ray free-electron lasers or by using non-rotating crystals together with a polychromatic X-ray beam, referred to as Laue crystallography (*1*, *2*). In the latter case more complete diffraction images are obtained. Due to the broader wavelength spectrum of the incoming beam a larger amount of reflections fulfill the diffraction condition at the same time and, in addition, the total flux of incoming X-ray photons is much higher. It has been shown that the Laue diffraction technique enables diffraction experiments with a time resolution of down to ~100 ps (*3*). Laue diffraction experiments have successfully been performed using both standard crystal-mounting methods (*4*) and multi-crystal methods (*5*).

In preparation for future microbeam Laue experiments at beamline P11 at PETRA III at DESY we have performed first test experiments by exposing insulin and lysozyme crystals to the first harmonic of the undulator spectrum (see Figure 1 left). Still images from a polychromatic beam were simulated by scanning the monochromator energy through the corresponding energy range. The monochromator scans were repeated for several angular positions of a single crystal so that for each orientation of the crystal a single diffraction image was obtained which is equivalent to a Laue diffraction image (see Figure 1 right). Diffraction images were processed using appropriate Laue software and the results will be compared to the corresponding monochromatic datasets.

Figure 1: Spectrum of a PETRA III undulator at 9.2 keV (left) and the corresponding diffraction image obtained from an insulin crystal (right).

References:

1. Z. Ren *et al.*, Laue crystallography: coming of age. *J. Synchrotron Radiat.* 6 (1999), pp. 891-917.
2. K. Moffat, D. Szebenyi, D. Bilderback, X-ray Laue Diffraction from Protein Crystals. *Science*. 223, 1423-1425 (1984).
3. T. Graber *et al.*, BioCARS: A synchrotron resource for time-resolved X-ray science. *J. Synchrotron Radiat.* 18, 658-670 (2011).
4. S. Cornaby *et al.*, Feasibility of one-shot-per-crystal structure determination using Laue diffraction. *Acta Crystallogr. Sect. D Biol. Crystallogr.* 66, 2-11 (2010).
5. S. L. Perry *et al.*, In situ serial Laue diffraction on a microfluidic crystallization device. *J. Appl. Crystallogr.* 47, 1975-1982 (2014).

Figure 1

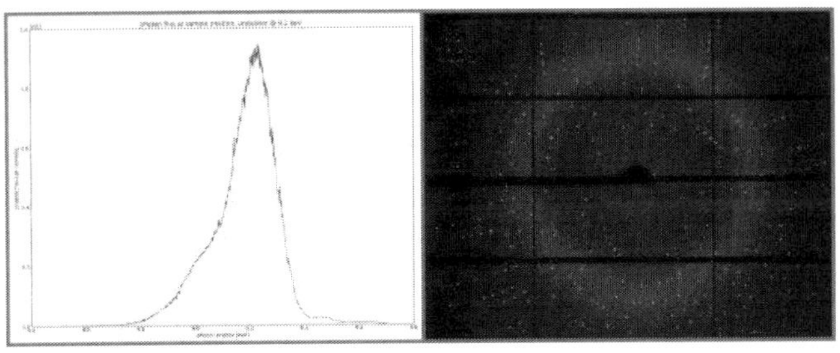

BIO-P21

The Crystallography Endstation at Beamline P11 at PETRA III

A. Burkhardt[1], P. Fischer[1], D. Göries[1], J. Meyer[1], T. Pakendorf[1], S. Panneerselvam[1], B. Reime[1], J. Röver[1], N. Stübe[1], M. Warmer[1], A. Meents[1]
[1]*Deutsches Elektronen-Synchrotron DESY, Photon Science, Hamburg, Germany*

The "Bio-Imaging and Diffraction Beamline" P11 at PETRA III is in operation since January 2013 and was designed for structure determinations on biological samples. For this purpose two state-of-the-art experimental endstations are available: an X-ray microscope and a crystallography experiment.

Basis of design was to make full use of the excellent source properties of the PETRA III storage ring. The flexible X-ray optics of beamline P11 allow for tailoring the beam properties to the needs of the experiment: A large parallel beam is available for the investigation of viruses and large molecular complexes. For structure determinations from microcrystals a micrometer sized beam can be generated.

The P11 X-ray optics comprise an LN_2 cooled double crystal monochromator, a first Kirkpatrick-Baez (KB) mirror system consisting of two horizontal deflecting and one vertical deflecting mirror located in the P11 optics hutch, and a second KB system very close to the crystallography endstation. The three mirrors of the first KB system are dynamically bendable and can be used to generate a secondary source at 65.5 m. With this beam sizes down to 300 × 300 μm^2 FWHM (v × h) can be realized at the sample position (73 m). The second KB system is also equipped with benders and can be used for further refocusing the X-ray beam into a spot of 4 × 9 μm^2 FWHM (v × h) with full flux from the source (> 10^{13} ph/s at 12 keV). Smaller beam sizes down to 1 × 1 μm^2 with more than 2 × 10^{11} ph/s in the focus can be obtained by slitting down the secondary source at the cost of flux.

The crystallography experiment is equipped with a high precision single axis goniostat with a combined sphere of confusion of less than 100 nm. X-ray energies are tunable between 5.5 and 30 keV in order to provide full SAD/MAD capability.

At P11 a full data set can be typically collected in less than two 2 minutes. The beamline is equipped with a Pilatus 6M-F detector for fast data collection with frame rates of up to 25 Hz. Crystals can be mounted in less than 10 s using an automatic sample changer. The large sample dewar can accommodate 368 samples. This makes P11 ideally suited for high-throughput crystallography, fast crystal screening and serial crystallography experiments.

BIO-P22

Same yet Different - Molecular Basis of $GABA_A$ versus Glycine Receptor Clustering through Gephyrin

V. B. Kasaragod[1], H. M. Maric[2], T. J. Hausrat[3], M. Kneussel[3], K. Strømgaard[1], H. Schindelin[1]
[1]*University of Würzburg, Rudolf Virchow Center for Experimental Biomedicine, Würzburg, Germany*
[2]*University of Copenhagen, Department of Drug Design and Pharmacology, Copenhagen, Denmark*
[3]*University Medical Center Hamburg-Eppendorf, Center for Molecular Neurobiology, ZMNH, Hamburg, Germany*

γ-Aminobutyric acid type A and glycine receptors ($GABA_ARs$, GlyRs) are the major inhibitory neurotransmitter receptors and contribute to many synaptic functions, dysfunctions and human diseases. $GABA_ARs$ are important drug targets, which are regulated by direct interactions with the scaffolding protein gephyrin. Gephyrin consists of an N terminal G domain and a C terminal E domain, the latter interacts with the intracellular loop between transmembrane α-helices 3 and 4 of the GlyR β-subunit and the GABA$_A$R a1-a3 subunits. The structure of the complex between GephE and the GlyR β-loop was solved nearly a decade ago but the interaction between GABA$_A$Rs and gephyrin remained enigmatic. Thus, the main aim was to structurally characterize and thermodynamically dissect the GephE-GABA$_A$R interaction. Initial efforts to crystallize the GephE-GABA$_A$R α3 complex were hindered due to aggregation and degradation problems of the GABA$_A$R α3-loop. To overcome this, we tailored peptides which contained the core binding motif. These peptides were used todeduce the molecular basis of the interaction between GABA$_A$R and gephyrin by X-ray crystallography and to study this interaction by isothermal titration calorimetry (ITC). Several crystal structures of GephE-GABA$_A$R α3 complexes revealed that the N-terminal region of the α3-peptide occupies the same binding site as the GlyR β-subunit, while the C-terminal end, which is conserved among all synaptic GABA$_A$R α-subunits, engages in subunit specific interactions. Thermodynamic dissections of the gephyrin-receptor interactions identified two residues, Asn369 and Thr374, in the GABA$_A$R α3 subunit as primary determinants for gephyrin's subunit preference. Thus, these first structural insight into GABA$_A$R clustering through gephyrin offer a framework for future investigations into the regulation of inhibitory synaptic strength and for the development of therapeutically relevant compounds targeting the gephyrin-GABA$_A$R interaction.

References :

Maric, H. M., Kasaragod, V. B. & Schindelin, H. Modulation of gephyrin-glycine receptor affinity by multivalency. *ACS Chem. Biol.* 9(11):2554-62 (2014).

Maric, H.M., Kasaragod, V.B., Kedstrom, L.H., Hausrat, T.J., Kneussel, M.,., Strømgaard, K., Schindelin, H. Design and synthesis of high-affinity dimeric inhibitors targeting the interaction between gephyrin and inhibitory neurotransmitter receptors. *Angew.Chem.Int.Ed.* (2014)

Maric, H.M., Kasaragod, V.B., Hausrat, T.J., Kneussel, M., Tretter, V., Strømgaard, K., Schindelin, H. Molecular basis of the alternative recruitment of GABA$_A$ versus glycine receptors through gephyrin. *Nat.Commun.* (in press).

BIO-P23

Expression, purification and crystallisation of MKK7

P. Wolle[1], S. Mayer-Wrangowski[1], D. Rauh[1]

[1]*Technische Universität Dortmund, Dortmund, Germany*

The mitogen-activated protein kinase kinase 7 (MKK7) is a dual-specific protein kinase and a member of the c-Jun N-terminal protein kinase (JNK) signalling pathway, which is involved in the regulation of numerous physiological processes during cellular development and in response to stress [1]. MKK7 is phosphorylated and thereby activated by MAP3K and phosphorylates JNK together with MKK4 on specific Tyr and Thr residues in the activation loop [2]. For a better understanding of the cellular function of MKK7 in disease states, we focus on the development of potent and selective inhibitors. As a starting point, we set out to solve the crystal structure of MKK7 in complex with classic ATP competitive inhibitors. For this, we cloned various constructs of MKK7, which differ in mutations in the activation loop and modifications at the N-terminus. All constructs were expressed in *E. coli* and purified by affinity chromatography, ion exchange chromatography and size exclusion chromatography. Suitable crystallisation conditions were selected by screening all constructs against 386 different conditions at 4 °C and 20 °C. With these conditions, we could obtain the first complex structures of MKK7. This complex gave key insights into the pharmacological perturbation of MKK7 and now serves as starting point for further compound development.

References:

[1] Asaoka, Y.; Nishina, H.: J. Biochem. 2010,148(4:) 393-401.

[2] Haeusgen, W.; Herdegen, T.; Waetzig,V.: European Journal of Cell Biology, 201190: 536-544.

CTC - Computational & Theoretical Crystallography

CTC-P01

PDF-refinement of crystalline & nanocrystalline Bi_2WO_6

M. Teck[1], H. K. Grossmann[2], T. Grieb[2], T. Hartmann[3], L. Mädler[2], T. M. Gesing[1]
[1]*Universität Bremen, Chemische Kristallographie fester Stoffe /FB02, Leobener Straße /NW2, 28359 Bremen, Germany*
[2]*Universität Bremen, Stiftung Institut für Werkstofftechnik, Mechanische Verfahrenstechnik, Badgasteiner Straße 3, 28359 Bremen, Germany*
[3]*Stoe&Cie GmbH, Hilpertstraße 10, 64295 Darmstadt, Germany*

As the simplest member of the Aurivillius family, Bi_2WO_6 has recently been investigated because of its superior performance as visible-light driven photocatalytic active material [1]. Considering the fact that the photocatalytic activity is strongly correlated to the surface area, average crystallite and particle size and also the morphology [2], the real structure of the nano-sized crystallites is fundamental to know. Therefore the structural differences between crystalline and nano-crystalline Bi_2WO_6 samples are reported here. Pair distribution function (PDF) data were refined with an evolutionary algorithm, taking the crystallite-shape into account for the refinement progress of the nano-crystalline sample. Whereas a well crystalline sample was prepared via hydrothermal synthesis (L_{Vol}(IB) = 17.09(8) nm), the nano-crystalline material was produced via flame spray pyrolysis with an average crystallite size around 8 nm in diameter. XRD data to prepare the PDFs were measured on a Stoe Stadi P diffractometer using monochromatic Ag $K_{\alpha 1}$ radiation together with a MYTHEN 1K PSD detector. The PDFs were calculated from the background corrected raw data with PDFgetX3 [3]. By using the DISCUS [4] and DIFFEV [5] approach for the whole-particle structure refinements of the Bi_2WO_6 structure, lattice parameters, scale-factor, pdf-factors, all atom positions and the B-values of the metal atoms were refined. For the nano-crystalline sample the particle size was additionally calculated. The resulting shape for the nano-crystallites (Fig. 1) was verified by HRTEM micrographs (Fig.2).

Fig. 1: resulting Bi_2WO_6 nano-particle Fig. 2: HRTEM-micrographs (FSP-Bi_2WO_6)

References

[1] D. Y. Kim, S. Kim, M. Kang, *Bull. Korean Chem. Soc.* 2009, *30*, 630.
[2] M. Shang, W. Wang, S. Sun, L. Zhou, L. Zhang, *J. Phys. Chem. C* 2008, *112*, 10407.
[3] P. Juhás, T. Davis, C. L. Farrow, S. J. L. Billinge, *J. Appl. Crystallogr.* 2013, *46*, 560.
[4] T. Proffen, R. B. Neder, *J. Appl. Crystallogr.* 1997, *30*, 171.
[5] K. Page, T. C. Hood, T. Proffen, R. B. Neder, *J. Appl. Crystallogr.* 2011, *44*, 327.

Figure 1

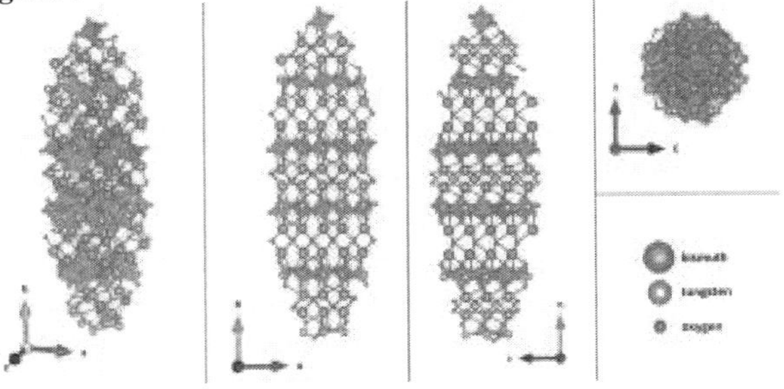

Figure 2

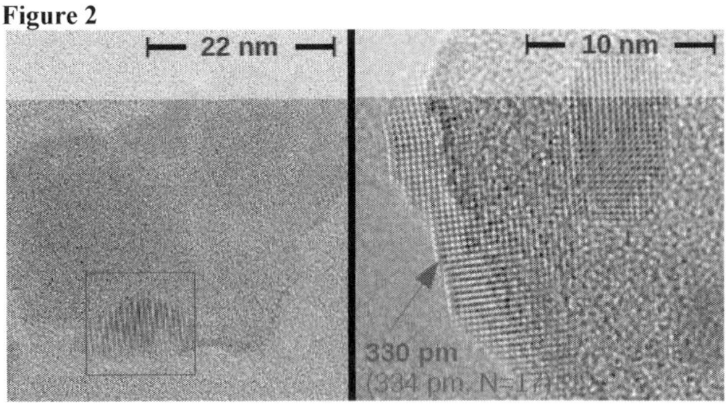

CTC-P02

An Empirical Correction for Low Energy Contamination of X-ray Data

L. Krause[1], R. Herbst-Irmer[1], D. Stalke[1]
[1]*University of Göttingen, Institute of Inorganic Chemistry, Göttingen, Germany*

Microfocus sealed-tube X-ray sources have become standard in many laboratories because of their low power consumption and minimal maintenance. The high X-ray flux density and the fine focus achieved by the total reflection on mirror optics allow for high performance experiments in house [1,2]. However, in 2011 Macchi et al. showed systematical defects with this technology as the marginal spectroscopic impurity of the beam [3]. A low energy radiation of approximately three times the wavelength of the incident beam bypasses the mirror optics and causes additional diffraction. Reflections with *3h3k3l* indices overlap with the reflections *hkl* from the low energy diffraction leading to systematically overestimated intensities for these particular reflections. We observed similar but weaker effects also for data sets collected on a TXS-Mo rotating anode equipped with mirror optics.

Macchi et al. described a solution by passing the X-ray beam through an aluminum foil into the X-ray beam. Alternatively we suggest a mathematical solution to avoid loss of X-ray intensity; furthermore, it is applicable for already measured data sets. To a good approximation, the following expression can be a description for the level of contamination [4]:

$$F_{obs}(3h3k3l) = F_{calc}(3h3k3l) + k \cdot F_{obs}(hkl)$$

The contamination coefficient k is then determined by linear regression and can be used for data correction with a routine implemented in SADABS [5].

This empirical approach is compared to the above mentioned experimental method on several model compounds.

References:

[1] S. J. Coles, P. A. Gale, *Chem. Sci.* 2012, *3*, 683-689.
[2] T. Schulz, K. Meindl, D. Leusser, D. Stern, J. Graf, C. Michaelsen, M. Ruf, G. M. Sheldrick, D. Stalke, *J. Appl. Cryst.* 2009, *42*, 885-891.
[3] P. Macchi, H-B. Bürgi, A. Chimpri, J. Hauser, Z. Gál, *J. Appl. Cryst.* 2011. *44*, 763-771.
[4] K. Kirschbaum, A. Martin, A. Pinkerton, *J. Appl. Cryst.* 1997, *30*, 514-516.
[5] L. Krause, R. Herbst-Irmer, G. M. Sheldrick, D. Stalke, *J. App. Cryst.* **2015**, *48*, in print.

CTC-P03

TDS: a First Step towards an Experimental Correction

B. Niepötter[1], R. Herbst-Irmer[1], D. Stalke[1]
[1]*University of Göttingen, Institute of Inorganic Chemistry, Göttingen, Germany*

Thermal diffuse scattering (TDS) mainly results in an underestimation of the atomic displacement parameters. However, smaller but nevertheless important errors occur in other parameters as well.[1] Hence the heights of the maxima in the electron density are changed without displacing them.[2,3] First order TDS leads to peak broadening in the diffraction experiment. As shown by Jennings [4] it is possible to estimate the TDS contribution by analyzing the peak profile. Blessing [5] as well as Stash and Zavodnik [6] extended this idea and developed programs to assess the TDS contribution to the measured intensities for data collected with point detectors. However, with area detectors these methods were no longer useable. Currently the effects of TDS on the model are more or less ignored referring to the low temperatures during the measurement. Nevertheless problems due to TDS can occur also at 100 K or even at 15 K. Residual density occurs close to atomic positions. Therefore it is essential for high quality single crystal X-ray experiments to correct for TDS.

By refining resolution dependent scale factors it is possible to minimize the residual density. In a simple approach the known ideas [7] are transferred to area detector techniques. By iteratively determining the correct size of the integration box the influence of TDS is minimized. An empirical TDS correction for the measured intensities can be derived from these methods. Model improvements are demonstrated for several structures.

[1] G. Alexandropoulos, M. J. Cooper, P. Suortti, B. T. M. Willis in *International Tables for Crystallography, Vol. C.* (Ed.: E. Prince), Kluwer Academic Publishers, Dordrecht, 2004, pp. 653-665
[2] V. G. Tsirelson, R. P. Ozerov, *Electron Density and Bonding in Crystals*, IOP Publishing, London, 1996, pp. 111-116
[3] R. B. Helmholdt, A. Vos, *Acta Crystallogr.* 1977, *A33*, 38-45
[4] L. D. Jennings, *Acta Crystallogr.* 1970, *A26*, 613-622
[5] R. H. Blessing, *J. Appl. Crystallogr.* 1986, *19*, 412
[6] A. I. Stash, V. E. Zavodnik, *Crystallogr. Rep.* 1996, *41,* 404
[7] V. E. Zavodnik, A. I. Stash, V. G. Tsirelson, R. de Vries, D. Feil, *Acta Crystallogr.* 1999, *B55*, 45-55.

CTC-P04

Charge Density Investigations on Low Valent Silicon Containing Compounds

D. Stalke[1]
[1]*University of Göttingen, Insitute of Inorganic Chemistry, Göttingen, Germany*

Silicon, the heavier congener of carbon, frequently shows different chemistry compared to its sister element. It has been realized that silicon prefers a positive charge as it is most frequently bonded to more electronegative atoms, i. e. carbon of substituents. Additionally, compounds with silicon in lower oxidation states are very important, since they can activate small organic molecules which otherwise are not available using transition metals. Allene type molecules can also be rationalized along a structure with the central C(0) atom to be bonded by the ligand L, where L perform as donor ligands. In the general formula $C(L:)_2$ which shows two lone pairs at the central carbon atom have been termed carbones whereas carbenes CR_2 accommodate only one lone pair.[1] When L is a carbene ligand in $C(L:)_2$, the compound is termed a carbodicarbene. A related situation may be envisaged for the silicon homologue of allenes.[2] Theoretical studies by Frenking et al.[3] suggested that the trisilaallene that was synthesized by Kira et al.[4] should rather be considered as a silylone $Si(L:)_2$ which explains why the compound has a rather acute Si-Si-Si angle and why the substituents at the terminal silicon atoms are not orthogonal to each other which is typical for an allene structure.[5] In addition two valence shell charge concentrations were located in the non-bonding region at the central silicon atom. The same is valid for germylones.[6]

[1] K. C. Mondal, H. W. Roesky, M. C. Schwarzer, G. Frenking, S. Neudeck, I. Tkach, H. Wolf, D. Kratzert, R. Herbst-Irmer, B. Niepötter, D. Stalke, *Angew. Chem.* 2013, *125*, 3036-3040; *Angew. Chem. Int. Ed.* 2013, *52*, 2963-2967.
[2] S. Ishida, T. Iwamoto, C. Kabuto, M. A. Kira, *Nature* 2008, *421,* 725 - 727.
[3] N. Takagi, R. Tonner, G. Frenking, *Chem.-Eur. J.* 2012, *18*, 1772 - 1780.
[4]M. Kira, *Chem. Comunm.* 2010, *46*, 2893 - 2903.
[5]B. Niepötter, R. Herbst-Irmer, D. Kratzert, P. P. Samuel, K. C. Mondal, H. W. Roesky, P. Jerabek, G. Frenking, D. Stalke, *Angew. Chem.* 2014, *126*, 2806-2811; *Angew. Chem. Int. Ed.* 2014, *53*, 2766-2770.
[6] Y. Li, K. C. Mondal, H. W. Roesky, H. Zhu, P. Stollberg, R. Herbst-Irmer, D. Stalke, D. M. Andrada, *J. Am. Chem. Soc.* 2013, *135,* 12422–12428.

CTC-P05

Validation of Charge Density Refinement Strategies

F. Engelhardt[1], L. Krause[1], B. Niepötter[1], R. Herbst-Irmer[1], D. Stalke[1]
[1]*University of Göttingen, Insitute of Inorganic Chemistry, Göttingen, Germany*

Modern experimental charge density investigations require the most accurate data, the highest possible resolution and an optimized refinement strategy [1,2]. As stated in the XD2006 manual [3] the complexity of the model should be increased in a stepwise manner and the final data to parameter ratio should not fall below a value of 10. However, fulfilling this rule does not guarantee that all refined parameters are reasonable. Common criteria such as residuals of the linear least squares refinement tend to decrease with increasing number of parameters. The statistical method of cross-validation helps to decide about the addition of parameters. This method is well established in macromolecular crystallography known as the R_{free} concept [4]. A fraction of the measured data is excluded from the whole refinement process and an R value is calculated for the refined model against the unused data.

Two problems need to be addressed in charge density investigations: First, the changes in *R* values are small compared to the precision of the R_{free} value. Second, the omission of particular reflection could have a high impact on the refinement. Therefore, k-fold cross-validation is used. Here the data set is divided into k subsets. One set acts as validation set, while the remaining sets are used for the refinement. The refinement is repeated k times with each of the subset serving once as R_{free} set [5]. Therefore every single reflection is used for validation. This approach is automated for charge density refinement procedures using the XD2006 program suite [3].

This developed procedure is demonstrated on the example of a high resolution low temperature data set of an alkaline organometallic complex.

[1] *Electron density and chemical bonding I experimental charge density studies*, ed. D. Stalke, Springer, Berlin; New York, 2012.

[2] P. Coppens, *X-ray charge densities and chemical bonding*, International Union of Crystallography ; Oxford University Press, [Chester, England]; Oxford; New York, 1997.

[3] A. Volkov, P. Macchi, L. J. Farrugia, C. Gatti, P. R. Mallinson, T. Richter, T. Koritsanszky, in *XD2006*, 2006.
[4] A. T. Brünger, *Nature* 1992, *355*, 472-475.
[5] A. Paul, M. Kubicki, C. Jelsch, P. Durand, C. Lecomte, *Acta Crystallogr.* 2011, B67, 365-378.

CTC-P06

Data Collection Strategies for High Resolution Datasets

C. J. Schürmann[1], F. Engelhardt[1], R. Herbst-Irmer[1], D. Stalke[1]
[1]*Georg-August University, Instiute for Inorganic Chemistry, Göttingen, Germany*

The precise determination of the low order reflections is of highest importance for charge density studies as they mainly contribute to the valence shell density.[1] In order to find an optimal collection strategy for these reflections, we collected data from a crystal of S_4N_4 at different experimental setups.

Diffraction data were collected at 100 K with a Bruker rotating anode (Mo K_α), equipped with an Al/Zr filter, multilayer optics and a APEX 2 CCD detector. The data were integrated with SAINT [2] and scaled with SADABS [3].

Data sets derived from different experimental setups were compared following the concept of paired refinement [4,5]. A multipole model refinement with XD2006 [6] is performed using the same refinement protocol for all data sets. The resulting models are evaluated by calculating R-values against each data set. Thereby the best model and subsequently the best data set and underlying experimental setup can be determined.

[1] H. Wolf, M. R. V. Jørgensen, Y. Chen, R. Herbst-Irmer, D. Stalke, *Acta Crystallogr. Sect. B* 2014, in production.
[2] Bruker SAINT 8.30c, Bruker ASX Inc., Madison, WI, USA, 2013.
[3] L. Krause, R. Herbst-Irmer, G. M. Sheldrick, D. Stalke, *J. Appl, Cryst*. 2015, in print.
[4] P. A. Karplus, K. Diederichs, *Science* 2012, *336*, 1030-1033.
[5] K. Diederichs, P.A. Karplus, *Acta Cryst.* 2013, D69, 1215-1222.
[6] A. Volkov, P. Macchi, L. J. Farrugia, C. Gatti, P. R. Mallinson, T. Richter, T. Koritsanszky, in *XD2006*, 2006.

GIN - General Interest

GIN-P01

Kinetics of ferroelastic domain switching in $SrTiO_3$ by pulsed electric fields

G. Eckold[1], J. Sidoruk[1], J. Leist[1], H. Gibhardt[1], O. Sobolev[1], B. Ouladiaff[2]
[1]*Georg-August-Universität Göttingen, Institut für Physikalische Chemie, Göttingen, Germany*
[2]*Institut Laue Langevin , Grenoble, France*

Strontium titanate ($SrTiO_3$) is a well-known member of the perovskite family.
It exhibits an antiferrodistortive phase transition at 105 K leading to a tetragonal paraelectric phase with three different structural domains.

The distribution of ferroelastic domains can be changed not only by applying uniaxial mechanical stresses but also by electric fields, although a direct linear coupling between the order parameter and the electric field is forbidden by symmetry [1]. In addition, strong enough electric fields can induce a transition into a polar ferroelectric phase at temperatures below about 40 K [2].

The basic mechanisms that govern this complex phase behaviour can very efficiently be investigated in real-time experiments under the influence of pulsed electric fields when the time evolution of superlattice reflections are used to monitor the kinetics of the domain redistribution. Stroboscopic data acquisition at the three axes spectrometer PUMA@FRM II and the single crystal diffractometer D10@ILL allowed the determination of the characteristic relaxation rates as functions of temperature, field amplitude and mechanical stress. As an example, Fig.1 shows the time evolution of the volume fraction of z-domains under the influence of a cycled electric field (500 Hz, 4 kV/cm) along the pseudocubic [001] direction for different mechanical stresses along [110]. The switching behaviour of ferroelastic domains does in fact strongly depend on the stress state of the sample. Moreover, two types of kinetic processes could be identified which occur on time scales of microseconds and seconds, respectively. This behaviour seems to be associated with the existence of soft and hard domain walls as predicted theoretically [3].

References:

[1] J. Sidoruk, J. Leist, H. Gibhardt, M. Meven, K. Hradil and G. Eckold,
J. Phys. Condens. Matter **22**, 235903 (2010).
[2] J. Hemberger, M. Nicklas, R. Viana, P. Lunkenheimer, A. Loidl and R. Böhmer
J. Phys. Condens. Matter **8**, 4673-4690 (1996).
[3] A.N. Morozovska, E.A. Eliseev, M.D. Glinchuk, L.-Q. Chen, V. Gopalan,
Phys. Rev. B **85**, 094107 (2012)

Fig.1: Variation of the z-domain fraction during a cycled electric field along [001] with amplitude 4 kV/cm and frequency 500 Hz at 20 K for different values of the applied mechanical stress along [110].

Figure 1

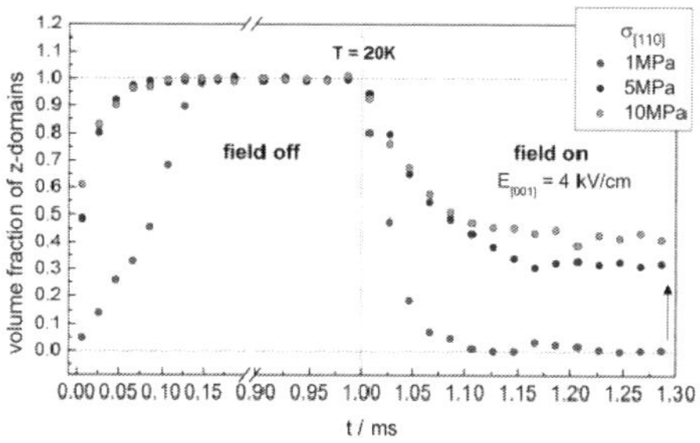

GIN-P02

Light-Atom Structures: Absolute Configuration Determination and Beyond.

J. Graf[1], M. Ruf[2], B. Noll[2], S. Freisz[3], A. Gerisch[3], H. Ott[3], B. Dittrich[4], C. Michaelsen[1]
[1]*Incoatec GmbH, Geesthacht, Germany*
[2]*Bruker AXS Inc., Madison, United States*
[3]*Bruker AXS GmbH, Karlsruhe, Germany*
[4]*Georg-August-Universität Göttingen, Göttingen, Germany*

The determination of the absolute configuration for light-atom structures is central to research in pharmaceuticals and natural-product synthesis [1]. In the absence of elements heavier than silicon, it is often problematic to make a significant assignment of absolute configuration. Traditionally, heavy-atom derivatives were prepared which have a stronger anomalous signal compared to the native compound. However, this is not always feasible.

The assignment of the absolute structure of pure organic compounds has become somewhat easier with the advent of high-intensity microfocus sources [2], as the increased flux density improves the anomalous signal through improvements in counting statistics. In order to maximize the anomalous signal, X-ray sources with Cu anodes are usually used for the absolute structure determination. However, these data are usually limited to a maximum resolution of about 0.80 Å. High-brilliance microfocus X-ray sources with Mo targets enable the collection of high quality data beyond 0.40 Å within a reasonable amount of time. This allows not only a more accurate modelling of the electron density by using aspherical scattering factors, but also enables a reliable determination of the absolute structure, despite the significantly lower anomalous signal obtained with Mo Kα radiation.

With the recently introduced liquid-Gallium-jet X-ray source unprecedented beam intensities can be achieved [3]. The shorter wavelength of Ga Kα compared to Cu Kα slightly weakens the anomalous signal of a typical light-atom structure. However, due to the shorter wavelength, the highest resolution for the liquid metal-jet source is typically at about 0.70 Å, compared to about 0.80 Å for Cu Kα. Hence, about 50% more unique reflections can be recorded. This clearly improves the structural model and the quality of the Flack parameter.

Selected results on the absolute structure and charge density determinations for light-atom structures will be presented.

[1] H. D. Flack, G. Bernardinelli, *Chriality*, 2008, *20*, 681 - 690.
[2] T. Schulz, K. Meindl, D. Leusser et al., *J. Appl. Cryst.*, 2009, *42*, 885 - 891.
[3] M. Otendal, T. Tuohimaa, H. M. Hertz et al., *Rev. Sci. Instrum.*, 2008, *79*, 016102.

GIN-P03

SYNTHESIS, CRYSTAL STRUCTURES , HYDROGEN BONDS STUDIES AND DFT CALCULATION OF NEW QUINOLINE DERIVATIVES

B. Nabil[1], M. Amel[1], Z. Bachir[1]
[1]*Abbés Laghrour University, SM, Khenchela, Algeria*

The quinoline ring system is a common structural component of a wide variety of natural or synthetically prepared products with highly desirable biological activity. It is a heterocyclic scaffold of paramount importance to human race. Indeed, quinoline derivatives are some of the oldest compounds which have been utilized for the treatment of a variety of diseases.

Compounds containing quinoline motif are most widely used as antimalarials, antibacterials, antifungals , anti VIH and antitumor agents. They have antiseptic, antipyretic and antiperiodic properties. Additionally, quinoline derivatives find use in the synthesis of fungicides, virucides, biocides, alkaloids, rubber chemicals and flavoring agents. They are also used as polymers, catalysts, corrosion inhibitors, preservatives, and as solvent for resins and terpenes. Furthermore, these compounds find applications in chemistry of transition-metal catalyst for uniform polymerization and luminescence chemistry. The synthesis of substituted quinolines has been a subject of great focus in organic chemistry.

Owing to the interesting of quinoline systems, this poster aims to describe the synthesis and the structure of two quinolines adducts obtained from the Baylis-Hillman derivates. Those adducts are the result of the acetylation reaction of Baylis-Hillman products in low temperature catalyzed by 1,4-diazabicyclo[2]octane (DABCO) and silicium oxid.

The crystal structure of **(1a)** crystallizes in orthorhombic space group pna2_1, a=7.7237 (12) Å, b = 20.2539 (28) Å, c = 8.7164 (12) Å, however the crystal structure of **(2a)** crystallizes in triclinic space group P-1, a=7.9145 (12) Å, b = 9.1412 (11) Å, c = 10.7286 (16) Å, α = 94.760 (11) °, β = 105.988 (13)° , γ = 101.818(11)° and its cohesion was assured by C-H···N, O-H···O and C-H···O hydrogen bonds. A theoretical studies was effected by DFT calculation.

figure legends:
figure 1: (1a)
Figure 2: (2a)

Figure 1

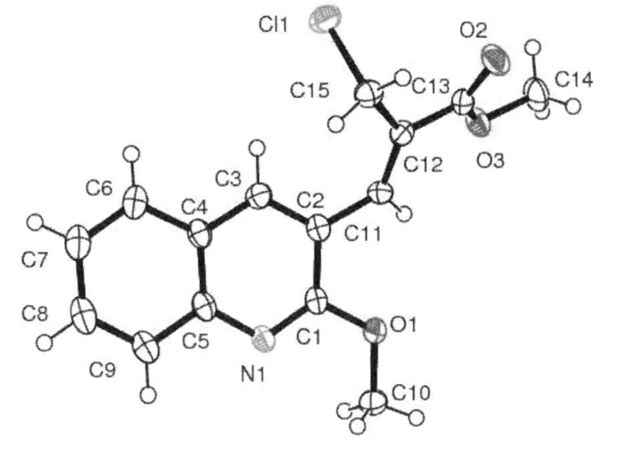

Figure 2

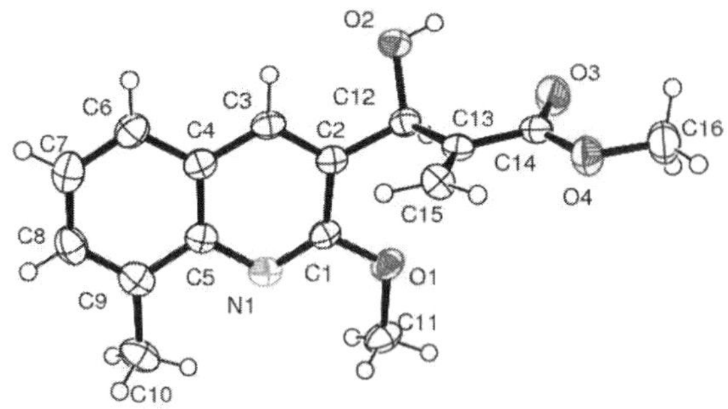

GIN-P04

Sulfur as hydrogen bond acceptor within the *ADA/DAD* synthon in cocrystals of 6-methyl-2-thiouracil

W. M. Hützler[1], E. Egert[1]
[1]*Goethe-University Frankfurt am Main, Institute of Organic Chemistry and Chemical Biology, Frankfurt am Main, Germany*

The synthesis of cocrystals provides an elegant option in order to obtain new crystalline forms of a drug with different physical properties.[1,2] A key step in the experimental procedure is the identification of a suitable coformer for each drug wherein the tools of crystal engineering are applied. Due to their high stability triply hydrogen-bonded synthons are potential building blocks.[3,4] The antithyroid drug 6-methyl-2-thiouracil[5] (MTU) features an *ADA* (*A* = acceptor, *D* = donor) hydrogen-bonding site. 2,4-Diaminopyrimidine (DAPY), 2,4,6-triaminopyrimidine (TAPY) and 6-amino-(3*H*)-isocytosine (AICT) show complementary *DAD* hydrogen-bonding sites. Thus they are suitable coformers for obtaining cocrystals of MTU containing the *ADA/DAD* N-H···O/N-H···N/N-H···S synthon. Cocrystallization experiments of MTU with these coformers yielded into one solvent-free cocrystal and six cocrystal solvates (Fig. 1). The cocrystals (II) to (VII) contain the desired N-H···O/N-H···N/N-H···S synthon. In crystal (I), on the other hand, an $R^2_2(8)$ interaction[6] similar to the Watson-Crick pattern of the adenine/uracil base-pair is formed. Each structure containing the *ADA/DAD* synthon shows a one-dimensional hydrogen-bonding network while in structure (I) a two-dimensional hydrogen-bonding network is provided.

Fig. 1: Structures of the compounds used and composition of the cocrystals.

[1] Schultheiss, N. & Newman, A. (2009). *Cryst. Growth Des.* 9, 2950-2967.
[2] Shan, N. & Zaworotko, M. J. (2008). *Drug Discovery Today*, 13, 440-446.
[3] Desiraju, G. R. (1995). *Angew. Chem. Int. Ed.* 34, 2311-2327.
[4] Prins, L. J., Reinhoudt, D. N. & Timmerman, P. (2001). *Angew. Chem. Int. Ed.* 40, 2382-2426.
[5] Hershman, J. M. & Van Middlesworth, L. (1962). *Endocrinology*, 71, 94-100.
[6] Bernstein, J., Davis, R. E., Shimoni, L. & Chang, N.-L. (1995). *Angew. Chem. Int. Ed. Engl.* 34, 1555-1573.

Figure 1

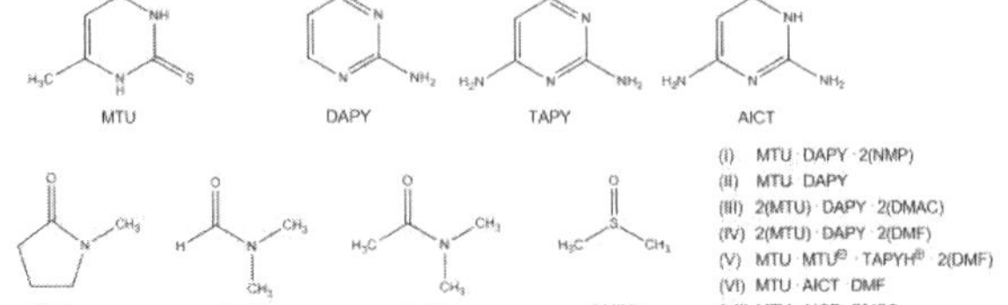

GIN-P05

Time-resolved X-ray diffraction reveals the hidden mechanism of high electromechanical activity in $Sr_{0.5}Ba_{0.5}Nb_2O_6$ uniaxial ferroelectric

S. Gorfman[1], H. Choe[1], V. Shvartsman[2], J. Dec[3], M. Ziolkowski[1], U. Pietsch[1]
[1] *University of Siegen, Department of Physics, Siegen, Germany*
[2] *University of Duisburg-Essen, Institute for Materials Science, Essen, Germany*
[3] *Institute of Materials Science, University of Silesia, Katowice, Poland*

High electromechanical activity (piezoelectricity) of many ferroelectrics has been the focus of numerous recent studies. The structural origin of this activity remains poorly understood due to a lack of appropriate experimental techniques and mixing of different mechanisms related to ferroelectricity and ferroelasticity. Our work reports on the study of a uniaxial $Sr_{0.5}Ba_{0.5}Nb_2O_6$ ferroelectric, where existence of the ferroelastic domains is ruled out for the symmetry reason and the interrelation between piezoelectricity and ferroelectricity can be inspected in the isolated form.

We performed X-ray diffraction experiment on a single crystalline sample under alternating electric field (using the stroboscopic technique described in e.g. [1],[2]) and observed unknown hidden-in-the-bulk electromechanical coupling mechanism. Benefiting from the ability to follow the bulk lattice parameter(s) under alternating electric field, we established that polarization may be reverted via two different routes: without and with nucleation of small ferroelectric domains. The first route passes over the strain minimum while the second route passes over the strain maximum (Figure 1). The second route shows a dramatic enhancement of the electromechanical activity in the form of the correlation between the domain sizes and their lattice parameter. This new mechanism creates a novel roadmap for designing materials with enhanced electromechanical properties.

[1] S. Gorfman, Crystallogr. Rev. 20, 210 (2014)
[2] S. Gorfman, O. Schmidt, M. Ziolkowski, M. Kozierowski, and U. Pietsch, J. Appl. Phys. 108, (2010).

Figure 1

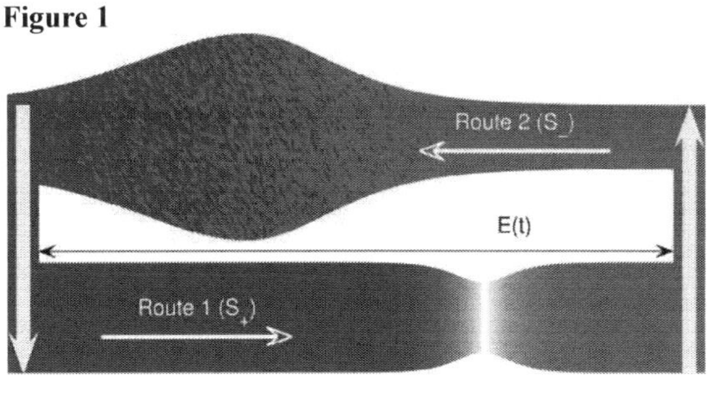

GIN-P06

Fexofenadinhydrochlorid, ein Antihistaminikum der dritten Generation

J. Brüning[1,2], M. U. Schmidt[1]
[1] *Goethe-Universität Frankfurt, Institut für Anorganische und Analytische Chemie, Frankfurt am Main, Germany*
[2] *Aeropharm GmbH, Francois-Mitterand-Allee 1, 07404 Rudolstadt, Germany*

Fexofenadinhydrochlorid ($C_{32}H_{39}NO_4$, **I**) ist ein Antihistaminwirkstoff der dritten Generation [1]. Der Wirkstoff wird bei der Behandlung von z.B. Heuschnupfen, chronischer idiopathischer Nesselsucht und ähnlicher Erkrankungen eingesetzt. Die Wirkung resultiert aus einer kompetitiven, reversiblen Inhibition des H1-Rezeptors [2,3].

Da für **I** keine Einkristalle erhalten werden konnten, wurde die Kristallstruktur aus den Röntgenpulverdaten mithilfe der Programme *DASH* [4] und *TOPAS* [5] bestimmt.

Die Verbindung **I** kristallisiert in der triklinen Raumgruppe *P-1* mit $Z = 2$ und $a = 6.36533(20)$, $b = 15.1310(5)$, $c = 15.5895(5)$Å, $\alpha = 106.245(1)$,

$\beta = 106.245(1)$, $\gamma = 106.245(1)°$. Das Molekül ist fast vollständig gestreckt. **I** zeigt an der zentralen n-Butylgruppe eine gauche-Anordnung und nicht die typischerweise vorliegende all-trans-Anordnung. Der C-C-C-C-Torsionswinkel beträgt 75°. Der Piperidinring zeigt die erwartete Sesselkonformation. Die endständigen Phenylringe stehen fast senkrecht zueinander. Die Elementarzelle beinhaltet zwei Fexofenadin-Kationen (1x S- und 1x R-Konfiguration) und zwei Chlorid-Anionen. Die Kat- und Anionen bilden starke Wasserstoffbrücken in einer trigonal-planaren Form aus (1x $^{+}$N-H···Cl^{-}, 2x O-H···Cl^{-} pro Fexofenadinmolekül). Die Carbonsäuregruppen zeigen Dimere des *graph set*-Typs $C^2_2(8)$ und vervollständigen ein dreidimensionales Wasserstoffbrückennetzwerk.

[1] Mutschler E. Mutschler Arzneimittelwirkungen. Pharmakologie, klinische Pharmakologie, Toxikologie. 10. Auflage. Stuttgart, 2013.
[2] Simons FE, Simons KJ. Peripheral H1-blockade effect of fexofenadine. Ann Allergy Asthma Immunol 1997; 79: 530-2.
[3] Simpson K, Jarvis B. Fexofenadine: a review of its use in the management of seasonal allergic rhinitis and chronic idiopathic urticaria. Drugs 2000; 59: 301-21.
[4] David WIF et al. DASH: a program for crystal structure determination from powder diffraction data. *J Appl Cryst* 2006; 39: 910-915.
[5] Coelho AA. TOPAS Academic, 2007.

Figure 1

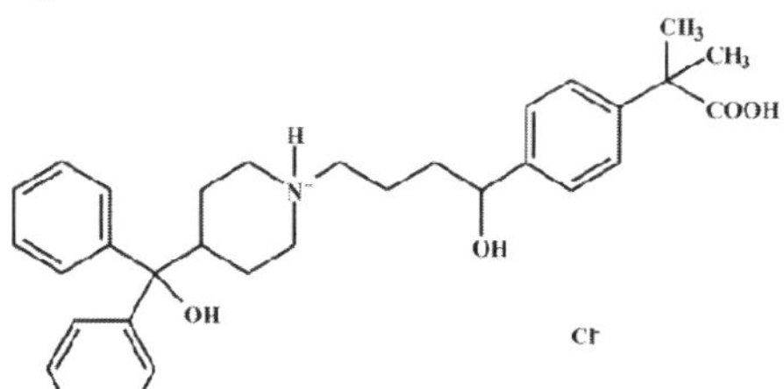

I

GIN-P07

High-resolution crystallographic investigations of organometallic compounds

D. Storozhuk[1], S. Techert[1,2], D. Raiser[2]
[1] *DESY, FS-SCS, Hamburg, Germany*
[2] *Max Planck Institute for Biophysical Chemistry, Go¨ttingen, Germany*

Our motivation is to shed more light into the ultrafast dynamics of some organometallic molecules and study the mechanism underlying, drawing the complete picture of the system under investigation. But since the important electronic changes due to optical excitation take place on very short ps- and sub-ps timescales it gives rise to the necessity of faster time-resolved measurements than those, which are currently possible at third generation synchrotrons. We performed high-resolution crystallographic measurements at various temperatures and study the dynamics of the system as the function of temperature and will be presented.

ISC - Inorganic structural chemistry

ISC-P01
Synthesis, Characterization and Biological Applications of Some Transition Metal Complexes of Doxycyline

O. O. Abosede[1], J. A. Obaleye[2], P. Smith[3], O. N. Majolagbe[4]
[1]*Federal University Otuoke, Department of Chemical Sciences, Otuoke-Yenagoa, Nigeria*
[2]*University of Ilorin, Chemistry, Ilorin, Nigeria*
[3]*University of Cape Town, Pharmacology, Cape Town, South Africa*
[4]*Ladoke Akintola University of Technology, Pure and Applied Biology, Ogbomoso, Nigeria*

We have synthesized some transition metal complexes of the antibiotic doxycylinewith polypyridyl ligands because of the pleiotropic properties of tetracyclines, thebiological activities of transition metals, the improved biological activities oftetracyclines upon chelation by metal ions as well as the biological importance ofpolypyridyl ligands. Mixed ligand complexes of copper (II), doxycycline and polypyridyl ligands were synthesized and well characterized in solid state and in solution by UV-Vis, FT-IR, EPR and ESI-MS. Doxycycline coordinates by the oxygen of the amide group and the carbonyl oxygen at ring A of doxycycline in all these complexes. This was confirmed by the X-ray crystal structure of complex **2** and **3** which are the first crystal structures to be reported for a metal complex of tetracycline antibiotics, as far as we know. DNA binding of the copper complexes were monitored by electronic titration of the complexes with CT-DNA, thermal melting of CT-DNA in the presence and absence of the complexes and viscosity measurements. The intrinsic binding constants of the complexes were 6.1 × 104, 7.2 × 104 and 6.7 × 104 for complexes **5**, **4** and **3** respectively. DNA cleavage experiment by gel electrophoresis revealed that **3** and **5** cleave DNA in the dark, at low concentrations and in the absence of external reducing agents.

Other mixed doxycycline complexes prepared include those of manganese (II), iron (III) and oxovanadium (II), and they were well characterized by elemental analysis, UV-Vis, FT-IR and ESI-MS. Antimicrobial activities of these complexes against pathogenic bacteria isolated from waste food, dung and soil as well as chloroquinesensitive and chloroquine-resistant *P. falciparum* were examined and compared with the parent antibiotic doxycyline. In all cases, the complexes are more potent than doxycycline, the parent antibiotic from which they were synthesized while some are more active than the chloroquine diphosphate and the drug Artesunate against CQ-resistant parasites *P. falciparum* Dd2 strain. The antibacterial activities of some of these complexes are comparable and in some cases higher than that of doxycycline. The cytotoxicity experiments of the copper complexes show that they can also be considered as potential anticancer agents and are effective as Matrix Metalloproteinase (MMP-2) inhibitors.

Figure 1

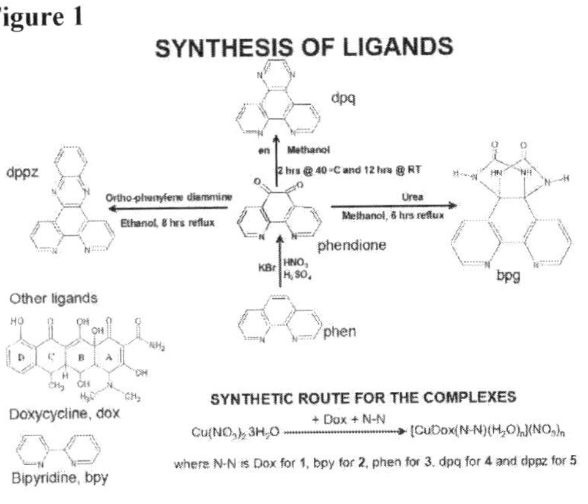

ISC-P02
Synthesis experiments on zeotype framework host substitution: Lithium ABW type materials and stable by-products

P. Törber[1], J.- C. Buhl[1]
[1]*Leibniz Universität Hannover, Hannover, Germany*

The relatively scarce amount of overall lithium bearing zeolite compounds leads to the assumption that lithium ABW type materials [1] could be of interest for future investigations. Determination of the diffusion and sorption properties, of electric conductivity, and thermal as well as hydrothermal stability would be desirable. In the present study syntheses within the ABW framework with the focus on $|Li_4(H_2O)_4|[T_4^1T_4^2O_{16}]$-ABW ($T^1$=Al, Ga; T^2=Si, Ge) are examined.

In contrast to the standard recipe [2], the focus of the present research lies in the modification of the synthesis conditions and methodology, respectively. These include:

A decrease of the synthesis temperature from 523 K to 473 K in all experiments in this study and an examination of the crystallization behavior without and with substitution of the T-atoms using several aluminosilicate, gallosilicate, aluminogermanate and gallogermanate source materials. These educts were transformed into the products under mild hydrothermal conditions at 473 K for 72 h (respectively, 24 h and 96 h).

Synthesis products are investigated structurally by X-ray powder diffraction. Further analyses were examined by Fourier transform infrared spectroscopy, scanning electron microscopy coupled with energy dispersive X-ray studies, as well as simultaneous thermogravimetric- and differential thermal analyses.

Lithium bearing ABW zeolites and phenakite-derivate structures have been successfully synthesized. The aluminosilicate Li-A(BW) zeolite products are orthorhombic (cell parameters of our optimal sample: a = 10.316(11) Å, b = 8.206(6) Å, and c = 4.998(6) Å), SG $Pna2_1$, in accordance with Barrer and White's very first synthesis product [3] and the sample of [1]. Tempering as-synthesized Li–A(BW) crystals up to 1273 K leads to a dehydrated and condensed ß-quartz structured phase $Li_2Al_2Si_3O_{10}$ (hexagonal, SG $P6_222$).

$|Li_4(H_2O)_4|[GaSiO_4]_4$-ABW were derived from gallosiliceous, wetted pastes pressed to pellets prior to the heating phase. Refinements revealed orthorhombic symmetry, SG $Pna2_1$, with a = 10.52(3) Å, b = 8.251(12) Å, and c = 5.056(14) Å.

In the system $Li_2O–Ga_2O_3–GeO_2–H_2O$ $LiGaGeO_4$, phenakite-derivative structure crystallizes from beryllonite analogous phase $NaGaGeO_4$, as well as from $|Na_6(H_2O)_8|[GaGeO_4]_6$–SOD, as the source materials.

Fig. 1: SEM images of selected products: T4: large crystal of aluminosilicate Li-A(BW), obtained from zeolite LTA at 473 K; S-4: $|Li_4(H_2O)_4|[GaSiO_4]_4$-ABW, derived from gallosiliceous, wetted pastes pressed to pellets prior to the heating phase. S-7: $LiGaGeO_4$ (phenakite-derivative structure) crystallized from beryllonite analogous phase $NaGaGeO_4$.

References: [1] Baerlocher, C., McCusker L. B., Olson, D. H.: "Atlas of Zeolite Framework Types". 6th ed., Elsevier, Amsterdam, 2001.

[2] Robson, E.: "Verified syntheses of zeolitic materials". 2nd ed., Elsevier, Amsterdam, 2001.

[3] Barrer, R. M., White, E. A. D.:J.: The Hydrothermal Chemistry of Silicates. Part 1. Synthetic lithium aluminosilicates. Chem. Soc. (1951)1267 - 1278.

Figure 1

ISC-P03
Zeolite crystallization experiments under insertion of laboratory ware waste material

T. M. Busche[1], A. Hartmann[1], V. Petrov[1], J.- C. Buhl[1], K. Rübner[2], M. Lindemann[2]
[1]Leibniz Universität Hannover, Institut für Mineralogie, Hannover, Germany
[2]BAM Bundesanstalt für Materialforschung und -prüfung, Fachbereich 7.4 Baustofftechnologie, Berlin, Germany

A study of leaching and crystallization behavior of laboratory ware (LW) was performed with the aim of an insertion of this waste material in zeolite Na-A (LTA) synthesis [1].

The chemical composition of LW was mainly SiO_2 (69 M %) and Al_2O_3 (29 M %) with mullite, quartz and minor amounts of tridymite as crystalline phases beside high amorphous parts. Firstly the transformation of LW without and with the additive $NaAlO_2$ was investigated. In a second series these products were transformed into LTA by an acid leaching step (1) followed by the alkaline synthesis step (2). Samples were characterized using XRD, SEM/EDX, FTIR as well as TG/DTA.

Without further addition of aluminate the mullite content of LW and the SiO_2 phases totally dissolve at 180°C in 5M NaOH and sodalite (SOD), cancrinite (CAN) and sodium aluminum silicate hydrate (PDF 28-1036) were formed. Addition of $NaAlO_2$ results in crystallization of analcime (ANA) in co-crystallization with SOD and few amounts of zeolite P at 160°C in 2M NaOH. Increasing NaOH concentrations (3M) yield zeolite JBW beside the mentioned phases. At 5M NaOH only SOD, CAN and phase PDF 28-1036 were found at 160°C (Fig. 1). Thus the 160°C-2M NaOH product was selected as suitable educt for transformation into the desired zeolite LTA. Two-step zeolite syntheses [2] were carried out for this sample at 90°C and 110°C. In this procedure acid solutions of the educts were performed to obtain a precursor gel for LTA synthesis followed under usually alkaline conditions as a second step. LTA was successfully obtained for both temperatures, but samples treated at 110°C contain a large amount of SOD. Best LTA crystals were synthesized at 90°C. Figure 2 shows a SEM photograph of this product.The LTA synthesis step 2 (alkaline process) was tested under insertion of: a) used NaOH solution of the LW treatment; b) fresh prepared NaOH solution and c) a 50% mix of fresh and used NaOH. In case a) no crystalline products were formed, whereas in cases b-c) LTA successfully crystallized.

Summarizing the results it can be concluded that the experiments describe interesting model reactions for a recycling process of laboratory ware waste material into valuable zeolites of LTA-Type.

Fig. 1: XRD-results of alkaline treatment of LW + $NaAlO_2$ at 160°C in: a) 2M NaOH, b) 3M NaOH and c) 5M NaOH (* = Sodium Aluminium Silicate Hydrate, PDF 28-1036).

Fig. 2: LTA from two step synthesis under insertion of the product of alkaline treatment of LW at 160°C in 2M NaOH and fresh prepared NaOH solution.

References:
[1]: T. Jakobczyk: Untersuchungen zur Zeolithsynthese unter Einsatz des industriellen Reststoffs Laborkeramik (LK). Masterarbeit im Studiengang Geowissenschaften, Leibniz Universität Hannover, 2014.
[2]: M.M. Nemati: Herstellung und Kristallisationsverhalten von Zeolith-Precursor Gelen aus Sodalith und Cancrinit. Masterarbeit im Studiengang Geowissenschaften, Leibniz Universität Hannover, 2013.

Figure 1

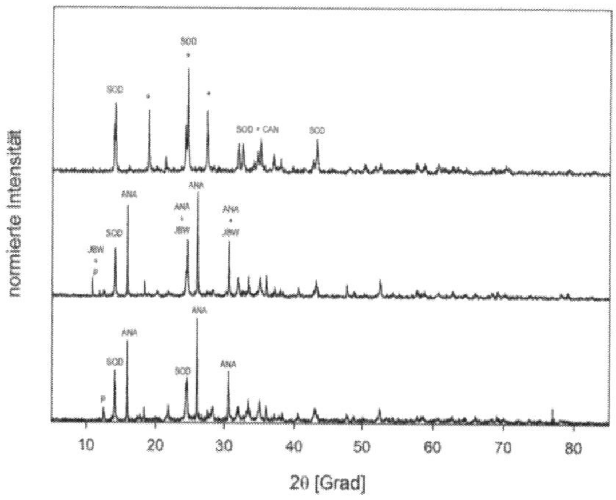

Figure 2

ISC-P04

Growth of M:$TiO_{2\pm\delta}$ (M = Ni, Fe, Co, Mn, Nb) Single Crystals by means of Induction Skull - Melting

M. Paun[1], M. Lerch[1]
[1]*Technische Universität Berlin, Institut für Chemie , Berlin, Germany*

Titanium dioxide (TiO_2) is an excellent photocatalyst due to its high activity, good stability, non-toxicity and low cost, which has many promising applications in the fields of renewable energy and environmental protection. However, only the ultraviolet part of the solar irradiation, amounting to ~4% of the solar energy on the earth's surface, could be absorbed by TiO_2 due to its high intrinsic band gap (~3.0 eV for rutile). In order to increase the usage of the solar output, TiO_2 can be doped with various transition metals to reduce the band gap.

Doped TiO_2 single crystals were grown with the Induction Skull - Melting method (ISM) which is a process for melting a material by applying a high frequency electromagnetic field (induction heating) to a water-cooled copper vessel (cold crucible). Advantages of ISM: growth of single crystals a) at temperatures higher than 2500 °C, b) with low defect density, and c) with high chemical purity.

Up to now, the following crystals have been grown: $Ni_{0.0075}Ti_{0.9925}O_{2-\delta}$, $Fe_{0.0075}Ti_{0.9925}O_{2-\delta}$, $Co_{0,0075}Ti_{0,9925}O_{2-\delta}$, $Mn_{0,0075}Ti_{0,9925}O_{2-\delta}$,$Mn_{0,0015}Ti_{0,9985}O_{2-\delta}$ and $Nb_{0,0015}Ti_{0,9985}O_{2\pm\delta}$.

Fig. 1. $Ni_{0.0075}Ti_{0.9925}O_2$ single crystals as grown (left) and after heating in O_2 atmosphere (right).

A. Fujishima, K. Honda, Nature 238 (1972) 37.

Kuz'minov Yu. S, Lomonova E.E, Osiko V. V, *Cubic Zirconia and Skull Melting*, Cambrige International Science Publishing, p. ix, 2008

Figure 1

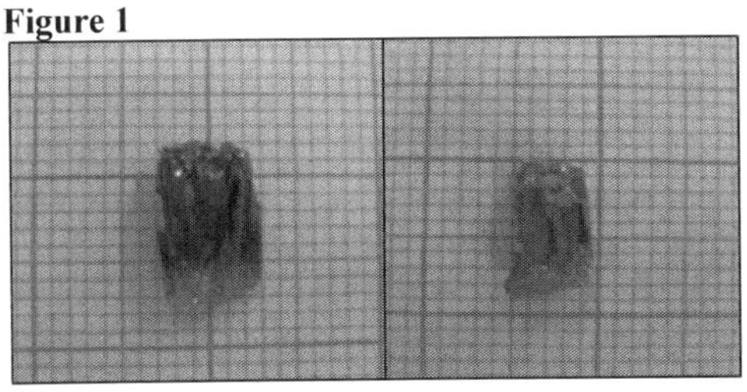

ISC-P05

Growth of Luminescent Zirconia-based Single Crystals by means of Induction Skull – Melting

M. Paun[1], E. Heppke[1], M. Lerch[1]
[1]*Technische Universität Berlin, Institut für Chemie , Berlin, Germany*

As a potential host for phosphor materials, yttrium-stabilized ZrO_2 (YSZ) is chemically and photochemically stable, has a high refractive index, and low phonon energy.

Zirconia-based single crystals were grown using the Induction Skull - Melting technique. Induction Skull - Melting (ISM) is a process for melting a material by applying a high frequency electromagnetic field (induction heating) to a water-cooled copper vessel (cold crucible).

Advantage of ISM: growth of single crystals a) at temperatures higher than 2500 °C, b) with low defect density, and c) with high chemical purity.

Up to now, the following crystals (from left to right in Fig.1) have been grown: $Zr_{0.76}Y_{0.2}Eu_{0.04}O_{1.89}$, $Zr_{0.78}Y_{0.2}Er0_{.015}Yb_{0.01}O_{1.89}$, $Zr_{0.78}Y_{0.2}Tm_{0.01}Yb_{0.01}O_{1.89}$, and $Zr_{0.78}Y_{0.2}Tb_{0.02}O_{1.89}$.

Fig.1: Zirconia-based single crystals under sunlight (top) and UV-radiation (bottom).

A. Patra, C.S. Friend, R. Kapoor, P.N. Prasad, J. Phys. Chem. B 106 (2002) 1909 - 1912.

Kuz'minov Yu. S, Lomonova E.E, Osiko V. V, *Cubic Zirconia and Skull Melting*, Cambridge International Science Publishing, p. ix, 2008

Figure 1

ISC-P06

Neutron scattering investigations on mechanochemically prepared CZTS

A. Ritscher[1], M. Lerch[1], O. Dolotko[2]
[1]*Technische Universität Berlin, Institut für Chemie, Berlin, Germany*
[2]*Heinz Maier-Leibnitz Zentrum FRM II, Garching b. München, Germany*

With respect to absorber materials in solar cells, Cu_2ZnSnS_4 (CZTS) has been a focus of interest in recent years. CZTS is a promising semiconductor material and a potential compound for future thin film photovoltaic applications. Cu_2ZnSnS_4 crystallizes in the kesterite type space group *I-4* . This space group is useful describing a complete or partial ordering of Cu and Zn. In addition, kesterite-type phases exhibiting a statistical distribution of Cu/Zn are also known. Such a structure can be described in space group *I-42m* .

Recently we developed a new route leading to single phase Cu_2ZnSnS_4 (CZTS) powders. A low-temperature formation process was created where the elements should be already mixed at room temperature at an atomic level. This can be realized by a mechanochemical treatment of the binary sulfides in a planetary ball mill. In a second step, the poorly crystalline product is annealed in H_2S-atmosphere.

For a deeper understanding of the correlation between structural parameters and physical properties the knowledge of the cation distribution is required. With standard X-ray techniques it is not possible to determine the Cu/Zn distribution. For a reliable structural characterization neutron diffraction measurements are absolutely necessary.

In Fig. 1 the XRD as well as the neutron scattering pattern of an apparently single phase sample is depicted. The obtained diffraction patterns show a kesterite-type phase with full Sn/(Cu/Zn) order. The results of our first Rietveld refinements using the neutron data (SPODI, FRM II) predict that Cu fully occupies the 2a (0,0,0) position. Zinc and the remaining copper do not show a complete order or a statistical distribution on the 2c (0,½,¼) and 2d (0,½,¾) positions, respectively, but rather reveal a partial disorder of the cations (approx.: 2c: 1/3 Cu 2/3 Zn, 2d: 2/3 Cu 1/3 Zn).

Fig.1: X-ray (left) and neutron (right) diffraction patterns of the final material with the results of the Rietveld refinements using a one-phase model (Kesterite, Cu_2ZnSnS_4, *I-4*).

Figure 1

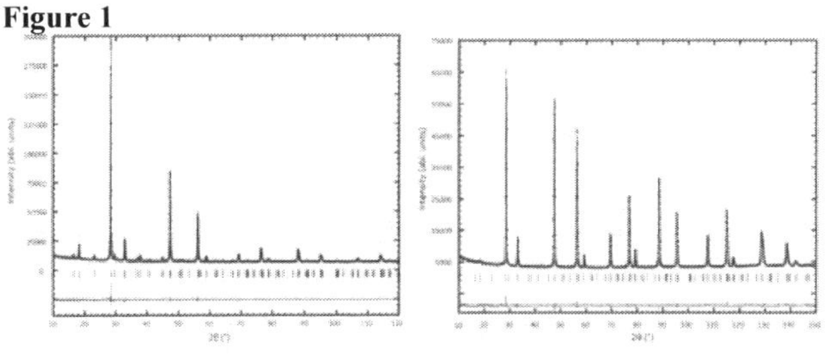

ISC-P07

Evaluation of amorphous, turbostratic and crystalline carbon membranes in the focus of bio-ethanol steam reforming.

A. Schulz[1], Y. Li[2], J. Caro[1]
[1]*Institute für Physikalische Chemie und Elektrochemie/ LUH, AK Prof. Caro, Hannover, Germany*
[2]*Institute of Chemical Physics/ Chinese Academy of Sciences, Dalian, China*

For bio-ethanol steam reforming the first choice of materials can be carbon with its hydrophobic character in general. Several types of carbon made membranes can be found in the literature today and are used for gas or steam separation (adsorption experiments) [1-8].

This work shows the influence of different structured carbon membranes and their separation behaviour in the presence of water steam. The most challenging problem in the separation of pure hydrogen using porous membranes for catalytic membrane reactors at ambient pressures and temperatures between 294 to 537 K ($C_2H_5OH + 3\ H_2O \leftrightarrow 2\ CO_2 + 6\ H_2$) is the presence of water [9].

Therefore we analyzed the separation characteristics of amorphous carbon on a porous α-Al_2O_3 support (⌀ 18 mm) coated with high vacuum sputter coater (Leica EM SCD500) and a carbon yarn without rotation in a static arrangement. Thereby the carbon yarn was heated up to the sublimation temperature of carbon in a high vacuum chamber at 10^{-4} mbar which was rinsed with argon for several times before. The turbostratic carbon membranes were prepared and provided by the "Fraunhofer Institute for Ceramic Technologies and Systems IKTS- Hermsdorf (Germany)" (see Fig. 1).

As crystalline carbon membrane we choose some graphite powder which could be easily pressed by a hydraulic press using about 100 kN for the preparation of a disc with a diameter of 18 mm and a thickness of 1 mm.

The experiments show mixed gas separation factors $\alpha(H_2/CO_2)$ of 4.6, typical for knudsen controlled diffusion in the case of the pressed graphite membranes, 3 for the amorphous carbon membrane and up to 14 for the turbostratic carbon membrane. Ethanol is separated from hydrogen in a equimolar mixture from 2.4 up to 26 depending on the membrane type. The comparison of the hydrogen permeabilities is shown Figure 2.

Figure 1: SEM cross section picture of the turbostratic carbon membran.

Figure 2: Comparison of hydrogen permeabilities (the thickness and pressure normalized)

of the three described membranes.

References:

[1] T. Kiyoshi, T. Namba, M. Yamawaki: J. Nuclear Materials. 155-157 (1988) 230-233.
[2] K. Yamaguchi, Y. Suzuki , S. Tanaka, M. Yamawaki: J. Nuclear Materials. 179-181 (1991) 242-245.
[3] H. Atsumi, K. Tauchi: J. Alloys Comp. 356-357 (2003) 705-709.
[4] A.Spitsyn, A. Pisarev, A. Skovoroda, V. Gureev, Yu Martynenko: J. Nuclear Materials. 363-365 (2007) 833-838
[5] A. V. Spitsyn, A.V. Golubeva, M. Mayer, A.A. Skovoroda: J. Nuclear Materials. 390-391 (2009) 701-704.
[6] A.F. Ismail, L.I.B. David, J. Membr. Science. 193 (2001) 1-18.
[7] T. A. Centeno, A. B. Fuertes, Carbon 38 (2000) 1067-1073.
[8] A. Schulz, F. Steinbach, J. Caro, J. Membr. Science. 469 (2014) 284-291.
[9] M. Ni, D.Y.C. Leung, M.K.H. Leung, Int. J. Hydrogen Energy 32 (2007) 3238.

Figure 1

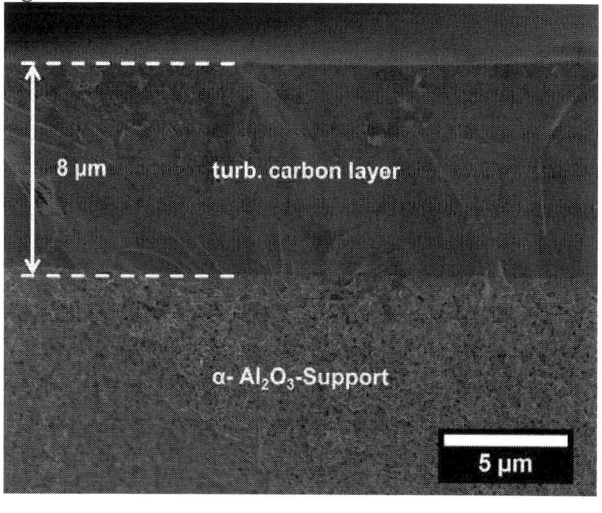

Figure 2

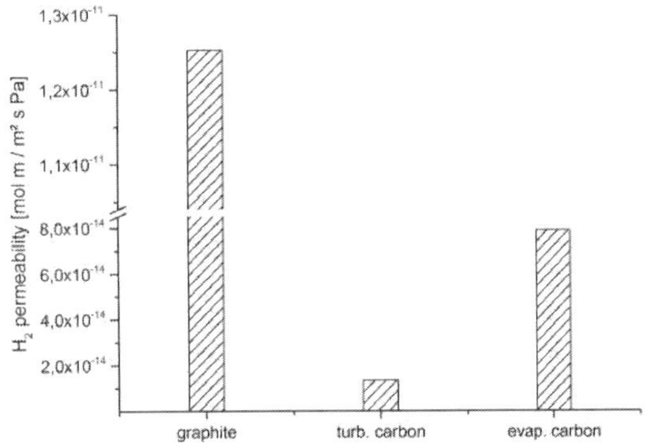

ISC-P08

$SrBa_2(NH_2)_6$: A new ternary amide from ammonothermal synthesis

J. Hertrampf[1], N. S. A. Alt[2], E. Schlücker[2], R. Niewa[1]
[1]*Universität Stuttgart, Institut für Anorganische Chemie, Stuttgart, Germany*
[2]*Friedrich-Alexander Universität Erlangen-Nürnberg, Lehrstuhl für Prozessmaschinen und Anlagentechnik, Erlangen, Germany*

Ammonothermal synthesis (critical point of ammonia: p = 11.3 MPa, T = 403.5 K) is a suitable method to synthesize a large range of materials, such as nitrides, amides, imides and ammoniates.[1] A high number of metal amides has been obtained using this technique. So far, there is no ternary inter-alkaline-earth metal amide known. [2-5]

We report the first ternary amide constituted exclusively from alkaline-earth metal ions, $SrBa_2(NH_2)_6$. Single crystals were obtained under ammonothermal conditions (p = 180 MPa, T = 573 K, 81.8% filling degree) in 97 mL autoclaves made of nickel-based alloy 718. [6]

$SrBa_2(NH_2)_6$ crystallizes in $P2_1/n$ with a = 6.395(2) Å, b = 7.195(2) Å, c = 10.471(2) Å, β = 97.99(2)°, Z = 2, V = 477.13(2) Å^3, and is build up from Sr in octahedral and Ba in capped-trigonal prismatic coordination, like similarly present in the respective binary amides [4,5,7]. The Ba polyhedra form a network with channels, in which the Sr is located. Two distinct amide species are present, one surrounded by one Sr and three Ba, the other coordinated by one Sr and two Ba. Those different coordination motifs cause the asymmetric and symmetric stretching modes in the Raman spectrum to split up into two signals each, located at 3484 cm^{-1}, 3325 cm^{-1} and 3252 cm^{-1}, 3168 cm^{-1}, respectively (Fig. 1). These characteristic modes in combination with elemental analysis confirm the presence of amide groups.

Figure 1. IR and Raman spectra of $SrBa_2(NH_2)_6$

[1] T. M. M. Richter, R. Niewa, *Inorganics* 2014, *2*, 29–78.
[2] H. Jacobs, D. Schmidt, *Curr. Top. Mater. Sci.* 1982, *8*, 387–427.
[3] R. Juza, H. Schumacher, *Z. Allg. Anorg. Chem.* 1963, *324*, 278–286.
[4] M. Nagib, H. Jacobs, H. Kistrup, *Atomkernenergie* 1979, *33*, 38–42.
[5] H. Jacobs, C. Hadenfeldt, *Z. Anorg. Allg. Chem.* 1975, *418*, 132–140.
[6] N. S. A. Alt, E. Meissner, E. Schlücker, *J. Cryst. Growth* 2008, *350*, 2.
[7] J. Hertrampf, N. S. A. Alt, E. Schlücker, R. Niewa *Z. allg. anorg. Chem.* 2014, *in press*.

Figure 1

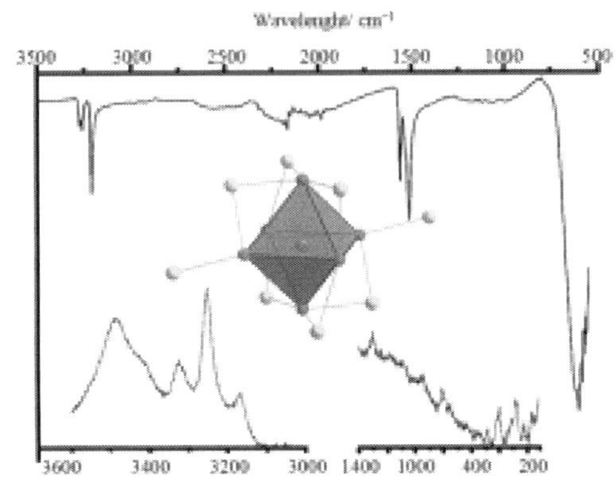

ISC-P09

Single crystal growth of Mn-doped melilites

M. Burianek[1], H. Lührs[1], R. X. Fischer[1]
[1]*Universität Bremen, Crystallography, GEO, Bremen, Germany*

The large family of tetragonal, noncentrosymmetric melilite-type compounds represents an interesting class of materials due to its magneto-electric and multiferroic properties. The chemical variability of the melilite family of crystal structures offers a wide range of possibilities for the design of new crystalline materials. The general composition is given by the formula $M_2^{[8]}T^{[4]}T'^{[4]}_2X_7$ with M = Ca, Na, Sr, Pb, Ba, RE; T = Be, Mg, Cu, Co, Zn, Fe^{2+}, Fe^{3+}, Mn^{2+}, Cd, Al, Ga; T' = Si, Ge, Al, Fe^{3+}, Ga, Be; and X = O, F, N, S. [1, 2] A comprehensive overview of the known melilite-type crystals is given by Kaminskii (2008) [1].

Interesting magnetic, optical, and dielectric properties have been reported for Mn- and Sr- containing melilites [3-6]. However, these results are mainly based on polycrystalline materials. In order to systematically study the magnetic structure as well as the physical and chemical properties, large single crystals of Mn- and Sr-doped melilites are needed. We have successfully grown large single crystals of the host mineral åkermanite ($Ca_2MgSi_2O_7$) using the Czochralski method. The crystals were grown at about 1500°C under nitrogen atmosphere in iridium crucibles yielding optically pure single crystals. In first experiments Mn-doped single crystals (Figure 1) were grown from a melt close to the åkermanite composition containing 1.5 mol % MnO. The powder diffraction patterns can be indexed with the åkermanite structure model and Mn incorporation was confirmed by EDX analyses. In subsequent experiments the amount of MnO will be gradually increased.

Figure 1: Mn-doped åkermanite single crystal. Length 88 mm, diameter 12.0-12.6 mm.

References:
1. Kaminskii, A.A., *et al.* (2008) **5**, 845-868.
2. Röthlisberger, F., *et al.* (1990) **2**, 585-594.
3. Hao, Y. and Y.H. Wang (2007) Mater Res Bull **42**, 2219-2223.
4. Joseph, T. and M.T. Sebastian (2010) J Am Ceram Soc **93**, 147-154.

5. Koo, H.J. (2012) Solid State Commun **152**, 1116-1118.
6. Koo, H.-J., *et al.* (2012) **324**, 3716-3718.

Figure 1

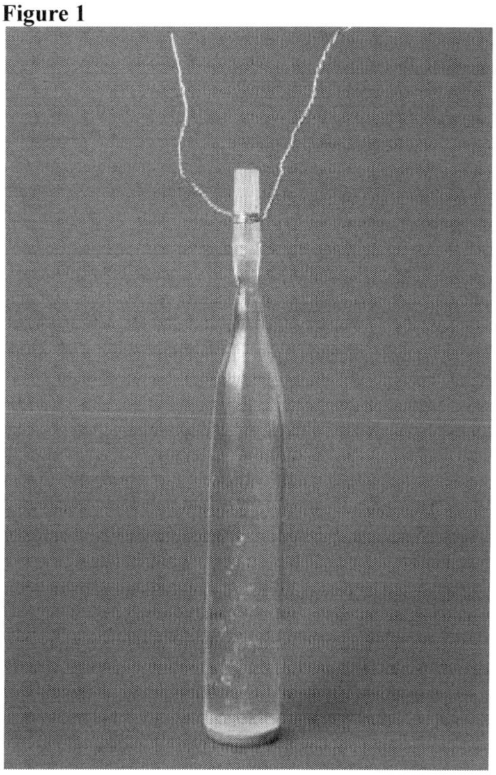

ISC-P10
High-Temperature Transformation from 1T- to 3R-Li_xTiS_2 Observed *in situ* with Neutron Diffraction

D. Wiedemann[1], S. Nakhal[1], A. Senyshyn[2], T. Bredow[3], M. Lerch[1]
[1]*Technische Universität Berlin, Institut für Chemie, Berlin, Germany*
[2]*Technische Universität München, FRM II, München, Germany*
[3]*Universität Bonn, Mulliken Center for Theoretical Chemistry, Bonn, Germany*

Layered titanium disulfide is used as lithium-ion intercalating electrode material in batteries. The room-temperature stable trigonal 1T polymorphs of the intercalates Li_xTiS_2 ($x \leq 1$) are widely-investigated. However, the rhombohedral 3R polymorphs, being stable at higher temperatures for large x, are less well known.

We recently reported on the synthesis of phase-pure 1T-Li_xTiS_2 (x = 0.7 and 0.9) and its reversible reconstructive transformation to the 3R phase between 673 and 873 K.[1] Neutron powder diffractograms, that we acquired at multiple temperature points, allowed for full Rietveld refinements resulting in good to very good fits. As proposed previously,[2] lithium ions were unambiguously located in octahedral voids at fractional coordinates 0, 0, ½ and no ordering was observed. (The same had been found for 1T-Li_xTiS_2 before.)[3] Known trends in the temperature evolution of lattice parameters were reproduced. Analyzing the Madelung parts of the lattice energies, we found the 3R polymorph to be electrostatically more stable than the 1T polymorph. However, periodic quantum-chemical calculations indicated that this approach is oversimplified due to neglect of polarization and electron correlation effects, at least for the relative stability of the polymorphs. The anomalous decrease of lattice parameters in 3R-$Li_{0.7}TiS_2$, when being heated from 873 to 973 K, is tentatively attributed to effects of lithium migration.

The insights gained do not only elucidate the structure of 3R-Li_xTiS_2, but also help to understand and control polymorphism in layered transition-metal sulfides.

This work is based upon experiments performed at the SPODI instrument operated by FRM II at the Forschungs-Neutronenquelle Heinz Maier-Leibnitz (FRM II), Garching, Germany. Financial support by the Deutsche Forschungsgemeinschaft (FOR 1277: "Mobilität von Lithiumionen in Festkörpern [molife]") is gratefully acknowledged.

[1] D. Wiedemann, S. Nakhal, A. Senyshyn, T. Bredow, M. Lerch, *Z. Phys. Chem.* 2015, *submitted.*
[2] S. Nakhal, M. Lerch, J. Koopman, M. M. Islam, T. Bredow, *Z. Anorg. Allg. Chem.* 2013, *639*, 2822-2825.
[3] J. R. Dahn, W. R. McKinnon, R. R. Haering, W. J. L. Buyers, B. M. Powell, *Can. J. Phys.* 1980, *58*, 207-213.

Fig. 1. Crystal structures of 1T- (left) and 3R-$LiTiS_2$ (right). View along [010], unit-cell edges in black, stacking sequences at the margins, arrows indicating the hypothetical movement during the transformation 1T → 3R.

Fig. 2. Stacking plot of temperature-dependent diffractograms of $Li_{0.7}TiS_2$ (normalized to the intensity of the niobium (011) reflection at *ca.* 38.5°). 1T phase in black, 3R phase in grey, container reflections cut off at high intensities for clarity.

Figure 1

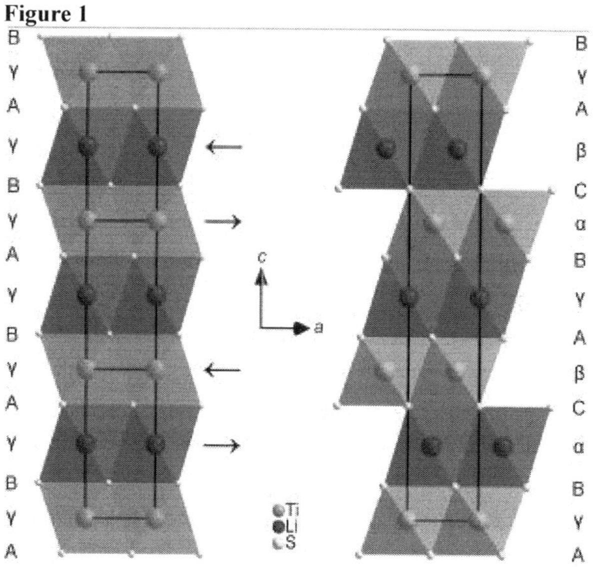

Figure 2

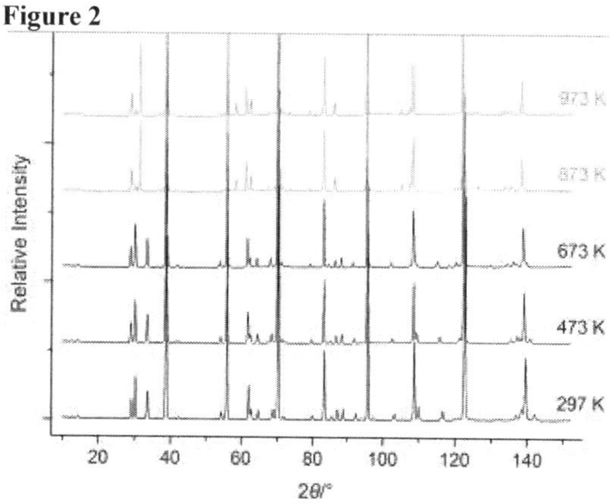

ISC-P11

Intra-cage oxidation and reaction kinetics of nitrite-sodalite to nitrate-sodalite

M. Sehovic[1], L. Robben[1], T. M. Gesing[1]
[1]*University Bremen, Chemistry, Bremen, Germany*

The three-dimensional highly flexible framework of sodalites forming *toc*-cages (better known as *β*-cages) is used as a model system for various zeolites, such as LTA, and the Faujasite-type ones. In most cases the *β*-cages are filled with different anions or neutral molecules (templates) and respective extra framework cations are needed for charge balance and structure stabilization. At higher temperatures those templates tend to interact and react with the surrounding atmosphere. This may lead to phase transformations like it is the case for the aluminosilicate nitrite sodalite ($|Na_8(NO_2)_2|[AlSiO_4]_6$) in carbondioxide atmosphere at higher temperatures, forming aluminosilicate carbonate nosean ($|Na_8(CO_3)\square|[AlSiO_4]_6$) [1-2]. In this study we have investigated the reaction processes of aluminosilicate nitrite sodalite at different temperatures under normal air and nitrogen atmosphere with in-situ temperature-dependent X-ray powder diffraction, simultaneous thermal analysis and spectroscopic methods. Measurements performed under inert nitrogen atmosphere were needed to separate pure thermal effects from those caused by the reactive atmosphere.

While heating NO_2^--sodalite in air an oxidation of NO_2^- to NO_3^- can be observed in temperature dependent X-ray diffraction measurements [3-4] from about 820 K on. At a temperature of 937(3) K the newly formed NO_3^--sodalite undergoes a phase transition from $P\bar{4}3n$ to $Pm\bar{3}n$ [5], which is explained by the full expansion of the framework, expressed in a tilt angle [6] of $\varphi = 0$. The oxidation process shows in the case of in-situ powder X-ray diffraction only small visible changes in the diffraction patterns. While the symmetry remains unchanged during the oxidation process the cubic lattice parameter increases slightly at time-dependent measurements (e.g. from 906.19(2) pm to 913.09(3) pm at 900(3) K). Data obtained from X-ray diffraction experiments are indicating no remaining NO_2^--sodalite within the sample, which is confirmed by IR-Spectroscopy showing only very minor traces of NO_2^- in the sample after the experiment.

References:

[1] Th. M. Gesing, J.-Ch. Buhl, *Eur. J. Mineral.* 10 (1998) 71-77. [2] M. Šehović, L. Robben, Th. M. Gesing, *Z. Kristallogr. accepted* [3] M. T. Weller, G.Wong, C. L. Adamson, S. M. Dodd, J. J. B. Roe, *J. Chem. Soc. Dalton Trans.* (1990) 593-597. [4] J.-Ch. Buhl, *React. Kinet. Catal. Lett.* 43, (1991) 577-582 [5] C. H. Rüscher, Th. M. Gesing, J.-Ch. Buhl, *Z. Kristallogr.* 218 (2003) 332-344. [6] W. Depmeier, *Acta Crystallogr.* B40 (1984) 434-436.

ISC-P12

The Defect *Scheelite*-Type Europium(III) *Ortho*-Oxidomolybdate(VI) $Eu_{0.667}[MoO_4]$

T. Schustereit[1], T. Schleid[1], I. Hartenbach[1]
[1]*University of Stuttgart, Institute for Inorganic Chemistry, Stuttgart, Germany*

An unsuccessful attempt to synthesize $[Eu_2(H_2O)_{12}Mo_8O_{27}] \cdot n\ H_2O$ as precursor for two different europium(III) oxidomolybdates(VI) led to colorless, coarse single crystals of $Eu_{0.667}[MoO_4]$. The defect *scheelite*-type lanthanoid(III) *ortho*-oxidomolybdates(VI) with the general formula $Ln_{0.667}[MoO_4]$ and statistically under-occupied cation sites are already known for *Ln* = La [1], Ce, Pr, Nd, and Sm [2]. The title compound crystallizes isotypically to the aforementioned examples in the tetragonal space group $I4_1/a$ with a = 522.61(3) and c = 1151.96(7) pm serving as a direct extension of this row. The coordination polyhedra around the crystallographically unique Eu^{3+} cations at *Wyckoff* position 4*b* with the site symmetry -4 consist of eight oxygen atoms (d(Eu-O) = 246.6 and 248.0 pm, 4 x each) arranged as trigonal dodecahedra (Figure 1). In order to maintain electroneutrality this Eu^{3+} site shows an occupation of merely $^2/_3$. The same site symmetry is found for the Mo^{6+} cations at *Wyckoff* position 4*a*. They represent the centers of isolated bisphenoidally distorted *ortho*-oxidomolybdate(VI) tetrahedra $[MoO_4]^{2-}$ (d(Mo-O) = 176.6 pm, 4 x ; ∢(O-Mo-O) = 4 x 107.0° and 2 x 114.5°). The $[EuO_8]^{13-}$ polyhedra share common vertices with eight $[MoO_4]^{2-}$ tetrahedra, while they are interconnected with each other via four common edges to build up the three-dimensional *scheelite*-type crystal structure (Figure 2). The phase purity of the title compound was analyzed by X-ray powder diffractometry. Furthermore, $Eu_{0.667}[MoO_4]$ exhibits the Eu^{3+}-typical red bulk luminescence upon excitation with UV light (λ = 254 nm).

[1] T. Schustereit, Th. Schleid, I. Hartenbach, *Acta Crystallogr.* 2013, *E 69*, i7.
[2] T. Schustereit, S. L. Müller, Th. Schleid, I. Hartenbach, *Crystals* 2011, *1*, 244.

Figure 1. Coordination polyhedron around the Eu^{3+} cations (ellipsoid representation at 80 %) in the defect *scheelite*-type structure of $Eu_{0.667}[MoO_4]$.

Figure 2. View at the defect *scheelite*-type structure of $Eu_{0.667}[MoO_4]$ along [100].

Figure 1

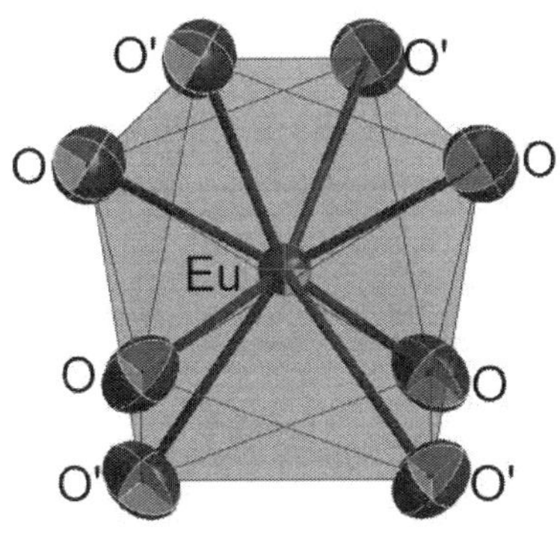

Figure 2

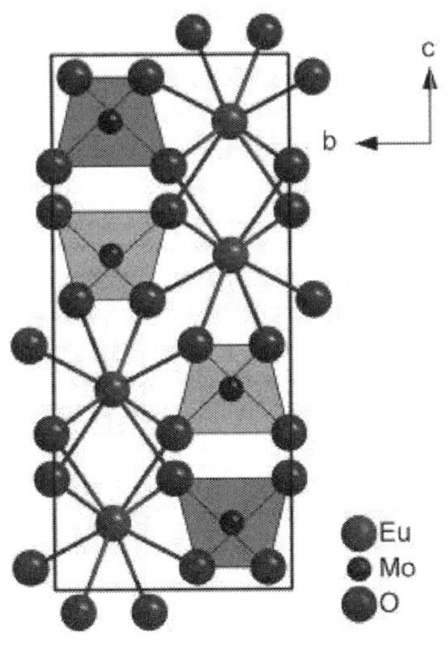

ISC-P13

Structural Distortion in the Compounds $(A_3N)As$ (A = Mg, Ca, Sr, Ba)

D. Stoiber[1], R. Niewa[1]
[1]*Universität Stuttgart, Institut für Anorganische Chemie, Stuttgart, Germany*

We present two new distorted inverse perovskites, namely $(Sr_3N)As$ and $(Ba_3N)As$, which crystallize in the spacegroup *Pnma* ($GdFeO_3$ structure type) like the already known $(Ca_3N)As$ [1] and analyse the structures and degree of distortion with respect to the undistorted cubic inverse perovskite $(Mg_3N)As$ [2] with spacegroup *Pm*-3*m*.

$(Sr_3N)As$ and $(Ba_3N)As$ were obtained via the reaction of arsenic with Sr_2N or Ba_2N. The substances were ground into powders, mixed, pressed into pellets and sealed into tantalum ampoules. The samples were heated to 800 °C for 48 hours and resulted in dark grey to black microcrystalline powders. The structures were refined using the Rietveld-method.

In the cubic inverse perovskite $(Mg_3N)As$ the size of the arsenic is sufficient to fill the cuboctahedral hole formed by eight (Mg_6N) octahedra. Due to the increasing size of the alkaline earth metal in $(A_3N)As$ with A = Mg, Ca, Sr, Ba, the space available for arsenic increases until it is no longer capable to effectively filling the cuboctahedral hole. As a result the (A_6N) octahedra tilt and the arsenic shifts away from the central position (see figure 1) [3], increasing in magnitude from calcium to barium. Remarkably, the distortion is realized by tilting the octahedra, with only minor deviations from ideal geometry. Due to the distortion the coordination number of the arsenic changes from 12 in the aristotype to about 4+4+2+2 in the strongest distorted $(Ba_3N)As$. Thus, such semiconducting inverse perovskites can be understood with the same simple geometric relations as known for oxide-perovskites ABO_3.

Figure 1: Crystal structure of $(A_3N)As$ with view along [010]. The tilting of the (A_6N) octahedra increases from A = Mg to A = Ba.

[1] M. Y. Chern, D. A. Vennos, F. J. DiSalvo, *J. Solid State Chem.* 1992, *96*, 415.

[2] E. O. Chi, W. S. Kim, N. H. Hur, D. Jung, *Solid State Commun.* 2002, *121*, 309.

[3] M. Y. Chern, F. J. DiSalvo, J. B. Parise, J. A. Goldstone, *J. Solid State Chem.* 1992, *96*, 426.

Figure 1

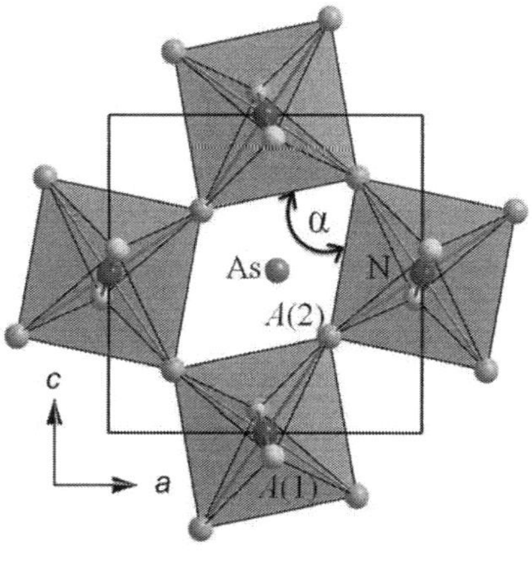

A	α
Mg	90°
Ca	76.2°
Sr	69.8°
Ba	68.5°

ISC-P14

Silicates and Phosphates with complex hydrogen bonds: Studies with neutron single crystal diffraction

M. Meven[1,2], G. D. Gatta[3]
[1]*RWTH Aachen University, Institut für Kristallographie, Aachen, Germany*
[2]*Forschungezentrum Jülich GmbH, Jülich Centre for Neutron Science, Jülich, Germany*
[3]*Università degli Studi di Milano, Dipartimento di Scienze Della Terra, Milano, Italy*

Minerals of the silicate and phosphate families provide a broad field of research especially because of their diverse framework structures. Although many of these minerals were studied extensively in the past with X-ray methods, the important role of hydrogen bonds on their structures could not always resolved clearly. Recently various silicates and phosphates with very different crystallographic space groups were studied on the hot neutron single crystal diffractometer HEIDI at the Heinz Maier-Leibnitz Zentrum in Garching (MLZ) . The combination of neutron and X-ray diffraction together with other methods reveal for all studied minerals - zoisite, elbaite, pezzottaite, brazillianite and paravauxite - valuable information about their structural details, not only the different roles of hydrogen bonds but also local disorder and mixed occupancies [1, 2, 3, 4, 5].

[1] F. Cámara, G.D. Gatta, M. Meven and D. Pasqual; Thermal expansion and high temperature structure evolution of zoisite by single-crystal X-ray and neutron diffraction; Physics and Chemistry of Minerals, 39(1):27-45 (2011). [2] G.D. Gatta, R.M. Danisi, I. Adamo, M. Meven and V. Diella; A single-crystal neutron and X-ray diffraction study of elbaite; Physics and Chemistry of Minerals, 39(7):577-588 (2012). [3]G.D. Gatta, I. Adamo, M. Meven and E. Lambruschi; A single-crystal neutron and X-ray diffraction study of pezzottaite, Cs(Be2Li)Al2Si6O18; Physics and Chemistry of Minerals, (39):829-840 (2012). [4] G.D. Gatta,P. Vignola,M. Meven and R. Rinaldi; Neutron diffraction in gemology: Single-crystal diffraction study of brazilianite, NaAl3(PO4)2(OH)4;American Mineralogist,98(8-9):1624-1630 (2013). [5]. G.D. Gatta, P. Vignola and M. Meven; On the complex H-bonding network in paravauxite, Fe2+Al2(PO4)2(OH)2•8H2O: A single-crystal neutron diffraction study; Mineralogical Magazine, Vol 78(4): 841-850 (2014)

ISC-P15

$Cs_2Eu[Nb_6Cl_{18}]$: A Sixteen Valence Electron Cluster Compound

D. Zimmermann[1], T. Schleid[1]
[1]*University of Stuttgart, Institute for Inorganic Chemistry, Stuttgart, Germany*

The standard method to synthesize rare-earth metal(III) oxide fluoride sufides (*e.g.* M_3OF_5S [1] and $M_6O_2F_8S_3$ [2]) utilizes the binary components M_2O_3 and MF_3, along with the elements M (= Sc, Y, La; Ce - Lu) and sulfur. Because of the high concentration of fluoride in the target compounds refractory metals (niobium or tantalum) are used as container material known for their chemically inert behavior. As support for both reaction and crystal growth fluxing agents like alkali-metal halides have proven well. By changing the reaction conditions with trial-and-error methods to close the gap of missing links for the europium(III) representatives, $Cs_2Eu[Nb_6Cl_{18}]$ was obtained instead from niobium capsules.

$Cs_2Eu[Nb_6Cl_{18}]$ crystallizes in the trigonal space group *R*-3 (no. 148) with the unit-cell parameters a = 957.71(2) and c = 2652.38(9) pm (Z = 3). It has to be classified as compound with sixteen valence electrons per octahedral Nb_6 cluster resulting in eight pairs at all triangular faces as 2e3c bonds. Single crystals emerge black and plate-shaped with strong lustre (Figure 1) indicating at least semiconductor properties.

Three different cations can be found, with Cs^+ as first one coordinated by twelve chloride anions as cuboctahedron (d(Cs–Cl) = 350 – 373 pm). Secondly, Eu^{2+} resides in an octahedral void surrounded by six chloride anions (d(Eu–Cl) = 292 pm). Six niobium atoms finally form a Nb_6 cluster unit (d(Nb–Nb) = 293 – 294 pm) with octahedral shape. A second sphere of altogether eighteen (better: twelve plus six) chloride anions encages the Nb_6 octahedron (d(Nb–Cl) = 248 – 249 plus 267 pm) as closer by edge-capping (Cl2 and Cl3) and far away terminal ligands (Cl1) (Figure 2).

Isostructurally with $K_2Sr[Nb_6Cl_{18}]$ [3] those cluster halides are known for their broad structural variety and intriguing physical behavior like to be insulators or semiconductors. $Cs_2Eu[Nb_6Cl_{18}]$ was until now only mentioned in the literature [4] without any crystal structure data.

[1] H. Grossholz, O. Janka, Th. Schleid, *Z. Naturforsch.* 2011, *66 b*, 213 - 220.
[2] H. Grossholz, D. D. Zimmermann, O. Janka, Th. Schleid, *Z. Naturforsch.* 2013, *68 b*, 751 -760.
[3] T. Duraisamy, A. Lachgar, *Acta Crystallogr.* 2002, *C 24*, i85.
[4] C. Perrin, S. Ihmaine, M. Sergent, *New J. Chem.* 1988, *12*, 321 - 328.

Figure 1 : Scanning electron microscope (SEM) picture of $Cs_2Eu[Nb_6Cl_{18}]$.

Figure 2 : Octahedral Nb_6 cluster surrounded by a chloride sphere of eighteen anions.

Figure 1

Figure 2

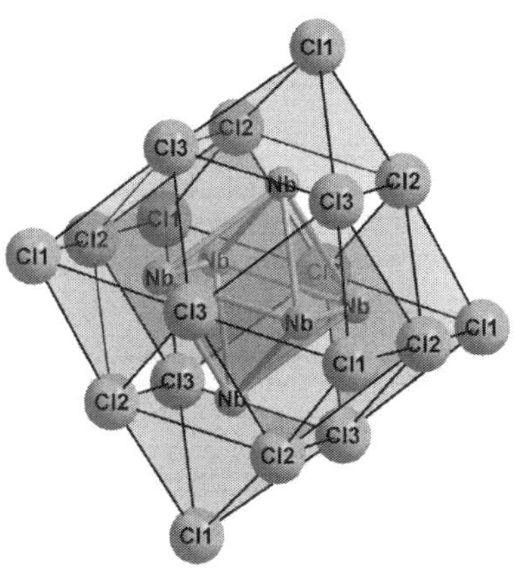

ISC-P16

Single Crystals of Matlockite-Type YbHCl

D. Rudolph[1], T. Schleid[1]
[1]*University of Stuttgart, Institute for Inorganic Chemistry, Stuttgart, Germany*

Syntheses of hydride chlorides with the divalent lanthanoids europium and ytterbium have already been reported in the literature [1, 2], but their characterization by means of X-ray diffraction was solely carried out on polycrystalline samples. Yellow, plate-shaped single crystals of ytterbium(II) hydride chloride (YbHCl) were now obtained by the reaction of equimolar amounts of ytterbium powder, sodium chloride and sodium hydride or alternatively by the reaction of ytterbium powder and ammonium chloride in a molar ratio of 4 : 1.

Like all other hydride and fluoride halogenides of the divalent lanthanoids, YbHCl crystallizes in the PbFCl-type structure [3] of the mineral matlockite in the tetragonal space group $P4/nmm$ (no. 129) with two formula units per unit cell (Figure 1). The unit-cell parameters are a = 385.61(3) and c = 680.59(5) pm, thus significantly smaller as compared with the corresponding fluoride chloride YbFCl [4] (a = 394.0(1) and c = 682.5(2) pm) caused by either the higher polarizability or the smaller size of the hydride juxtaposed to the fluoride anion.

In the crystal structure of YbHCl (Figure 2), the crystallographically unique Yb^{2+} cations are surrounded by four hydride and five chloride anions forming a square antiprism that is capped by one of the chloride anions. The Yb-Cl distance to this capping Cl^- anion is 308 pm representing a considerably larger bond length than for the other four equally long Yb-Cl contacts at 292 pm. The hydride anions are coordinated tetrahedrally by four Yb^{2+} cations (d(H-Yb) = 235 pm) and the chloride anions have five Yb^{2+} cations as next neighbors forming a square pyramid. Both types of anions build up square layers that can be clearly seen in Figure 2. Above a layer of hydride anions a chloride layer is turned by 45° against the hydride one. A second layer of chloride anions is located above the first chloride layer in a way that the anions of this sheet reside above the voids of the first chloride anion layer. This results in a capping of the square antiprismatic coordination of the Yb^{2+} cations by a further Cl^- anion.

Figure 1: Crystal data, structure refinement, fractional atomic coordinates and displacement parameters for YbHCl ($U_{22} = U_{11}$ and $U_{12} = U_{13} = U_{23} = 0$ due to the space-group symmetry).

Figure 2: View at the crystal structure of YbHCl approximately along [100].

[1] B. Tanguy, M. Pezat, C. Fontenit, J. Portier, *Compt. Rend. Acad. Sci. Paris* 1975, *C 280*, 1019-1020.
[2] H. P. Beck, A. Limmer, *Z. Naturforsch.* 1982, *37 b*, 574-578.
[3] a) W. Nieuwenkamp, J. M. Bijvoet, *Z. Kristallogr.* 1932, *81*, 469-473; b) F. A. Bannister, M. H. Hey, *Min. Mag.* 1934, *23*, 587-597.
[4] a) H. P. Beck, *J. Solid State Chem.* 1978, *23*, 213-217; b) H. P. Beck, *Z. Anorg. Allg. Chem.* 1979, *451*, 73-81.

Figure 1

Chemical formula	YbHCl		
Space group	*P4/nmm*		
Crystal system	tetragonal		
a / pm	385.61(3)		
b / pm	680.59(5)		
c/a	1.765		
Z	2		
ρ_{calc} / g·cm^{-3}	6.875		
Molar volume, V_m / cm^{-3}·mol^{-1}	30.47		
Diffractometer	κ-CCD		
F(000)	176		
Range in $\pm h$, $\pm k$, $\pm l$	5, 5, 9		
μ / mm^{-1}	46.99		
Total number of reflections	1512		
Number of unique reflections	97		
R_{int}, R_σ	0.091, 0.030		
Structure solution and refinement	SHELX-97		
R_1, wR_2	0.025, 0.050		
Goodness of Fit	1.133		
Residual electron densities / e·Å^{-3}	1.47, –1.89		
CSD number	428163		
	Yb	H	Cl
site	2*a*	2*c*	2*a*
x/a	1/4	3/4	1/4
y/b	1/4	1/4	1/4
z/c	0.19706(9)	0	0.6493(6)
U_{11} / pm^2	133(3)	–	181(11)
U_{33} / pm^2	218(4)	–	203(16)
U_{eq}[a] / pm^2	161(3)	658[b]	188(8)

[a] $U_{eq} = 1/4(U_{11} + U_{22} + U_{33})$, [b] The isotropic displacement factor of the hydrogen atom was constrained by a factor of 3.5 to the equivalent displacement factor of chlorine as the last non-metal atom being unconstrained.

Figure 2

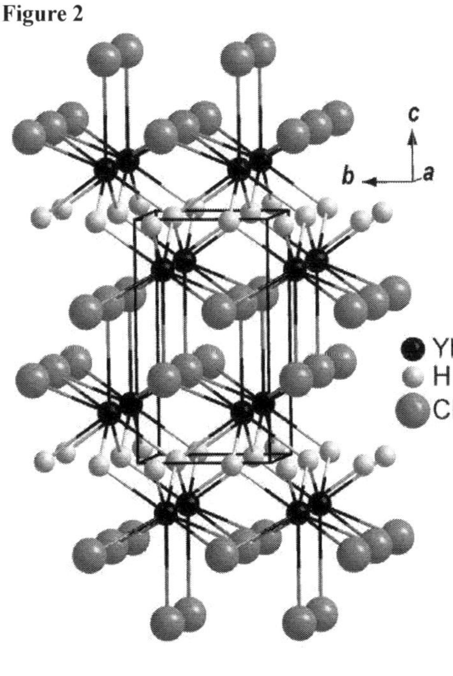

ISC-P17

UPTe, ThPTe and U_2PTe_2O - Actinide Pnictide Chalcogenides with Diphosphide Anions

K. Stolze[1], A. Isaeva[1], U. Schwarz[2], T. Doert[1]
[1]*TU Dresden, Chemistry and Food Chemistry, Dresden, Germany*
[2]*Max Planck Institute for Chemical Physics of Solids, Dresden, Germany*

Crystal structure re-determination of UPTe reveal a UGeTe derivative structure with a pseudo-tetragonal unit cell with lattice parameters a = 410.1(1) pm and c = 1697.1(1) pm at 120(2) K. The uranium atom is coordinated by four phosphorus and four tellurium atoms in a distorted square-antiprismatic environment; the U-P distances of 282(3) - 287(5) pm are considerably shorter than the U-Te distance with 315.5(1) pm. One more tellurium atom caps the Te-square of the antiprism at a distance of 335.8(1) pm from the uranium atom. The tellurium atoms, on the other hand, are located within a tetragonal-pyramidal arrangement of five uranium atoms (Figure 1a).

The new compound U_2PTe_2O can be considered as a partially oxidised phosphide telluride with a pseudo-tetragonal unit cell as well, with a = 403.7(1) pm, and c = 3206.8(10) pm at 120 K.

Single crystal X-ray diffraction data of UPTe and U_2PTe_2O exhibit — in contrast to the UGeTe-type structure[1] — phosphorus atoms on *Wyckoff* site 8*j*, which exclude a simultaneous occupation of both P-positions (x, ½, 0 and -x, ½, 0), as the distance between neighbouring atoms would amount to only ca. 70 pm. In addition, the Fourier maps indicate occupancies of 50 % of the individual phosphorus positions only.

The displacements result in an ordered arrangement of parallel P_2-dumbbells with intramolecular P-P distances of 242(5) pm for UPTe and 242(2) pm for U_2PTe_2O going along with a monoclinic distortion to space group *I*11*m*. The respective symmetry decrease from the UGeTe type (space group *I4/mmm*) is traceable by a *Bärnighausen* type group-subgroup relation.

The relatively long intramolecular P-P distances motivated real-space topological analysis of the chemical bonding which clearly evidences significant covalent bonding in the P_2 dumbbells.

For ThPTe, X-ray powder data indicate a structural pattern analogous to UPTe, its pseudo-tetragonal lattice parameters are a = 425.05(1) pm and c = 1726.8(1) pm at 296(1) K. The existence of diphosphide anions in UPTe and ThPTe is also supported by Raman data on single-crystals and powder samples. Considering the alternatives, the both compounds can best be rationalised by a *Zintl*-type approach as $2\,A^{4+} + P_2^{4-} + 2\,Te^{2-}$ with A = Th, U.

The structures of UPTe, ThPTe and U_2PTe_2O can be described by a stacking sequence of [A_2P_2] and [Te_2]-slabs (with A = Th, U), and [U_2P_2], [U_2O_2] and [Te_2]-slabs, respectively, whereby the [A_2P_2]-slabs contain the planar layers of P_2-dianions in a parallel arrangement (Figure 1c).

[1] A. Zygmunt, A. Murasik, S. Ligenza, J. Leciejewicz, Phys. Status Solidi A **1974**, 22, 75-79.

Figure 1: Projections of the unit cells of a) UPTe, b) U_2PTe_2O and c) the planar phosphide layer in c = 0 with P_2 dumbbells, the shortest P-P distances indicated as solid lines.

Figure 1

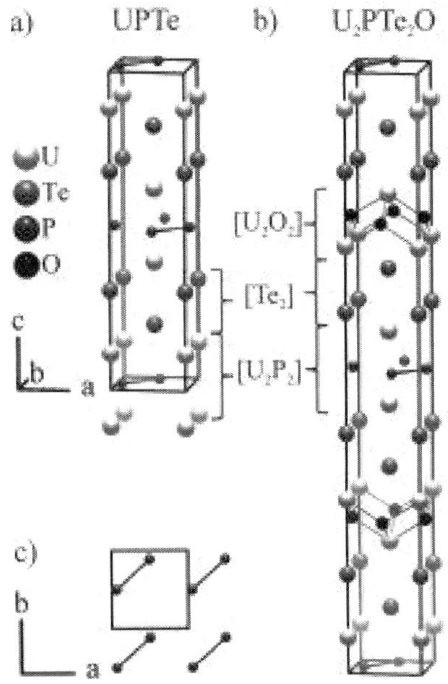

ISC-P18

Synthesis, Crystal Chemistry and Electronic Structure of New Mixed Strontium Trielides

C. Meyer[1], K. Köhler[1], C. Röhr[1]
[1] *Institut für anorganische und analytische Chemie, Albert-Ludwigs-Universität Freiburg, Freiburg, Germany*

Depending on the triel element M (M= Al, Ga, In) the binary and mixed ternary ditrielides SrM_2 form three different structure types, which are electron precise Zintl phases [1]. The systematic synthetic and single crystal investigations have been extended to the M-richer compounds SrM_4. In this case, the binary border phases $SrAl_4$ and $SrGa_4$ [2] crystallize in the electron precise $BaAl_4$-type structure [3] (fig. 1 a), whereas the hypoelectronic compound $SrIn_4$ exhibits the rare monoclinic $EuIn_4$-type [4,5] (fig. 1 c). In contrast to the latter indide, which shows no phase width at all, the stability range of the $BaAl_4$-type is quite large ($Sr(Al_{1-x}Ga_x)4$, x= 0-1; $Sr(Al_{1-x}In_x)_4$, x= 0-0.46; $Sr(Ga_{1-x}In_x)_4$, x= 0-0.25; cf. squares and gray line in fig. 2). On the triel exchange, the more electronegative elements Ga and In prefere the occupation of the tip of the pyramid (M(1) in fig. 1 a). M-poorer compounds of general composition Sr_3M_5 (Ca_3Ga_5-type, fig. 1 d) are formed in the likewise broad In-rich part of the ternary system (diamonds in fig. 2). For $Sr_3(M_{1-x}In_x)_5$ ternary phases could be obtained in the range of x= 0.35-1 for the substitution with Al and x= 0.59-1 for Ga, repectively. Unexpectively, the 1:1 aluminium/gallium compound exists altough the binary border phases are unknown. The phase width of this singular trielide is a work in progress. Coloring effect, i.e. the triel distribution among the different crystallographic sites in the three structure types are discussed in relation to geometric and electronic aspects, also taking results of DFT bandstructure calculations into account.

[1] W. Harms, M. Wendorff, C. Röhr, Z. Naturforsch. 62b, 177-194 (2007).
[2] M. L. Fornasini, F. Merlo, Z. Kristallogr. 22, 382-390 (2006).
[3] Ch. Zheng, R. Hoffmann, Z. Naturforsch. 41b, 292-320 (1986).
[4] D.-K. Seo, J. D. Corbett, J. Am. Chem. Soc. 122, 9621-9627 (2000).
[5] M. Wendorff, C. Röhr, Z. Naturforsch. 69b, 388-408 (2014).

Fig. 1: Projection of the four relevant structure types (a) $BaAl_4$, (b) La_3Al_{11}, (c) $EuIn_4$ and (d) Ca_3Ga_5 along the shortest axis.

Fig. 2: Ternary diagram showing the stabilitiy ranges of the mixed trielides.

Figure 1

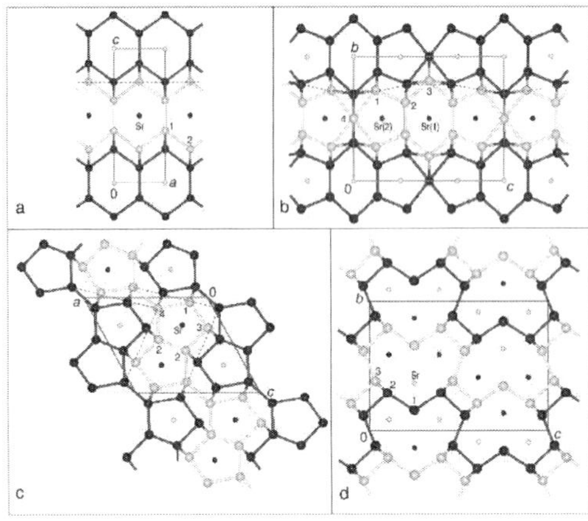

Figure 2

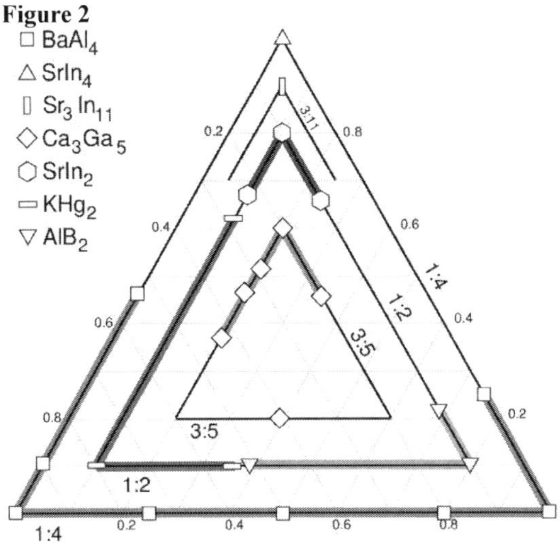

ISC-P19

Composition dependent transitions from cubic to tetragonal in $Li_{7+x}Te_xI_{1-x}N_2$

C. Ney[1], R. Niewa[1]
[1] *Universität Stuttgart, Institut für Anorganische Chemie, Stuttgart, Germany*

Powder samples of the compositions $Li_{7+x}Te_xI_{1-x}N_2$ (x = 0, 0.1, ..., 0.9, 1) have been synthesized from the ternary compounds Li_7IN_2 [1] and Li_8TeN_2 [2]. Both compounds, the cubic nitride halide (F-$43m$, a = 1038.8 pm) and the the tetragonal nitride chalcogenide ($I4_1md$, a = 706.2 pm, c = 996.9 pm), share the same anionic substructure with the structural motif of the cubic Laves phase. As such the anionic substructures of the both compounds can be brought into a group-subgroup relation with the common supergroup Fd-$3m$, the actual space group of the cubic Laves phase $MgCu_2$. Due to the complexity of the cationic substructure and the absense of lithium positions in the cubic Laves phase, a similar comparison of the cationic substructures is rather difficult. Therefore, a more direct mathematical approach to compare the lithium positions has been chosen, suggesting two third of the lithium positions to be closely related, the remaining to be significantly different.

In order to determine the composition for the transition from the cubic to the tetragonal structure and to investigate the development of the lithium positions, the samples have been characterized by both X-ray and neutron powder diffraction. Both techniques imply the transition to occur between $Li_{7.6}Te_{0.6}I_{0.4}N_2$ and $Li_{7.7}Te_{0.7}I_{0.3}N_2$. Additionally, a monotone decrease of the cell volume can be observed with increasing x in accordance with Vegard's Law. The occurrence of a potential orthorhombic distortion, as observed for the structurally related argyrodites [3], can so far not be supported.

Figure 1: Structural relation between Li_7IN_2 and Li_8TeN_2

Figure 1

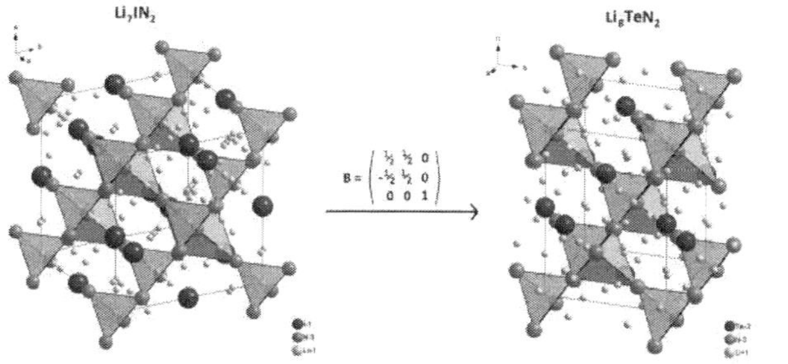

ISC-P20
Crystal Structure of the Cerium(III) Sulfide Chloride Thioarsenate(III) $Ce_5S_4Cl[AsS_3]_2$

F. Ledderboge[1], T. Schleid[1]
[1]*Institut für Anorganische Chemie der Universität Stuttgart, Stuttgart, Germany*

Water- and air-stable red single crystals of pentacerium(III) tetrasulfide chloride bis(thioarsenate(III)) $Ce_5S_4Cl[AsS_3]_2$ were prepared by the reaction between elemental cerium (Ce) and sulfur (S) with arsenic sesquisulfide (As_2S_3) and cesium chloride (CsCl) as both fluxing agent and Cl^- source in evacuated silica ampoules at 750 °C for 3 days. $Ce_5S_4Cl[AsS_3]_2$ crystallizes orthorhombically in the space group $Cmc2_1$ (no. 36) with the lattice constants a = 3557.06(19), b = 704.25(4) and c = 706.12(4) pm for Z = 4. The known structure of $Nd_5O_4Cl[AsO_3]_2$ [1] and $Pr_5O_4Cl[AsO_3]_2$ [2] and the new structure of $Ce_5S_4Cl[AsS_3]_2$ may show the same formal composition, but they are structurally not related to each other. The main structure elements of the title compound can rather be found in the crystal structures of $La_3S_2Cl_2[AsS_3]$ [3] or $Pr_3S_2Cl_2[AsS_3]$ [4] and $Pr_4S_3[AsS_3]_2$ [5].

The structure of $Ce_5S_4Cl[AsS_3]_2$ contains three crystallographically different Ce^{3+} cations. $(Ce1)^{3+}$ is eightfold coordinated as trigonal prism of sulfide (d(Ce1-S2) = 285 (2x) and 295 pm (2x)) and chloride anions (d(Ce1-Cl) = 284 and 303 pm), which is capped twice by the sulfur atoms of thioarsenate(III) group (d(Ce1-S3) = 311 pm (2x)). The sevenfold coordination of $(Ce2)^{3+}$ shows a capped trigonal prism of sulfur (d(Ce2-S) = 281 - 301 pm + 308 pm). The also sevenfold coordinated $(Ce3)^{3+}$ cation shows two faces. On the one hand there is a triangle of thioarsenate(III) sulfur atoms (d(Ce3-S) = 297 - 303 pm) and on the other hand a square of regular sulfide anions (d(Ce3-S1) = 283 - 287 pm). The As^{3+} cations form discrete ψ^1-tetrahedra $[AsS_3]^{3-}$ (d(As-S) = 225 - 228 pm), which are attached to six Ce^{3+} cations via all edges and corners. All S^{2-} anions show coordination numbers of four with more or less tetrahedral coordination spheres for S1 and S2, building up layers in the case of the $[(S1)(Ce3)_4]^{10+}$ tetrahedra and double chains for the $[(S2)(Ce1)_2(Ce2)_2]^{10+}$ tetrahedra (Figure 1). The sulfur atoms of the thioarsenate (III) anions (S3, S4 and S5) exhibit almost trigonal planar coordination spheres of Ce^{3+} cations. All stereochemically active lone pairs at the $[AsS_3]^{3-}$ groups point into the same direction along [001] owing to the polar space group $Cmc2_1$ (Figure 2). The Cl^- anions are only connected to two $(Ce1)^{3+}$ cations (d(Cl-Ce1) = 284 and 304 pm) in a saw-blade type chain (∢(Cl-Ce1-Cl) = 83°;∢(Ce1-Cl-Ce1) = 136°).

Figure 1: View at the crystal structure of $Ce_5S_4Cl[AsS_3]_2$ along [001].

Figure 2: View at the crystal structure of $Ce_5S_4Cl[AsS_3]_2$ along [010].

References
[1] D.-H. Kang, J. Wontcheu, Th. Schleid, *Solid State Sci.* 2009, *11*, 299-304.
[2] H. Ben Yahia, A. Villesuzanne, U. C. Rodewald, Th. Schleid, R. Pöttgen, *Z. Naturforsch.* 2010, *65 b*, 549-555.
[3] D.-H. Kang, Th. Schleid, *Z. Kristallogr.* 2008, *Supple. 28*, 46.
[4] D.-H. Kang, F. Ledderboge, H. Kleinke, Th. Schleid, *Z. Anorg. Allg. Chem.* 2015, *in preparation.*
[5] D.-H. Kang, Th. Schleid, *Z. Anorg. Allg. Chem.* 2009, *635*, 2170-2176.

Figure 1

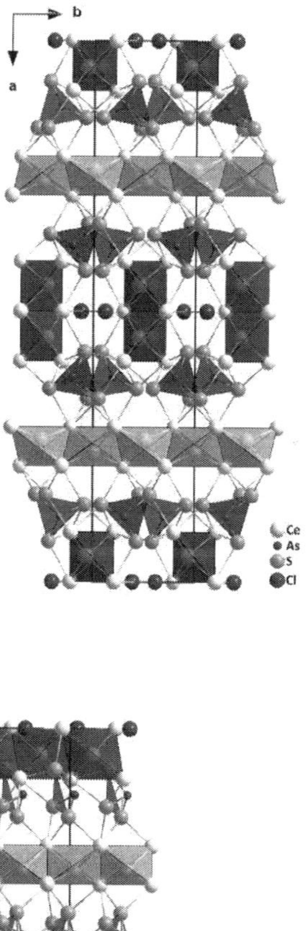

Figure 2

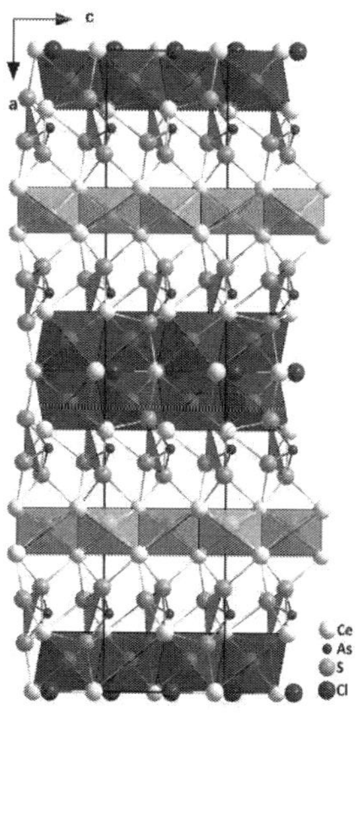

ISC-P21

$[Mg(H_2O)_6]_2Cl_2[B_{12}Cl_{12}]$: The First Mixed-Anionic Dodecachloro-*closo*-Dodecaborate Hydrate

F. M. Kleeberg[1], T. Schleid[1]
[1] *Universität Stuttgart, Institut für Anorganische Chemie, Stuttgart, Germany*

Reports on mixed-anionic *closo*-dodecaborates are quite rare in literature, but compounds like $Cs_2[B_{12}H_{12}]$ · CsCl (≡ $Cs_3Cl[B_{12}H_{12}]$) first described by *Muetterties et al.* [1] and $A_3X[B_{12}H_{12}]$ (A = K - Cs, X = Cl - I) from *Kuznetsov et al.* [2] were known for a long time. It took thirty years, however, until each of these salt-like *closo*-dodecaborates were fully characterized by single-crystal X-ray diffraction, all indicating the same trigonal perovskite structure with *anti*-$LaAlO_3$-type arrangement [3,4]. We now report on the synthesis and crystal structure of a new mixed-anionic *closo*-dodecaborate consisting of dianionic perchlorinated B_{12} icosahedra, water coordinated Mg^{2+} cations and isolated chloride anions.

The new compound with the composition $[Mg(H_2O)_6]_2Cl_2[B_{12}Cl_{12}]$ crystallizes in the orthorhombic space group *Pbca* (no. 61) with a = 1187.39(5), b = 2342.74(9) and c = 1213.16(5) pm for Z = 4. It has no analogies to a perovskite-related structure, since its crystal structure contains two $[Mg(H_2O)_6]^{2+}$ cations and two Cl^- anions per $[B_{12}Cl_{12}]^{2-}$ cluster (Figure 1). Six crystallographically different boron and chlorine atoms each build up *quasi*-icosahedral $[B_{12}Cl_{12}]^{2-}$ anions with typical interatomic distances (d(B-B) = 177-179 pm, d(B-Cl) = 178-179 pm). The Mg^{2+} cations occupy the *Wyckoff* position *8c* and are surrounded by six oxygen atoms from their corresponding water molecules forming nearly perfect octahedra (d(Mg-O) = 204-207 pm). Between four of these $[Mg(H_2O)_6]^{2+}$ octahedra, the secluded chloride anions reside (d(Mg···Cl) = 433-481 pm) and are fixed there by strong hydrogen bond interactions via six contacts to the nearest water molecules. The O···Cl distances range from 315 to 344 pm and are thus related to other compounds containing Mg^{2+}, Cl^- and H_2O such as $MgCl_2$ · 6 H_2O (d(O···Cl) = 318-356 pm) [5]. Even the chlorine atoms of the $[B_{12}Cl_{12}]^{2-}$ icosahedra seem to have hydrogen bonds to the water molecules of hydration at the Mg^{2+} cations ($d(O\cdots Cl_{Cluster})$ = 328-354 pm).

Figure 1. View at the unit cell of $[Mg(H_2O)_6]_2Cl_2[B_{12}Cl_{12}]$ along [001].

Figure 1

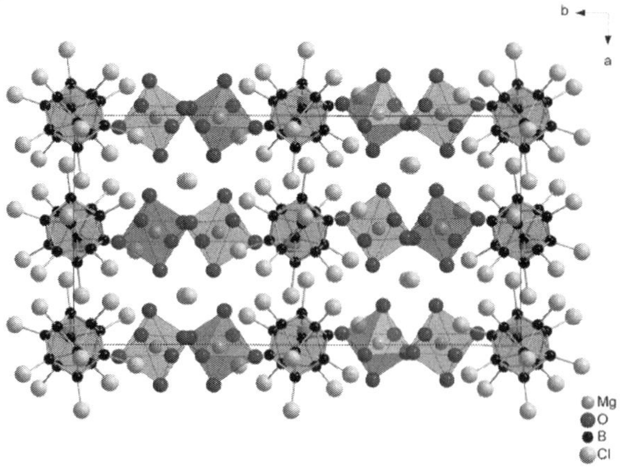

References

[1] E. L. Muetterties, J. H. Balthis, Y. T. Chia, W. H. Knoth, H. C. Miller, *Inorg. Chem.* 3 (1964) 444-451.
[2] N. T. Kuznetsov, G. S. Klimchuk, O. A. Kanaeva, K. A. Solntsev, *Russ. J. Inorg. Chem.* 21 (1976) 505-506.
[3] I. Tiritiris, Th. Schleid, *Z. Anorg. Allg. Chem.* 629 (2003) 1390-1402.
[4] I. Tiritiris, J. Weidlein, Th. Schleid, *Z. Naturforsch.* 60 b (2005) 627-639.
[5] K. R. Andress, J. Gundermann, *Z. Kristallogr.* 87 (1934) 345-369.

ISC-P22

Structural Investigations on $CsAlD_4$ by Time-of-Flight Neutron Powder Diffraction and DFT calculations

T. Bernert[1], D. Krech[1], M. Felderhoff[1], W. Kockelmann[2], T. J. Frankcombe[3], C. Weidenthaler[1]
[1] *Max-Planck-Institut für Kohlenforschung, Heterogeneous Catalysis, Mülheim an der Ruhr, Germany*
[2] *ISIS, Rutherford Appleton Laboratory, Didcot, OX11 0QX, United Kingdom*
[3] *Australian National University, Research School of Chemistry, Canberra, Australia*

In the light of identifying new possible candidates for hydrogen storage materials a lot of effort was spent to determine structure-property relations of metal tetrahydridoaluminate compounds with the general formula $MAlH_4$ (M = Li, Na, K, Rb, Cs) [1,2,3]. Recently, the crystal structures of two polymorphs of $CsAlH_4$ were determined by X-ray powder diffraction [4]. α-$CsAlH_4$ [$CsAlH_4(o)$] crystallizes in the barite structure type (space group *Pnma*) at ambient conditions. Furthermore, α-$CsAlH_4$ shows a mechanochemically induced phase transition to a tetragonal phase, δ-$CsAlH_4$ [$CsAlH_4(t)$], which was assumed to adopt the $NaAlH_4$ (scheelite) structure type (space group $I4_1/a$). By heating the tetragonal phase up to 473 K it reverts back to α-$CsAlH_4$ [4].

Since in this case X-ray powder diffraction cannot provide accurate hydrogen positions, time-of-flight neutron powder diffraction was performed on both polymorphs of deuterated $CsAlH_4$ samples on GEM at ISIS. The simultaneous Rietveld refinement of the diffraction data from the detector banks at 35°, 63.6° and 91.3° in 2θ was carried out using the structural model based on the results from Krech et al. [4] as well as calculated hydrogen positions, provided by DFT calculations using the program abinit.

The Rietveld refinement of the neutron diffraction data confirmed the predicted hydrogen positions for the orthorhombic α-$CsAlH_4$, according to the barite structure type (Figure 1). However, the positions of the Cs- and Al-atoms as well as the predicted hydrogen positions in δ-$CsAlH_4$ fulfill approximately the symmetry conditions for space group $I4_1/amd$. The Rietveld refinement of the neutron diffraction data of δ-$CsAlD_4$ in space group $I4_1/a$ revealed that the deuterium atoms are statically disordered with half of the deuterium atoms occupying the positions according to the $NaAlH_4$ structure type whereas the other half are located on positions predicted by the calculations. From these results the structural change during the phase transition from the tetragonal to the orthorhombic structure may be deduced.

References:

[1] T. J. Frankcombe, O. L. Løvvik, *J. Phys. Chem. B*, 2006, 110, 622 - 630.
[2] P. Vajeeston, P. Ravindran, R. Vidya, H. Fjellvåg, A. Kjekshus, *Cryst. Growth Des.*, 2003, 4, 471 - 477.
[3] B. C. Hauback, H. W. Brinks, C. M. Jensen, K. Murphy, A. J. Maeland, *J. Alloys Compd.*, 2003, 358, 142 - 145.
[4] D. Krech, B. Zibrowius, C. Weidenthaler, M. Felderhoff, *Eur. J. Inorg. Chem.*, 2014, 114, 5683 - 5688.

Figure 1: Crystal structures of the two polymorphs of α-$CsAlD_4$, green: Cs, blue: Al and grey: hydrogen. On the left: crystal structure of α-$CsAlD_4$ and on the right: δ-$CsAlD_4$, viewing along [001].

Figure 1

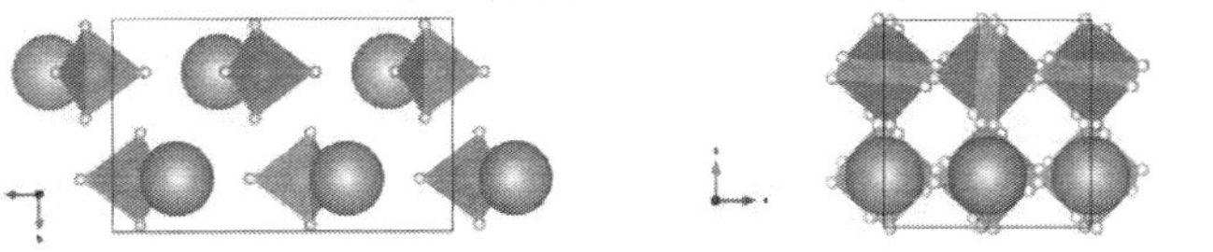

figure legends:

Figure 1: Ortep of $Ni(Ox)_2$.

Figure 1

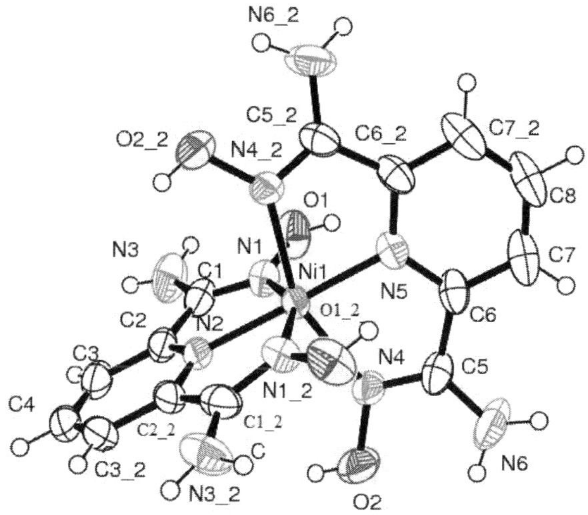

ISC-P23
SYNTHESIS, CRYSTAL STRUCTURES , HYDROGEN BONDS STUDIES AND DFT CALCULATION OF NEW COMPLEXE BASED ON Ni^{+II}

A. Messai[1,2], B. Nabil[1]
[1]*Abbés Laghrour University, SM, Khenchela, Algeria*
[2]*Abbés Laghrour University, Chemestry, Khenchela, Algeria*

The literature contains reports of the structure of amide oximes (amidoximes) which can co-ordinate as bi-,tri-,tetra-and hexa-dentate ligands. The majority of these amidoximes are aliphatic compounds which form five-membered rings utilizing the diimine moiety, -N=C-C=N-, involving oxime and imine nitrogens as the co-ordination sites with transition-metal atoms. The oxime $C_7H_9N_5O_2$ **(L)**, possesses the structural requirements to react as a tridentate ligand with the heterocyclic and two oxime nitrogen atoms forming five-membered rings with transition-metal ions.

This research was carried out with the specific purpose of determining the structure of **(L)** as well as his complexe formed with nickel(II).

We report here the single crystal X-ray structure for the complex **L** with nickel(II) , The structure of the ligand has been studied with 1H NMR and FTIR spectroscopy however the complexe has been studied experimentally by X-ray single crystal and theoretically by DFT calculation.

References:
1) G. A. Pearse, P. R. Raithby and J. Lewis, *Polyhedron*, 1989, 8, 301.
2) G. A. Pearse, P. R. Raithby, C. M. Hay and J. Lewis, *Polyhedron*, 1989, 8, 305.
3) M. Nasakkala, H. Saarinem, J. Korvenranta and M. Oranrua, *Acta Crystallogr., Sect. C*, 1989, 45, 1514.
4) D. L. Cullen and E. C. Liugafelten, *Inorg. Chem.*, 1970, **9**, 1865.
5) G. A. Pearse, P. R. Raithby and M. M. J. Maughan, *Polyhedron*,1994, 13, 553.
6) G. A. Pearse and C. E. Pfluger, *Inorg. Chim. Acta*, 1994, **227**, 171.
7) R. Banks and R. F. Brookes, *Chem. Ind.*, 1974, 617.

ISC-P24
New Alkaline-earth mercurides $AHg_{6-x}M_x$ (A=Ca, Sr, Ba; M = Sn, Ga, In, Zn, Cd)

M. Wendorff[1], M. Schwarz[1], C. Röhr[1]
[1]*Institut für Anorganische und Analytische Chemie, Freiburg, Germany*

Starting from the recently reported binary mercuride $BaHg_6$, Hg can be substituted by up to 2.25 In/f.u. [1]. A further decreased In content of the samples causes the formation of related 1:7 compounds [2]. Ga-substitued Ba compounds like $BaGa_{0.8}Hg_{5.2}$ form a likewise related 1:6-type structure [1]. The analogous exchange of Hg in $BaHg_6$ by the electron-richer element tin reaches up to x=1.23, i.e. to the same v.e.c. as in the In mercuride.

Despite the lack of a binary phase $SrHg_6$, ternary Sr-M mercurides $SrHg_{6-x}M_x$, which additionally contain In (x=0.30<xx=0.30) or Sn (0.30<x6-type. Also in this case, the maximum Sn content is approx. half as large as the highest In proportion observed.

The main feature of the $BaHg_6$ structure are planar nets of 5- and 8-membered Hg/M rings stacked perpendicular to the short b axis (fig. 1). The A cations are located in the centers of the octagons of these nets, thus exhibiting a 5:8:5 coordination by Hg/M. Typical for Hg-rich mercurides, the Hg-Hg bond lengths are distributed into two groups: Strong, short bonds (I, 285-300 pm) inside the nets, and longer distances (II, 320-350 pm), which serve to complete the coordination spheres of Hg/M (fig. 2). On substituting Hg by more electron-rich but less electronegative elements, the six sites are preferentially occupied according to the M electronegativity: The Bader charges obtained from bandstructure calculations of AHg_6 and ordered mixed Hg-rich model compounds [2,3,4] show, that the M elements are the positively charged bonding partners in the polyanions. In agreement, the M-content of the specific sites nicely corresponds to the Bader charge of the respective Hg position in (Sr/Ba)Hg_6 (fig. 2).

In the Ba systems, the substitution of Hg by isoelectronic Zn and Cd resulted in the formation of ternary members of the $Ba_{20}Hg_{103}$ family only [4]. For A=Sr and A=Ca, the well known YCd_6-type [5,6] with a considerable high Hg content ($SrCd_6$-$SrHg_{4.5}Cd_{1.5}$; $CaCd_6$-$CaHg_{5.4}Cd_{0.6}$) appears in this case.

[1] M. Wendorff, C. Röhr, J. Alloys Compd. 546, 320 (2013)
[2] M. Wendorff, M. Schwarz, C. Röhr, J. Solid State Chem. 203, 297 (2013)
[3] M. Wendorff, C. Röhr, Z. Naturforsch. 68b, 307 (2013)
[4] M. Wendorff, C. Röhr, Z. Naturforsch. 67b, 893 (2012)
[5] G. Bruzzone, Gazz. Chim. Ital. 102, 234 (1972)
[6] C. P. Gomez, S. Lidin, Phys. Rev. B, 68, 024203-1 (2003)

Fig 1: The $BaHg_6$-type structure.
Fig 2: *M* distribution in the ternary representatives.

Figure 1

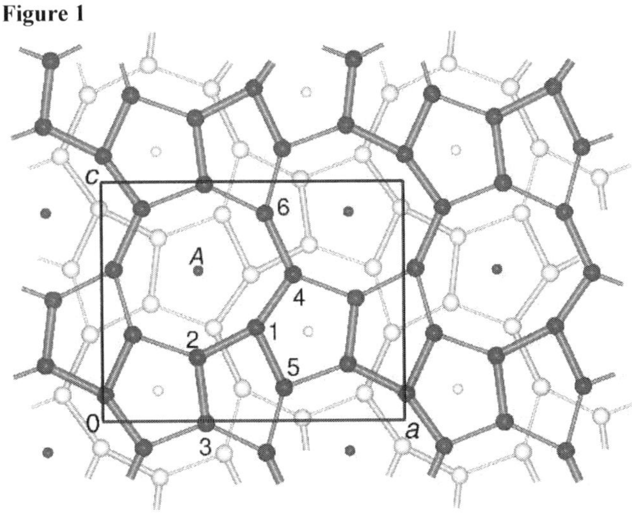

Figure 2

A:*M* combination	*M* content *x*	*M* content [%] of the crystallographic site no. 1	2	3	4	5	6
Ba:In	1.23	···	···	12	16	23	72
	2.25	···	24	29	21	49	82
Ba:Sn	1.23	···	···	4	18	12	21
Sr:In	0.30	···	···	···	···	···	31
	0.77	···	···	5	5	11	57
	1.55	···	10	16	22	22	85
Sr:Sn	0.30	···	···	···	···	8	25
	0.62	···	···	···	11	11	39
Sr:Ga	0.30	···	14	···	17	···	···
Bader Charge →		−0.257	−0.253	−0.208	−0.255	−0.183	−0.143
CN ($n_I+n_{II}+n_A$) →		3+4+3	5+2+3	5+2+3	5+3+4	5+3+3	3+6+2
$V_{Bader-Bassin}$ [$10^6 pm^3$] →		25.1	25.8	26.0	25.9	25.6	25.9

ISC-P25

Formation and structural characterization of the first basic tin(II)-fluoride, $Sn_9F_{13}O(OH)_3 \cdot 2H_2O$.

H. Reuter[1], E. Uglova[1], M. Reichelt[1]
[1] *University of Osnabrück, Institute of Chemistry of New Materials, Osnabrück, Germany*

Since some decades, small quantities of SnF_2 are added to some tooth pastes in order to prevent caries. On one side, the fluorine ions reconstruct and replace hydroxyapatite, $Ca_5(PO_4)_3(OH)$, that constitutes more than 70% of dentin, by fluorapatite, $Ca_5(PO_4)_3F$. On the other side, Berndt could show that tin(II) fluoride reacts with hydroxyapatite to form $Sn_3F_3(PO_4)$ [1], the structure of which was published in 1972 [2]. In rerunning this experiment, we have studied the reaction of concentrated aqueous solutions of tin(II) fluoride on $Ca_3(PO_4)_2$-pellets in more detail. For this purpose, the pellet was placed in a beaker containing the tin(II) fluoride solution. After some days at room temperature, the pellet was transferred without drying to a petri dish and examined under a microscope. Among the crystals which covered the surface of the pellet, a larger one was identified by single crystal X-ray diffraction to be $Sn_3F_3(PO_4)$. For further investigations and to prevent the pellet to fall dry it was covered by a second, smaller petri dish. After the weekend, this second petri dish was firmly connected with the first one by a lot of large single crystals exhibiting to different crystal shapes. One form was the well-known compound Sn_4OF_6 [3], while the second one represents the up to now unknown basic tin(II) fluoride $Sn_9F_{13}O(OH)_3 \cdot 2H_2O$. On the poster, we will describe this structure consisting of 9 crystallographic independent tin sites with special respect to the different coordination polyhedrons at tin and the dominating role of a Sn_4O_4-heterocuban framework.

[1] A.F. Berndt, *Reaction of Stannous Fluoride with Hydroxyapatite.* J. Dent. Res., 49 (1970) 638.
[2] A.F. Berndt, *Reaction of Stannous Fluoride with Hydroxyapatite: The Crystal Structure of $Sn_3PO_4F_3$.* J. Dent. Res., 51 (1972) 53-57.
[3] I. Abrahams, S.J. Clark, J.D. Donaldson, Z.I. Khan and J.T. Southern, *Hydrolysis of Tin(II) Fluorid and Crystal Structure of Sn_4OF_6.* J. Chem. Soc. Dalton Trans., (1994) 2581-2583.

Figure 1

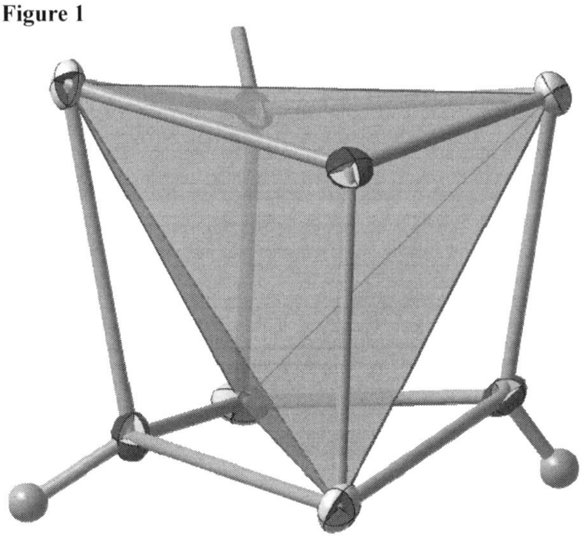

ISC-P26

Accurate structural parameters of the tin(II) halides from well-resolved, low-temperature single crystal data.

H. Reuter[1], M. Reichelt[1]
[1] *University of Osnabrück, Institute of Chemistry of New Materials, Osnabrück, Germany*

Although belonging to the basic chemicals of tin(II), the structure determinations of the tin(II) halides are of less quality and far away from modern requirements. Thus, only from SnF_2 that was measured several times an adequate single crystal structure determination exists [1]. The normal pressure modification of $SnCl_2$ was first investigated in 1961 [2] by use of two layers of Weissenberg photographs. These data were confirmed without significant improvements in 1996 [3] when the results of a powder data refinement were published. A similar situation is found in case of $SnBr_2$: a first structure determination was published in 1975 [4] followed by a re-determination in 2000 by means of powder data [5]. From SnI_2 also only a powder diffraction structure from 1972 exists [6]. In this case, bond lengths and angles calculated from deposited data are inconsistent with those published.

In order to raise the precision of the structure determinations on a higher level we have grown single crystals of all tin(II) halides and determined their structures in a similar way with high data redundancy and the same resolution (d_{max} = MoK_α) at 100 K. This allows for the first time to compare the structures and to elucidate differences and similarities between the different tin(II) halides with high precision. One main feature, common to nearly all tin sites, is a pseudo-tetrahedral coordination which can be extended to a square-pyramidal coordination by two significantly longer tin-halide bonds.

[1] G. Denes, J. Pannetier, J. Lucas and J.Y. Le Marouille, *About SnF_2 stannous fluoride. I. Crystallochemistry of alpha-SnF_2.* J. Sol. State Chem., 30 (1979) 335-343.
[2] J.M. van den Berg, *The crystal structure of $SnCl_2$.* Acta Cryst., 14 (1961) 1002-1003.
[3] J.M. Leger, J. Haines and A. Atoug, *The high pressure behaviour of the cotunnite and post-cotunnite phases of $PbCl_2$ and $SnCl_2$.* J. Phys. Chem. Solids, 1 (1996) 7-16.
[4] J. Andersson, *On the Crystal Structure of Tin(II) Bromide.* Acta Chem. Scand., A29 (1975) 956.
[5] I. Abrahams and D.Z. Demetriou, *Inert Pair Effects in Tin and Lead Dihalides: Crystal Structure of Tin(II) Bromide.* J. Solid. State Chem., 149 (2000) 28-32.
[6] R.A. Howie, W. Moser and I.C. Trevena, *The Crystal Structure of Tin(II) Iodide.* Acta Cryst., B28 (1972) 2965-2971.

Figure 1

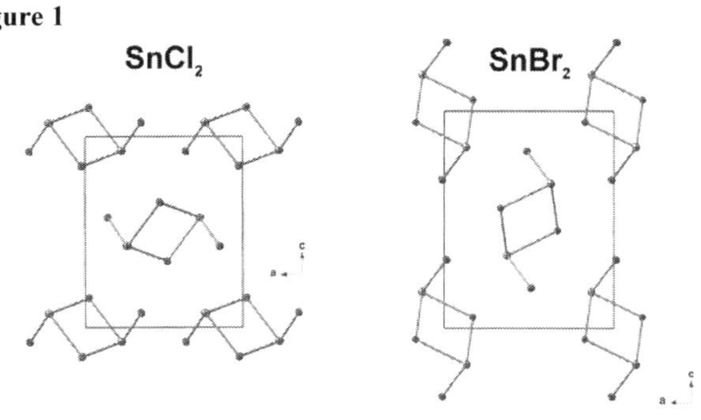

ISC-P27

[$^iPrSnI_2(pyNO)_3$][SnI_3] · $CHCl_3$: an organic-inorganic mixed-valence tin(IV)-tin(II) compound

H. Reuter[1], D. Schwarte[1], M. Reichelt[1]
[1] *University of Osnabrück, Institute of Chemistry of New Materials, Osnabrück, Germany*

In literature there are only some few examples of crystal structure determinations on coordination compounds of monoorganotin(IV) trihalides. In order to close this gap we have started a broad study with different mono- and bidentate nitrogen and oxygen Lewis-Bases including not only the trichlorides but also the tribromides and triiodides which in part we synthesized for the first time. The compounds were routinely characterized by IR- and Raman-spectroscopy, ^1H and ^{13}C solution NMR-spectroscopy and whenever possible by single crystal X-ray diffraction. During this study we were able to isolate a Lewis-base adduct of isopropyltin(IV) triiodide, $^iPrSnI_3$, with pyridine-N-oxide, PyNO, of composition $^iPrSnI_3$ · 3PyNO. Because crystals obtained from $CHCl_3$-solutions used for NMR experiments on this compound differed in shape from the original ones we determined this structure, too, suggesting another crystal packing. Structure solution, however, revealed an unexpected result: an organic-inorganic mixed-valence tin(IV)-tin(II) compound of composition [$^iPrSnI_2(pyNO)_3$][SnI_3] · $CHCl_3$. The poster will summarize the structural parameters of both compounds and highlight the coordination of the different tin atoms with respect to their different oxidation states.

Figure 1

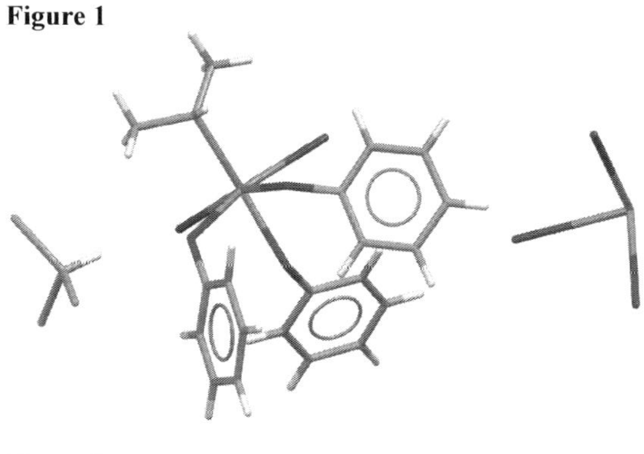

Figure 2

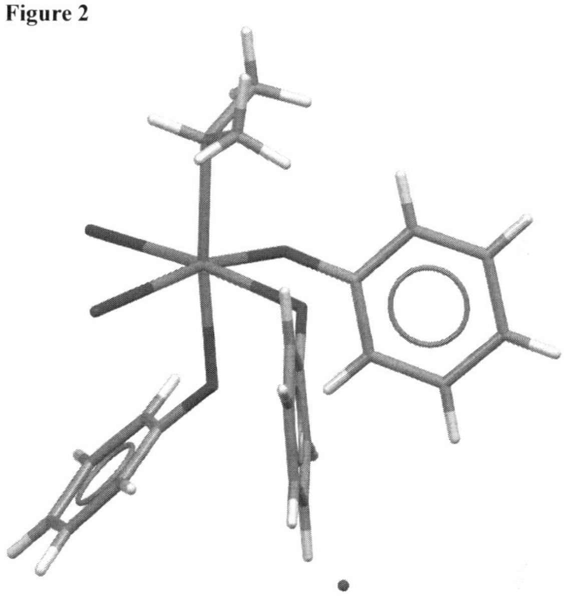

ISC-P28

Synthesis and Structural Characterization of the Europium(II,III) Oxide *Ortho*-Oxoaluminate $Eu_3O[AlO_4]$

M. E. Bohem[1], O. Reckeweg[2], T. Schleid[2]
[1] *University of Stuttgart, Dept. Inorganic Chemistry in coop. with Fraunhofer IPA, Stuttgart, Germany*
[2] *University of Stuttgart, Dept. Inorganic Chemistry, Stuttgart, Germany*

Strontium-aluminate materials doped with europium have been of interest for several decades [1, 2]. We were now able to synthetize the europium(II,III) oxide oxoaluminate $Eu_3O[AlO_4]$ as a thermodynamically stable tetragonal phase in a solvent-free process. The crystals were obtained from solid-state reactions of elemental europium, EuF_3 and alumosilicate glass flakes by the use of CsBr as a fluxing agent. The reactions were performed in evacuated niobium capsules by applying a stepwise temperature program with T_{max} = 1100 K for 1.5 days. The title compound crystallizes tetragonally in the space group *I*4/*mcm* (no. 140) with two formula units per unit cell and forms strongly anisotropic, red crystals with bar-like shape and the cell parameters a = 676.41(4) and c = 1103.06(7) pm. The composition requires both +II and +III as oxidation states of europium for charge neutrality and thus results in two crystallographically distinct cation sites for europium (*Wyckoff* sites: 4*a* and 8*h*). The position 4*a* (Eu1) is tenfold coordinated in the shape of a bicapped tetragonal antiprism, built up by $(O2)^{2-}$ anions at 16*l* and capped by $(O1)^{2-}$ at 4*c*. The Eu2 atoms at 8*h* have an eightfold coordination forming a bicapped trigonal prism, again built up by the $(O2)^{2-}$ anions with $(O1)^{2-}$ as capping species. The $[AlO_4]^{5-}$ tetrahedra are composed of $(O2)^{2-}$ anions, whereas $(O1)^{2-}$ represents the so-called "free" oxygen with no connection to Al^{3+}. With respect to larger interatomic distances, the Eu1 site at *Wyckoff* position 4*a* is preferentially occupied by Eu^{2+}, whereas the smaller Eu^{3+} cations are primarily situated at the Eu2 position (8*h*).

The structure of $Eu_3O[AlO_4]$ shares similarities with some isotypic alkaline-earth metal *ortho*-oxosilicates $AE_3O[SiO_4]$ (*AE* = Ca - Ba) [3] or with *anti*-isotypic Li_5OCl_3 (≡ $Cl_3Li[OLi_4]$) [4]. The overall structure shows chains of vertex-connected $[(O1)Eu_6]^{14+}$ octahedra that are stacked parallel to the *c* axis in a rod-like fashion with a slight twist of the single octahedra. These strands are further arranged in tetragonal rod packing along the (001) plane. The $[AlO_4]^{5-}$ tetrahedra reside within the octahedral holes of the *perovskite*-like 3D-$[(O1)Eu_{6/2}]^{5+}$ network and their edges lie strictly parallel to the face diagonals of (001).

Several application fields like photoluminescence emission properties of this compound might be of interest and will further be investigated, after having optimized the accessibility of $Eu_3O[AlO_4]$ in larger quantities as phase-pure material.

[1] M. Drofenik, L. Golič, *Acta Crystallogr.* 1979, *B 35*, 1059-1062.
[2] W. Carrillo-Cabrera, J. O. Thomas, G. C. Farrington, *Solid State Ion.* 1988, *28-30*, 317-323.
[3] E. Tillmanns, H. P. Grosse, *Acta Crystallogr.* 1978, *B 34*, 649-651.
[4] O. Reckeweg, B. Blaschkowski, Th. Schleid, *Z. Anorg. Allg. Chem.* 2012, *638*, 2081-2086.

Figure 1. View at the crystal structure of $Eu_3O[AlO_4]$ along the *c* axis.
Figure 2. Anionic environment the two crystallographically different europium cations.

Figure 1

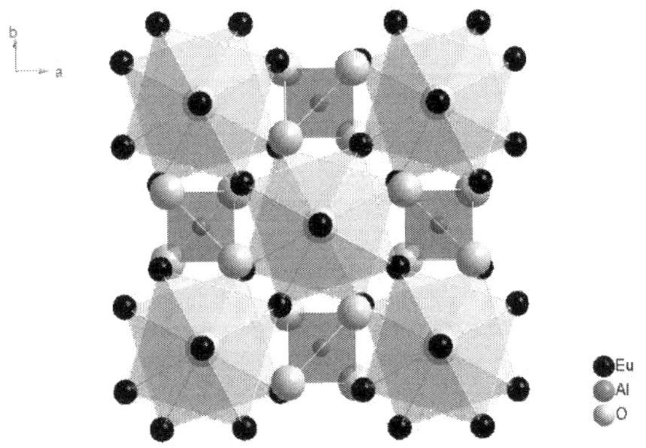

Figure 2

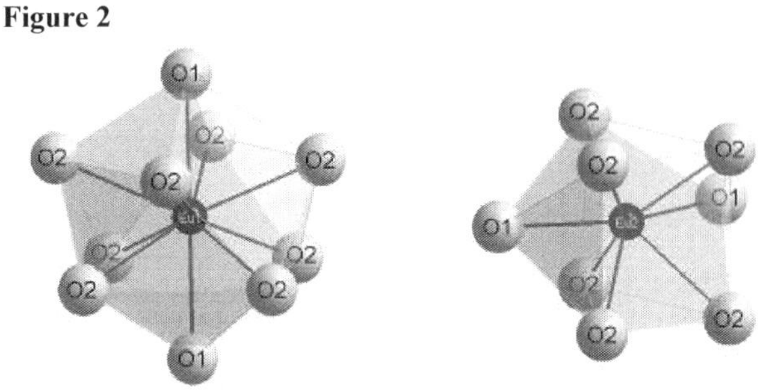

ISC-P29

Serendipitous Formation of Cesium Mercury(II) Selenide $Cs_2Hg_3Se_4$ as High-Pressure Phase

R. Stromsky[1], L. W. Zimmermann[1], I. Hartenbach[1], T. Schleid[1]
[1]*Universität Stuttgart, Institut für Anorganische Chemie, Stuttgart, Germany*

An attempt to synthesize $Cs_4Hg_2[B_{12}(BSe_3)_6]$ [1, 2] by fusing together elemental cesium, boron and selenium along with mercury(II) selenide in evacuated silica ampoules at 700 °C for 12 h did not result in the desired compound, but among other by-products, red, platelet-shaped single crystals of $Cs_2Hg_3Se_4$ emerged from this reaction mixture. The title compound crystallizes orthorhombically in space group *Ibam* (*a* = 643.45(4), *b* = 1200.26(7), and *c* = 1468.74(9) pm; CSD-428897) with four formula units per unit cell. The crystallographically unique Cs^+ cations reside at *Wyckoff* position 8*j* within a slightly distorted cubic coordination environment, built of also crystallographically unique Se^{2-} anions at the general *Wyckoff* position (16*k*). The interatomic distances between cesium and selenium range between 362 and 390 pm. Furthermore, the crystal structure contains two crystallographically distinguishable mercury dications, with $(Hg1)^{2+}$ at the *Wyckoff* position 4*a* being surrounded bisphenoidally by four Se^{2-} anions, while $(Hg2)^{2+}$ resides at the *Wyckoff* position 8*g* coordinated by only two Se^{2-} anions in a slightly bent fashion. The interatomic distances between Hg^{2+} and Se^{2-} were determined to 267 pm within the $[(Hg1)Se_4]^{6-}$ tetrahedra and 246 pm for the $[(Hg2)Se_2]^{2-}$ units. These $[(Hg2)Se_2]^{2-}$ boomerangs (Se–Hg–Se angle: 160°) and the $[(Hg1)Se_4]^{6-}$ tetrahedra are fused together by common selenide vertices to form chains of the composition 1D-$[Hg_3Se_4]^{2-}$ (Figure 1) parallel to [100]. The aforementioned chains are piled primitively to form a sheet perpendicular to [001], however no interconnection between the single chains in [010] direction can be detected. In order to form the three-dimensional structural setup, these negatively charged $[Hg_3Se_4]^{2-}$ sheets are separated from each other along the *c* direction with layers consisting of Cs^+ cations (Figure 2). The title compound represents the high-pressure variation for the composition $Cs_2Hg_3Se_4$, with the low-pressure phase crystallizing in space group *Pbcn* [3], exhibiting a slight tilting of the 1D-$[Hg_3Se_4]^{2-}$ chains towards each other in its crystal structure. The presented high-pressure phase can be considered as isostructural to the formula-analogous compound $Cs_2Zn_3S_4$ [4], but in the structure of the latter both crystallographically different Zn^{2+} cations exhibit a tetrahedral S^{2-} coordination, hence a layered partial structure according to 2D-$[Zn_3S_4]^{2-}$ is found.

[1] A. Lindemann, J. Kuchinke, B. Krebs, *Z. Anorg. Allg. Chem.* 1999, *625*, 1165.
[2] A. Hammerschmidt, A. Lindemann, M. Döch, B. Krebs, *Z. Anorg. Allg. Chem.* 2003, *629*, 1249.
[3] E. A. Axtell, Y. Park, K. Chondrouis, M. G. Kanatzidis, *J. Am. Chem. Soc.* 1998, *120*, 124.
[4] W. Bronger, U. Hendriks, P. Müller, *Z. Anorg. Allg. Chem.* 1988, *559*, 95.

Figure 1. $[Hg_3Se_4]^{2-}$ chains along [100] in the crystal structure of $Cs_2Hg_3Se_4$

Figure 2. View at the crystal structure of $Cs_2Hg_3Se_4$ along the *a* axis.

Figure 1

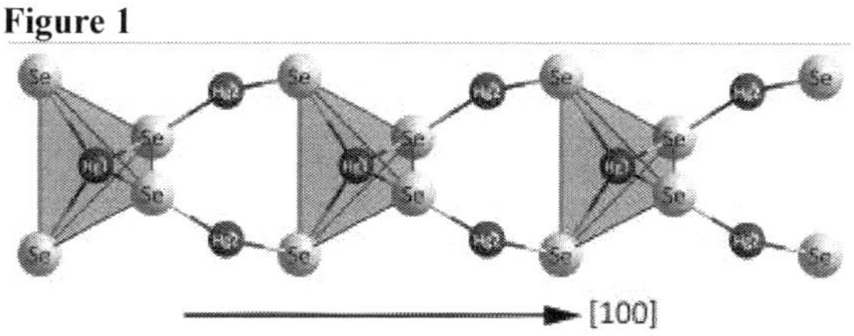

Figure 2

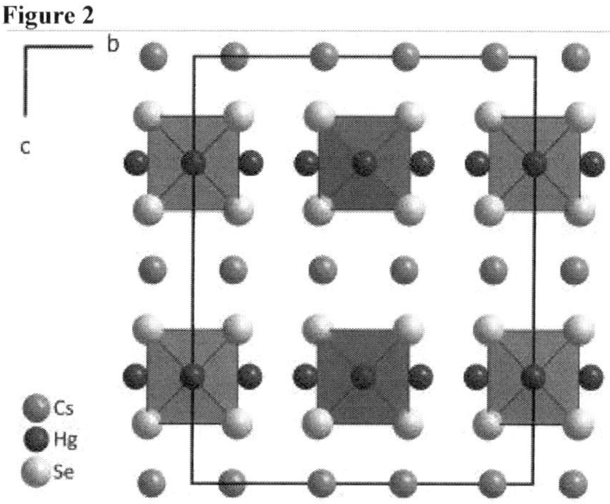

ISC-P30

The New Dysprosium(III) Oxide Selenide Dy_2OSe_2

A. H. Geyer[1], C. M. Schurz[1], T. Schleid[1]
[1]*University of Stuttgart, Inorganic Chemistry, Stuttgart, Germany*

Rare-earth metal(III) oxide selenides with the formula RE_2O_2Se (RE = Y, La - Nd, Sm - Lu) [1-3] are known for decades, but very few lanthanoid(III) oxide selenides of the inverted composition Ln_2OSe_2 (Ln = Pr and Gd) have found their way into literature so far. Gd_2OSe_2 crystallizes orthorhombically in the space group *Pnma* (a = 1605.0(3), b = 393.75(8), c = 703.09(14) pm, Z = 4) [4], whereas Pr_2OSe_2 forms monoclinic crystals with the space group $P2_1/c$ (a = 882.05(6), b = 732.89(5), c = 732.94(5) pm, β = 100.288(7)°, Z = 4) [3] as does the new compound Dy_2OSe_2 with the unit-cell parameters a = 854.53(3), b = 704.80(3), c = 707.71(3) pm and β = 100.669(2)°. One is startled by the fact that two isotypic structures for an element of the lighter (Pr) and the heavier lanthanoids (Dy) exist, but for an element in between (Gd) dramatic changes happen. This has to be investigated in the future.

The crystal structure of monoclinic Dy_2OSe_2 consists of layers erected by dimers of edge-sharing $[ODy_4]^{10+}$ tetrahedra, which are further connected via three vertices to form two-dimensional layers 2D-$\{[O(Dy1)_{3/3}(Dy2)_{1/1}]^{4+}\}$. Their three-dimensional interconnection is achieved by Se^{2-} anions. The Dy^{3+} cations are surrounded by each 7 or 7+1 chalcogenide anions. In the case of $(Dy1)^{3+}$ these are 4+1 selenide (d(Dy1-Se) = 287-328 pm + 377 pm) and 3 oxide anions (d(Dy1-O) = 228-233 pm). For $(Dy2)^{3+}$ it is only one oxide anion (d(Dy2-O) = 226 pm) along with six selenide anions (d(Dy2-Se) = 281-301 pm). This pattern can also be found in isostructural Pr_2OSe_2 [3]. In the case of Gd_2OSe_2 there are major differences in the condensation of the $[OGd_4]^{10+}$ tetrahedra, which are here connected to each other by two *cis*-oriented edges each, forming one-dimensional strands 1D-$\{[O(Gd1)_{3/3}(Gd2)_{1/1}]^{4+}\}$, and the coordination of the Gd^{3+} cations, which have coordination numbers of 6 and 8.

[1] H. A. Eick, *Acta Crystallogr.* 1960, *13*, 161-161.
[2] F. A. Weber, *Doctoral Thesis* 1999, University of Stuttgart.
[3] F. A. Weber, Th. Schleid, *Z. Anorg. Allg. Chem.* 2001, *627*, 1383-1388.
[4] O. Tougait, J. A. Ibers, *Acta Crystallogr.* 2000, C *56*, 623-624.

Figure 1. Coordination polyhedra of both Dy^{3+} cations, both Se^{2-} anions and planes of vertex- and edge-sharing $[ODy_4]^{10+}$ tetrahedra in the monoclinic crystal structure of Dy_2OSe_2.

Table 1. Atomic coordinates of monoclinic Dy_2OSe_2 (all atoms occupy the general *Wyckoff* site 4*e*).

Atom	*x/a*	*y/b*	*z/c*	U_{eq}/pm²
Dy1	0.57141(5)	0.65431(6)	0.82928(6)	72(1)
Dy2	0.14396(5)	0.55165(5)	0.75213(6)	72(1)
O1	0.3861(7)	0.0839(8)	0.3502(9)	77(12)
Se1	0.68391(11)	0.12167(12)	0.07742(13)	91(2)

Se2	0.07598(10)	0.27976(12)	0.03064(12)	75(2)

Figure 1

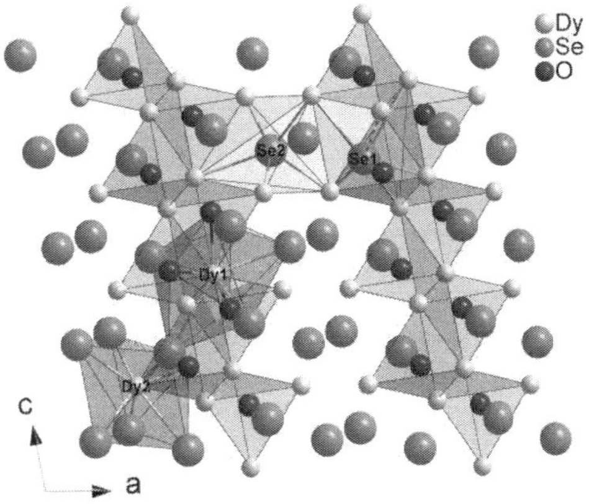

ISC-P31

$Y_2S[SiO_4]$: A New Yttrium Sulfide *Ortho*-Oxosilicate

M. Eberle[1], T. Schleid[1]
[1]*University of Stuttgart, Stuttgart, Germany*

Sulfide derivatives of lanthanoid(III) *ortho*-oxosilicates with the formula $M_2S[SiO_4]$ have been synthesized and structurally characterized for *M* = Ho [1] and Er [2], which occur in two different orthorhombic space groups. A similar yttrium analogue is just known with the more sulfide containing formula $Y_4S_3[Si_2O_7]$ [3] so far. The crystal structure of this compound consists of two crystallographically different Y^{3+} cations centering tricapped trigonal prisms of chalcogenide anions. The S^{2-} anions have two different coordination spheres of adjacent Y^{3+} cations with coordination numbers of six and four, resulting in the formation of distorted octahedra or squares, respectively. Discrete pyroanionic $[Si_2O_7]^{6-}$ units complete this crystal structure.

A new representative of rare-earth metal(III) sulfide *ortho*-oxosilicates, isotypic to $Ho_2S[SiO_4]$, is now presented with $Y_2S[SiO_4]$ (Table 1). Water- and air-resistant, pale yellow, lath shaped single crystals emerged from a mixture of Tl_2S, Cu, Y and S in a molar ratio of 1 : 4 : 2 : 5 along with CsBr as fluxing agent designed to produce $TlCu_2YS_3$ [4] as target product. The reaction took place for 8 days at 800 °C in evacuated silica ampoules, which recruit the source for the anionic $[SiO_4]^{4-}$ tetrahedra in the product.

$Y_2S[SiO_4]$ crystallizes in the orthorhombic space group *Pbcm* with *a* = 604.71(5), *b* = 689.90(6) and *c* = 1064.78(9) pm for Z = 4. The two crystallographically distinguishable Y^{3+} cations both have eight adjacent chalcogen atoms (six O^{2-} and two S^{2-} anions), but in different arrangements. While the $(Y1)^{3+}$ cations are found in bicapped trigonal prisms (*d*(Y1-O) = 239 - 249 pm, *d*(Y1-S) = 271 - 280 pm), the $(Y2)^{3+}$ cations center trigonal dodecahedra (*d*(Y2-O) = 231 - 245 pm, *d*(Y2-S) = 287 pm, 2x). The coordination polyhedra around the S^{2-} anions can be described as distorted squares of four Y^{3+} cations (∢(Y-S-Y) = 80 - 94°), which are fused together by shared vertices to result in strongly corrugated 2D-$\{[SY_{4/4}]^{4+}\}$ layers parallel to the (100) plane with diamond-shaped holes in it (Figure 1). The isolated $[SiO_4]^{4-}$ anions (*d*(Si-O) = 162 - 163 pm, (∢(O-Si-O) = 104 - 119°) carry four terminal and four edge-bridging Y^{3+} cations. They are located above and underneath $(Y1)^{3+}$ cations when looking along the [100] direction.

[1] I. Hartenbach, S. F. Meier, J. Wontcheu, Th. Schleid, *Z. Anorg. Allg. Chem.* 2002, *628*, 2907-2913.
[2] I. Hartenbach, P. Lauxmann, Th. Schleid, *Z. Anorg. Allg. Chem.* 2004, *630*, 1408-1412.
[3] L. A. Koscielski, J. A. Ibers, *Acta Crystallogr.* 2011, *E 67*, i16-i16.
[4] J.-M. Babo, *Doctoral Thesis*, University of Stuttgart, 2014.

Figure 1: Crystal structure of $Y_2S[SiO_4]$ with outlined unit cell.

Table 1: Atomic coordinates and U_{eq} values for $Y_2S[SiO_4]$.

Figure 1

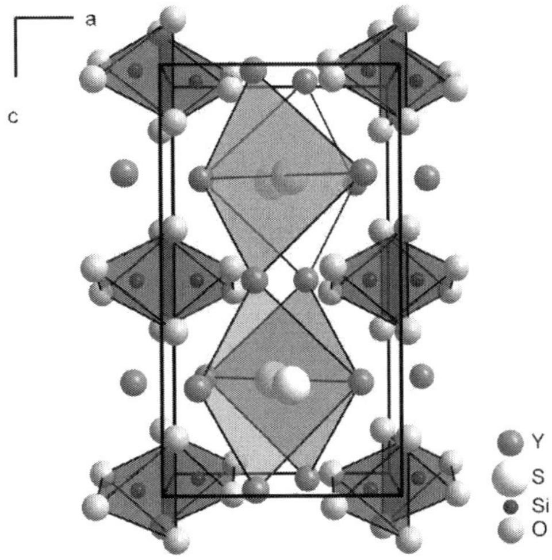

Figure 2

Atom	*Wyckoff* site	*x/a*	*y/b*	*z/c*	U_{eq}/pm^2
Y1	4*d*	0.15105(9)	0.47099(8)	1/4	96(2)
Y2	4*c*	0.61339(9)	1/4	0	94(2)
S	4*d*	0.4612(3)	0.1687(2)	1/4	134(3)
Si	4*c*	0.1189(3)	1/4	0	75(3)
O1	8*e*	0.0439(5)	0.7392(4)	0.3778(2)	92(6)
O2	8*e*	0.2814(5)	0.4318(4)	0.4706(2)	105(6)

ISC-P32

Crystal structure and phase transition of $BaCaBO_3F$

M. Burianek[1], C. Niekamp[1], M. Teck[1], I. Spieß[1], O. Medenbach[2], L. A. Fischer[3], P. E. Wolff[3], J. Neumann[4], R. X. Fischer[1]
[1]*Universität Bremen, Fachbereich Geowissenschaften, Bremen, Germany*
[2]*Ruhr-Universität Bochum, Institut für Geologie, Mineralogie und Geophysik, Bochum, Germany*
[3]*Leibniz Universität Hannover, Institut für Mineralogie, Hannover, Germany*
[4]*Fraunhofer Institute for Manufacturing Technology and Advanced Materials, Oldenburg, Germany*

$BaCaBO_3F$ (BCBF) is an important material with interesting nonlinear optical properties. It was first synthesized and described by Keszler et al. [1] in the hexagonal space group P-62m, confirmed by Xu et al. [2] who refined the crystal structure of a Czochralski grown crystal by powder diffraction methods. A first single-crystal study was performed by us [3] showing that BCBF is monoclinic (space group C2) at room temperature representing a superstructure of the hexagonal BCBF. Li and Zeng [4] reported a reversible phase transition at 242 °C with the hexagonal high-temperature structure and a trigonal structure in space group R3 at room temperature with lattice parameters tripled in all three dimensions.

Meanwhile we have grown new crystals of BCBF by the Czochralski method. Starting with a platinum rod as replacement for a seed crystal single crystals were developed along a preferred growth direction. BCBF was grown both under air and under nitrogen atmosphere. While the crystals grown in air are colorless, the crystals grown under nitrogen atmosphere exhibit a light blue/turquoise color. Pulled with a rate of 0.25-0.5 mm/h crystals reach lengths of 45 mm with a diameter of 12 mm.

The grown crystals (Fig. 1a) were characterized by thermal analyses, electron microprobe analysis (EPMA), optical microscopy, single-crystal X-ray diffraction, and high-temperature powder diffraction. A weak endothermic reaction was detected at 242.7 °C for the crystal grown in air indicating the phase transition, in perfect agreement with the observation in [4]. Upon cooling, the melt-grown crystal decomposes forming a lamellar structure shown in Fig. 1b. EPMA showed a homogeneous distribution of the elements with a slight excess of BaO (60.12 ± 0.32 wt%) and a slight deficiency of B_2O_3 (12.25 ± 0.05 wt%). Investigations with the petrographic microscope with conoscopic illumination confirmed the monoclinic character with a biaxial interference figure (Fig. 1c).

[1] Keszler, D.A. et al. (1994) Mat. Res. Soc. Symp. Proc. 329, 15-22.
[2] Xu, K. et al. (2009) J. Cryst. Growth 311, 2508-2512
[3] Teck, M. et al. (2013) 21st Annual Conference of the German Crystallographic Society, Abstracts p. 89-90, Freiberg, Germany.
[4] Li, R.K., Zeng, Q.D. (2013) J. Cryst. Growth 382, 47-51

Fig. 1 a) Left: BCBF (l=40 mm) grown in air; right: BCBF (l=40 mm) grown under nitrogen atmosphere. b) Lamellar structure under crossed nicols c) biaxial interference figure

Figure 1

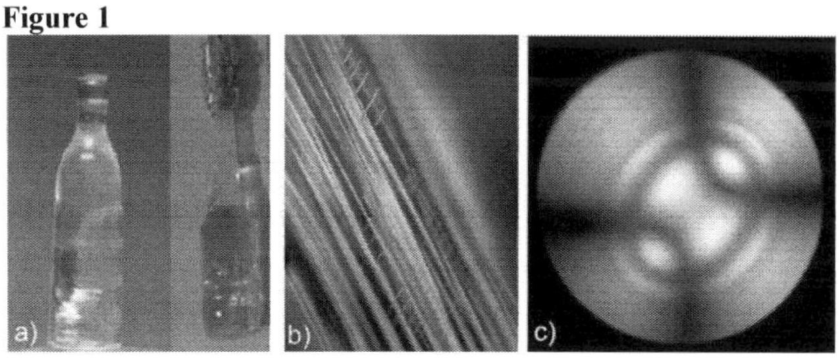

ISC-P33
Thermal expansion, bulk and atomistic Debye-temperatures of $|Na_8I_2|[AlSiO_4]_6$ determined by temperature-dependent neutron TOF and X-ray scattering.

L. Robben[1], I. Abrahams[2], T. M. Gesing[1]
[1]Universität Bremen, Chemische Kristallographie fester Stoffe, Bremen, Germany
[2]Queen Mary University of London, Materials Research Institute, School of Biological and Chemical Sciences, London, United Kingdom

The temperature-dependent structural behavior of aluminosilicate iodide sodalite $|Na_8I_2|[AlSiO_4]_6$ was examined by TOF neutron and X-ray scattering, and temperature-dependent Raman-spectroscopy. Special attention is paid to the interaction between the template ions (Na^+, I^-) and the framework.

Temperature-dependent neutron TOF diffraction patterns from 5 to 290 K and X-ray diffraction patterns from 298 K to 1300 K are used to determine structural and lattice parameters by Rietveld refinement. The title compound shows a phase transition from space group P-43n to Pm-3n at 1023 K and decomposes at 1173 K [1]. A Debye-Einstein model [2] was fitted against the temperature-dependent volume behavior and the thermal expansion coefficients, giving a Debye-temperature of Θ_D = 510 K and an Einstein temperature of Θ_E = 2450 K (see figure). The temperature dependent development of the Debye-Waller factors obtained from the scattering data was fitted using a Debye/Einstein/anharmonicity-model according to:

$$iso>^2 = c + D + A$$

With c being a constant describing contributions resulting from zero-point-energies and atomic disorder, D is the Debye-Waller-factor and A describes contributions from the Einstein approximation and the high-temperature anharmonicity. The Debye-Waller factor is calculated by

$$D = (424/M_A) * (T/\Theta)*[(\Theta/(4T)+(T/\Theta)*\int(y/(e^y-1))dy]$$

Where T is the temperature and M_A the atomic mass (taken from p. 191 in [3], with the lower integration boundary being 0 and the upper one Θ/T). The high-temperature Einstein/anharmonicity contribution is taken into account by a quartic function with a constant c_A [4]:

$$A = c_A*(T/\Theta)^4$$

The atomistic Debye temperatures for each species are determined by this model. The mean atomistic Debye-temperature ($\Theta_{D,atom}$), determined on the basis of the TOF data alone, of 654 K is in reasonable agreement with the one obtained from the lattice expansion fit. The reliability of the Debye-Waller factors obtained from X-ray diffraction data in this context is discussed. Temperature-dependent Raman spectra from 80 K to 800 K as well as force-field simulations are discussed.

[1] M.J. Dempsey, D. Taylor, Distance Least Squares Modelling of the Cubic Sodalite Structure and of the Thermal Expansion of Na8(Al6Si6O24)I2, Phys. Chem. Miner. 208 (1980) 197-208.
[2] M.M. Murshed, C.B. Mendive, M. Curti, G. Nenert, P. Kalita, A. Cornelius, et al., Anisotropic lattice thermal expansion of PbFeBO4: a study by X-ray and neutron diffraction, Raman spectroscopy and DFT calculations, Mater. Res. Bull. 59 (2014) 170-178.
[3] A. Guinier, X-Ray Diffraction in Crystals, Imperfect Crystals, and Amorphous Bodies, San Francisco, 1963.
[4] A. Maradudin, P. Flinn, Anharmonic contributions to the Debye-Waller Factor, Phys. Rev. 129 (1963) 2529-2547.

Figure: Lattice volume expansion of $|Na_8I_2|[AlSiO_4]_6$ fitted with a Debye-Einstein model.

Figure 1

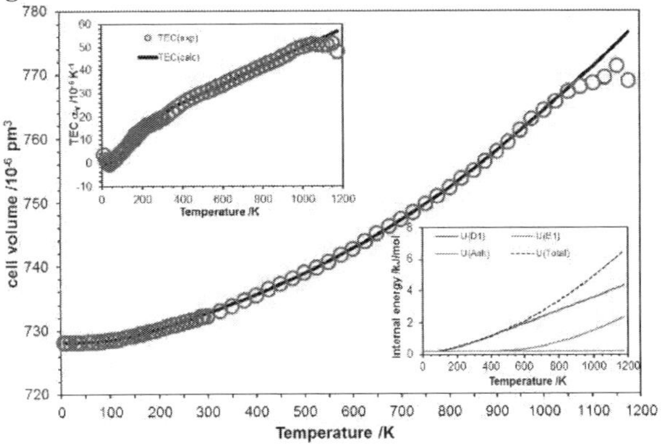

cell volume /10⁻⁶ pm³
Temperature /K
TEC(exp)
TEC(calc)
Temperature /K
U(D1)
U(E1)
U(Anh)
U(Total)
Internal energy /kJ/mol
Temperature /K

ISC-P34

Impossible H...H Contacts in Published Carboxylic Acid and Carboxylic Acid Complex Structures

C. Schwalbe[1]
[1]*Cambridge Crystallographic Data Centre, Cambridge, United Kingdom*

Because of their weak scattering power, hydrogen atoms are difficult to locate accurately from X-ray diffraction data. The difficulty is compounded in the presence of metal atoms. Therefore H atoms are often placed in calculated positions. However, the SHEXL manual warns about the possibility that "the automatic placing of hydrogen atoms has assigned the hydrogens of two different O-H or N-H groups to the same hydrogen bond" [1], and CheckCIF includes a test [2] of D-H...H-D contacts that gives a PLAT417 Class A alert if the H atoms are closer than 1.8 Å.

Nevertheless, a survey of the Cambridge Structural Database (version 5.35 with updates to February 2014) revealed 82 independent carboxylic acid structures with $R \leq 0.1$ and without gross (non-hydrogen) disorder that appeared to violate this criterion. The O=C-OH...HO-C=O arrangement in 61 of these did not form a pair-bonded dimer: although the great majority of their carboxyl groups had the *syn* conformation, adjacent molecules were translated or rotated so as to keep the OH groups together and the C=O groups apart, precluding such dimer formation. While close proximity of partially occupied H atom sites is feasible, as is sharing of a single fully occupied site between two O atoms, 27 non-dimeric structures (all published in 2000 or later, and 15 with metal ions) had an unacceptable clash between fully occupied H sites. Two papers even characterised such interactions as an alternative hydrogen bonding motif, $O^{-H}\ldots{}_{H-}O$ with graph set $R_2^2(4)$. In a number of such cases it is possible to spot locations where additional H atoms could be placed to compensate for the necessary reduction in occupancy of the clashing sites. An additional 5 structures (4 with metal ions) have correct occupancy at existing sites but lack designated alternative sites, yielding fewer H atoms in the unit cell than the molecular formula demands.

It is evident that, even in this era of validation software, authors, referees and readers must be on guard against COOH...HOOC clashes.

[1] http://shelx.uni-ac.gwdg.de/SHELX/shelx97.pdf, page 2-7. Accessed 11 December 2014.
[2] http://journals.iucr.org/services/cif/datavalidation.html . Accessed 11 December 2014.

ISC-P35

Crystal structure of the luminescent oxonitridosilicates $BaLa_3Si_5O_2N_9:Ce^{3+}$ and $La_{3-x}Ba_xSi_6N_{11-x}O_x:Ce^{3+}$ ($x \approx 0.1$)

L. Neudert[1], D. Durach[1], O. Oeckler[2], W. Schnick[1]
[1]*LMU Munich, Munich, Germany*
[2]*University Leipzig, IMKM, Leipzig, Germany*

The luminescence properties of rare earth doped oxonitridosilicates correlate with their crystal structures. Thus, precise structure determination of such phosphors is important, but challenging when syntheses lead to inhomogeneous microcrystalline samples. Single-crystal X-ray datasets of oxonitridosilicate crystallites pre-characterized by transmission electron microscopy (TEM), electron diffraction and X-ray spectroscopy were collected using a microfocused synchrotron beam. The refined structure models were confirmed with Z-contrast (STEM-HAADF) and high-resolution TEM images. The element distribution O/N or Ba/La on mixed occupied sites were further analyzed by bond-valence sum calculations.

$BaLa_3Si_5O_2N_9:Ce^{3+}$, with yellow luminescence, exhibits an orthorhombic (space group $Pmn2_1$ with a = 9.551 Å, b = 19.08 Å, c = 12.11 Å; R1 (all) = 0.036, wR2 (all) = 0.088) interrupted three-dimensional network built up of vertex-sharing Q^4-type and Q^2-type SiN_4 and SiN_2O_2 tetrahedra (cf. Figure, right). Condensation of these tetrahedra results in *dreier* and *zehner* rings (figure left: black and gray "bonds", respectively) whose topological arrangement [1,2] is represented by the point symbol $\{3.6^3.7^2\}_2\{3\}\{6^6\}_2$. There are nine crystallographically independent mixed Ba/La sites with coordination numbers of 7 - 9, three of which are split positions.

The green luminescent material $La_{3-x}Ba_xSi_6N_{11-x}O_x:Ce^{3+}$ ($x \approx 0.1$) crystallizes in the $La_3Si_6N_{11}$ structure type [3,4] (space group $P4bm$ with a = 10.193 Å, c = 4.838 Å; R1(all) = 0.021, wR2 (all) = 0.048). The non-centrosymmetric network is built up of $Si(O,N)_4$ tetrahedra that share all vertices and form layers parallel (001). These contain *vierer* and *achter* rings centered by mixed La/Ba sites and are interconnected via pairs of $Si(O,N)_4$ tetrahedra.

Figure: Topological representation of the network of $BaLa_3Si_5O_2N_9:Ce^{3+}$ (left, view along ~ [100], each line represents a Si-N,O-Si bond), and representation with polyhedra (right, projection along [100]).

[1] V. A. Blatov, M. O'Keeffe, D. M. Proserpio, *CrystEngComm* 2010, *12*, 44-48.
[2] V. A. Blatov, *IUCr CompComm Newsletter* 2006, *7*, 4-38.
[3] T. Seto, N. Kijima, N. Hirosaki, *ECS Transactions* 2009, 25, *9*, 247-252.
[4] M. Woike, W. Jeitschko, *Inorg. Chem.* 1995, *34*, 5105-5108.

Figure 1

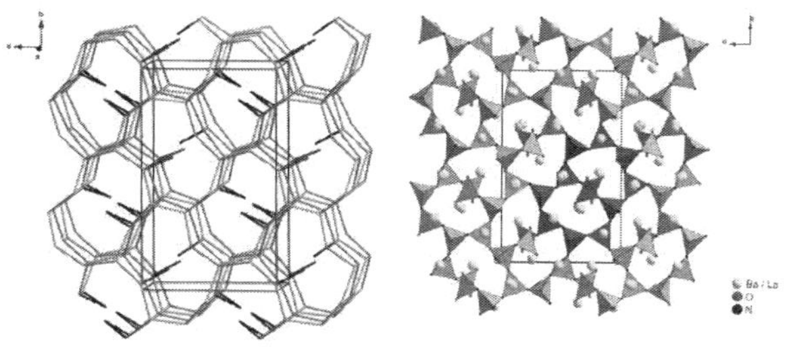

ISC-P36

A New Nitride Selenide Chloride of Lanthanum: $La_4NSe_3Cl_3$

M. Foltin[1], M. Meyer[1], F. Lissner[1], T. Schleid[1]
[1]*University of Stuttgart, Institute for Inorganic Chemistry, Stuttgart, Germany*

A huge variety of nitrogen-based compounds containing lanthanides, chalcogens and/or halogens in different compositions [1] share one structural feature, since lanthanido-ammonium tetrahedra $[NLn_4]^{9+}$ can be found in several connection motifs in each of these structures. Dimers, one-dimensional chains or even two-dimensional layers can be observed as prominent examples for those. In the case of the short quaternary series $Ln_4NS_3Cl_3$ (Ln = La - Nd) [2] the $[NLn_4]^{9+}$ tetrahedra appear in an unconnected form as isolated monomers, which are surrounded and charge-balanced by S^{2-} and Cl^- anions (Figure1).

$La_4NSe_3Cl_3$, which crystallizes isotypically with $La_4NS_3Cl_3$ [2] (Table1), adopts the polar hexagonal space group $P6_3mc$ (no. 186) with the lattice parameters a = 951.23(4) and c = 722.06(3) pm for Z = 2. Two crystallographically different La^{3+} cations, occurring with coordination numbers of 10 for $(La1)^{3+}$ and 8 for $(La2)^{3+}$, are surrounded by all three types of anions in the shape of capped trigonal prisms. Due to the isoelectronic anion pair S^{2-}/Cl^{-}, which is almost impossible to differentiate by X-ray diffraction, it was quite difficult to solve the positional problems of both anions in the already known short $Ln_4NS_3Cl_3$ series [2]. For this kind of question both the isotypic $Ln_4NS_3Br_3$ series (*Ln* = La - Pr) [3] and the newly obtained $La_4NSe_3Cl_3$ example give a good opportunity to confirm the suggestions, which were already made for this structure type yielding interatomic distances such as *d*(La-N) = 233 - 241 pm, *d*(La-Se) = 301 - 362 pm and *d*(La-Cl) = 299 - 338 pm.

$La_4NSe_3Cl_3$ was obtained by reactions of lanthanum metal with selenium, sodium azide, lanthanum trichloride and sodium chloride as flux in evacuated silica tubes. Appropriate molar ratios were chosen for these reactions, which ran for seven days at 900 °C. After subsequent slow cooling, brown lath-shaped crystals could be detected and selected for single crystal X-ray diffraction with following structure analysis.

Figure 1. Isolated $[NLa_4]^{9+}$ tetrahedron in the crystal structure of $La_4NSe_3Cl_3$ with its full anionic surrounding.
Table 1. Fractional atomic coordinates and U_{eq} values for $La_4NSe_3Cl_3$.

References
[1] Th. Schleid, F. Lissner, *J. Alloys Compds.* 2008, *451*, 610-616.
[2] F. Lissner, Th. Schleid, *Z. Anorg. Allg. Chem.* 1994, *620*, 1998-2002.
[3] M. Meyer, Th. Schleid, *Z. Kristallogr.* 2000, *Suppl. 17*, 179-179.

Figure 1

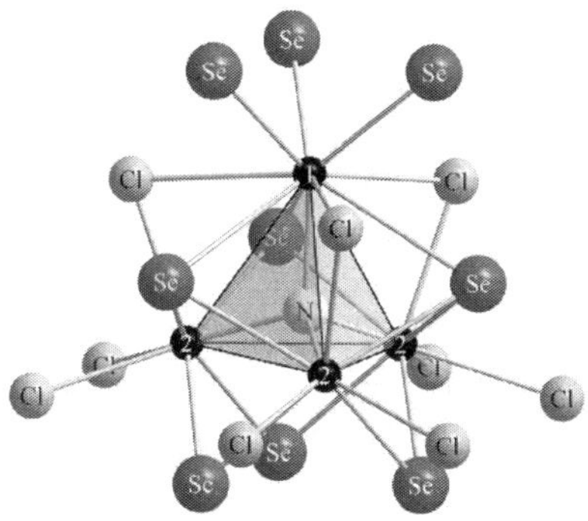

Figure 2

Atom	Site	*x/a*	*y/b*	*z/c*	U_{eq}/pm²
La1	2*b*	⅓	⅔	0.68715(15)	204(3)
La2	6*c*	0.19948(3)	−*x/a*	¼ a)	126(2)
N	2*b*	⅓	⅔	0.3531(19)	103(26)
Se	6*c*	0.47105(6)	−*x/a*	0.96033(14)	154(2)
Cl	6*c*	0.12963(17)	−*x/a*	0.6326(3)	214(6)

a) arbitrarily fixed for a proper definition of the origin.

ISC-P37
Synthesis and structure of two new crystalline silicic acids: H-Carletonite, $H_{32}Si_{64}O_{144}$ and Hydrated H-Apophyllite, $H_{16}Si_{16}O_{40} \cdot 8\ H_2O$

B. Marler[1]
[1]*Ruhr University Bochum, Dept. of Geology, Mineralogy and Geophysics, Bochum, Germany*

Crystalline silicic acids are layered silicates which are prepared by acid leaching of alkaline and alkaline earth layer silicates. Synthetic materials as well as minerals have been used as parent materials, e.g. Apophyllite, Magadiite, Kenyaite, α-$Na_2Si_2O_5$, etc. The ion exchange of Na^+, Ca^{2+}, etc. against H^+ leads to silicic acids like H-Magadiite, H-Kenyaite or Disilicic Acid I. The crystal structures of mostcrystalline silicic acids, including the ones ofH-Apophyllite and H-Carletonite [1], remained unsolved due to severe structural disorder. So far, the structures of only three silicic acids have been solved: H-LDS [2], H-RUB-18 [3] and Disilicic Acid I [4].

H-Carletonite and Hydr. H-Apophyllite were synthesized by leaching natural Carletonite, $KNa_4Ca_4[Si_8O_{18}(CO_3)_4(OH,F)]*H_2O$, or Apophyllite crystals, $KCa_4[Si_8O_{20}(OH/F)]*8H_2O$, in a large surplus of 1.2 molar hydrochloric acid. Physico-chemical characterization by TG/DTA, NMR and FTIR spectroscopy confirmed that H-Carletonite and Hydr. H-Apophyllite are hydrous layer silicates. The XRD powder patterns of both materials were indexed in tetragonal symmetry with a_0 = 8.49 Å, c_0 = 16.89 Å, space group P4/ncc (Hydr. H-Apophyllite) and with a_0 = 13.49 Å, c_0 = 20.47 Å, space group I4/mcm (H-Carletonite). The crystal structures were solved from model building. Rietveld refinements of the structures converged to residual values of R_{Bragg} = 0.033, χ^2 = 1.3 (Hydr. H-Apophyllite) and R_{Bragg} = 0.035, χ^2 = 2.3 (H-Carletonite) confirming the structure models. The structures of both silicic acids contain silicate layers of the same topology as present in their particular parent phases.

The silicate monolayer of Hydr. H-Apophyllite can be regarded as made up by interconnected 4-rings which point alternatively up and down and form a puckered silicate layer. A sheet of water molecules is intercalated between the silicate layers. Together with the silanol groups the water molecules form a complex system of hydrogen bonds. At room temperatureHydr. H-Apophyllitedecays within one day to form a nearly amorphous product.

The structure of H-Carletonite consists of much thicker silicate layers. These rigid and stable layers are interconnected by strong hydrogen bonds between the terminal O atoms of neighboring layers (d(O…O) = 2.4 and 2.7 Å). No structural water is present in H-Carletonite which is stable at room temperature for at least several years.

References
[1] G. Lagaly and R. Matouschek: *N. Jb. Miner. Abh.* 138 (1980) 81-93. [2] T. Ikeda et al., *Chem. Lett.* 39 (2010) 747-749. [3] M. Borowski et al., *Z. Kristallogr.* 217 (2002) 233-241. [4] F. Liebau, *Z. Kristallogr.* 120 (1964) 427-449.

Fig. 1: The structures of Hydrated H-Apophyllite (left) and H-Carletonite (right) projected on (100).

Figure 1

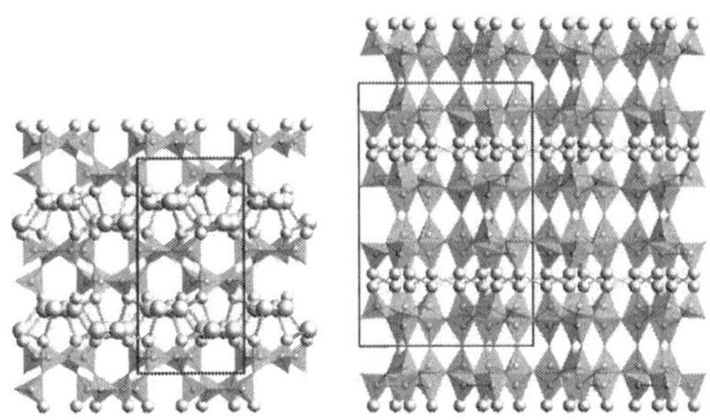

MSC - Material Science

MSC-P01

Halogengold(I)-Amine Complexes and their Oxidation Products

C. Döring[1], P. G. Jones[1]
[1]*Technische Universität Braunschweig, Institut für Anorg. und Analyt. Chemie, Braunschweig, Germany*

Halogengold(I)-amine complexes were synthesized by dissolving and crystallizing thtAuX (X = chlorine [1] or bromine) in a primary or secondary amine or azaaromatic. A molecular form (**2**) and three different ionic forms (**1, 3** and **4**) were obtained.

The products were oxidized, in the case of bromine, directly by the halogen and, in the case of chlorine, by phenyliodine(III) dichloride, which can be used stoichiometrically.

Most complexes were obtained as mixtures and/or in poor yields. They display a variety of forms, from molecular (**11**) to ionic (**5-9** and **12**).

Scheme 1. Possible forms of gold(I) and gold(III) complexes with amines and halogenide groups.

The structures of the products were analysed by X-ray structure determination and their packing was investigated with reference to hydrogen bonds, aurophilic and halogen contacts. In structures with N-H groups, classical hydrogen bonds between amine H atoms and the free halogenides or those bound to gold are generally formed. In some cases short gold-gold contacts around 3.3 Å, gold-halogen and halogen-halogen contacts between 3.2 Å and 3.9 Å occur, which often build chains and are therefore a central element of the packing. These are further linked by hydrogen bonds and occasionally by gold-hydrogen contacts to form ribbons, layers or three-dimensional networks.

[1] B. Ahrens, S. Friedrichs, R. Herbst-Irmer, P. G. Jones, *Eur. J. Inorg. Chem.* 2000, 2017 and refs. therein.

Figure 1

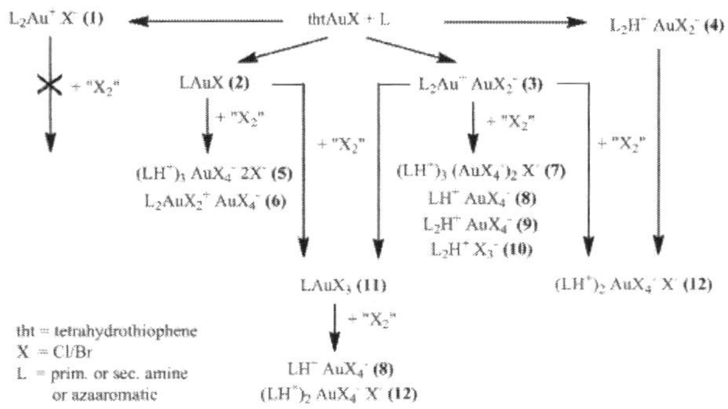

MSC-P02

Solvothermal synthesis, structure, and properties of New dioxorhenium(V) compound with aromatic amines

A. Kochel[1]
[1]*University of Wrocław, Faculty of Chemistry, Wrocław, Poland*

The synthesis, structure, and properties of $[ReO_2(py)_2bpy]Cl{\cdot}H_2O$ will be presented. The compound crystallizes as a monoclinic crystal system, $P2_1/c$ space group a=11.9948(2), b =11.9948(2) c=7.7630(3) Å, V=1138.29(6) Å^3, Z=2.

The geometry around the Re atom is octahedral, with four N atoms of pyridine and bpy

molecules and the Re atom defining the equatorial plane. The Re—O 1.768(3) Å distance is similar to those found in other oxo rhenium(V)-containing compounds[1].

The water molecules and Cl^- anions form intermolecular hydrogen bonds, giving a three-dimensional structure.

Figure 1.

View of the $[ReO_2(py)_2bpy]^+$ showing 30% displacement ellipsoids (arbitrary spheres for the H atoms), symmetry codes i (-x, y, -z,).

[1] Lock, C. J. L. & Turner, G. (1978). Acta Cryst. B34, 923-927.

Luck, R. L. & O'Neill, R. S. (2001). Polyhedron, 20, 773-782

Figure 1

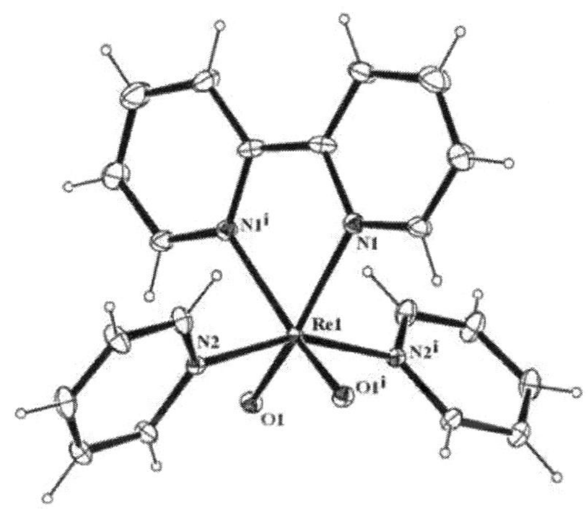

MSC-P03

Towards a crystal undulator

J. Härtwig[1,2], S. H. Connell[2], D. Mavunda[2,3], T. N. Tran Thi[2,4], W. Lauth[4]
[1]*ESRF - The European Synchrotron, Grenoble, France*
[2]*University of Johannesburg, Physics, Johannesburg, South Africa*
[3]*NECSA, PRETORIA, South Africa*
[4]*University of Mainz, Mainz, Germany*

A crystal undulator is similar to a normal undulator as typically found at a synchrotron to produce extremely brilliant X-ray beams. The difference is that the magnetic lattice is realized by the periodic electrostatic potential of a crystal lattice seen from the reference frame of the GeV range electron or positron beam. The extremely relativistic incident particle beam would need to be captured in a high index crystallographic channel of a crystal superlattice. The particle beam will then "see" a many Tesla range periodically varying magnetic field with a few micrometre pitch. This method could theoretically lead to a MeV range gamma ray laser by the FEL principal.

The most important problem for the moment is the creation of a "wiggling" lattice deformation in a suitable crystalline sample with an as high as possible degree of its crystalline quality. The first attempts were done with one sided and two sided scoring of silicon plates by mechanical kerfing. After this we used femtosecond laser ablation to producing statically bent diamonds. Both attempts were not satisfying. The first reason was that too much lattice damage was produced. The second one was that the strain sources on the crystal surfaces resulted in a resulting strain field that decreased as a function crystal depth, resulting in a distribution of undulator amplitudes within the crystal. So the idea appeared to use a crystal superlattice.

Because the "very robust" diamond is the best suited material for the planned application in a harsh environment (particle accelerator for the powerful incoming electron or positron beam), this material was preferred and selected to go in the direction of a diamond superlattice. We started to prepare the necessary technology for manufacturing of a prototype diamond superlattice as well as for the necessary X-ray techniques for the characterisation of its crystalline quality, which is by very strain sensitive X-ray diffraction topography (double crystal X-ray topography, plane wave topography).

The undulator fabrication principle involved CVD growth of diamond on a diamond substrate while varying the concentration of boron in the gas phase during growth. This should lead to the periodic variation of the lattice dilatation by the varying concentration of the single substitutional boron impurity atoms. The validation via X-ray diffraction topography proved non-trivial but was eventually promising. The first step was to produce a 50μm thick diamond graded layer on a type Ib diamond substrate by Element Six. A single graded boron doped layer was produced by varying the boron concentration up to $\sim 7 \times 10^{20}$ at cm^{-3} (0.3% or 3000 ppm). A second sample was capped with an "intrinsic" layer. Some first X-ray topography results will be presented.

MSC-P04

Formation of Ni(II) Oxime-Bridged Basket-Like Complexes and Their Structural Aspects

M. Holynska[1]
[1]*Philipps-Universität Marburg, Chemistry, Marburg, Germany*

Metal complexes with pyridyl oximes as versatile N,O-donor ligands have been studied extensively during the recent years. Nevertheless, they do continue surprising with new possibilities and can be expected to reveal even more interesting features in the future.

Phenyl 2-pyridyl ketoxime (ppkoH) is a representative example of pyridyl oxime ligands for which a variety of complexes with transition metal / main group metals or lanthanides are known. Particular tendency to form oxime complexes is observed in Ni(II) compounds [1].

In this contribution two new Ni(II) complexes with phenyl 2-pyridyl ketoxime ligand are reported, displaying unprecedented basket-like topologies of their complex cores (Figure 1). The compounds are studied by X-ray diffraction, EDX and IR spectroscopy, confirming uncommon coordination of a DMF ligand in one of them. The reaction system is described in detail and some of the products are analyzed by EDX/SEM, including EDX mapping analyses of the elemental distribution.

Figure 1 The basket-like cores in the title compounds. Part of the cores is shown with different transparency, for clarity.

Acknowledgment
Financial support from DFG and Philipps University Marburg.

[1] Milios, C.J.; Stamatatos, T.C.; Perlepes, S.P. The coordination chemistry of pyridyl oximes. Polyhedron, 2006, 25, 134-194.

Figure 2

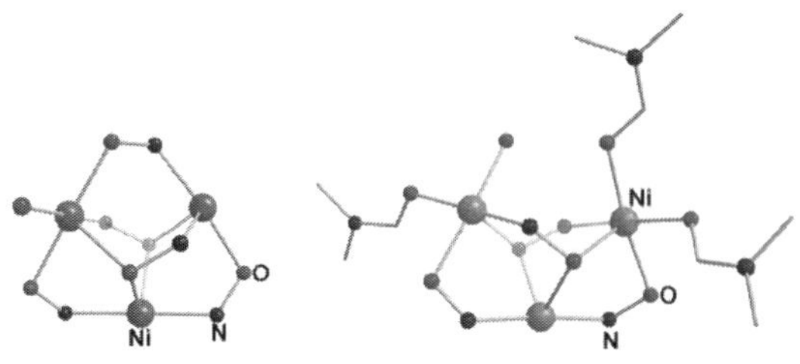

MSC-P05

Structural Complexity of Th Compounds Built from Tellurite Polyhedra and MO_4^{2-}(M = S^{6+}/W^{6+}/Mo^{6+}) Tetrahedra

B. Xiao[1], E. Alekseev[1]
[1]*Forschungszentrum Jülich GmbH, Jülich, Germany*

Solid-state materials accommodating lone pairs such as Te^{4+} and Se^{4+} is of current interest and great importance attributable to their rich structural chemistry. Up to now, a vast number of molybdotellurites and tungstotellurites containing alkali, alkaline earth, transition and rare earth metals have been isolated.[1-5] In contrast, actinide molybdotellurites and tungstotellurites remain in deep freeze, and until now there is no crystal structure reported in this family. It is believed that inclusion of actinide elements has great chance of obtaining unusual crystal structural because of the versatility of coordination geometries and a wide variety of oxidation state of actinide element.[6] Based on this, a series of novel thorium compounds containing mixed oxo-anions (S^{6+}/W^{6+}/Mo^{6+} with Te^{4+}) were synthesized and characterized. We found that the MoO_4^{2-} tetrahedra play the role of interlayer linkers in thorium molybdotellurite system. For the tungsotellurites system, the structures of thorium with novel Th spiral chains show a high level of complexity. The structures of thorium tellurium sulfates are as similar as those in thorium selenite/selenates system with small changes. $Na_2Th(MO_4)(Te_2O_5)_2$ (M = S, Mo, W) consists of $[Na_2Th(Te_2O_5)_2]^{2+}$ cationic framework with a system of channels (≈ 5.5 Å × 5.5 Å). The highly disordered MO_4^{2-} entities are resided in channels and they do not covalently bond to any framework atoms. The structural configurations of these compounds suggest the potential for anion exchange behavior.

Figure 1. View of structure of $Na_2Th(MO_4)(Te_2O_5)_2$ (M = S, Mo, W). Th polyhedra are in yellow, Te polyhedra are in blue and M polyhedra are in green. The sodium atoms are in black nodes.

References

1. Chi, E. O.; Ok, K. M.; Porter, Y.; Halasyamani, P. S., *Chem. Mater.* 2006, *18*, 2070.
2. Vidyavathy; Vidyasagar, K., *Inorg. Chem.* 1999, *38*, 1394.
3. Vidyavathy; Vidyasagar, K., *Inorg. Chem.* 1999, *38*, 5809.
4. Dussack, L. L.; Harrison, W. T. A.; Jacobson, A. J., *Mater. Res. Bull.* 1996, *31*, 249.
5. Oh, S.-J.; Lee, D. W.; Ok, K. M., *Inorg. Chem.* 2012, *51*, 5393.
6. Burns, P. C.; Ewing, R. C.; Hawthorne, F. C., *Canadian Mineralogist* 1997, *35*, 1551.

Figure 1

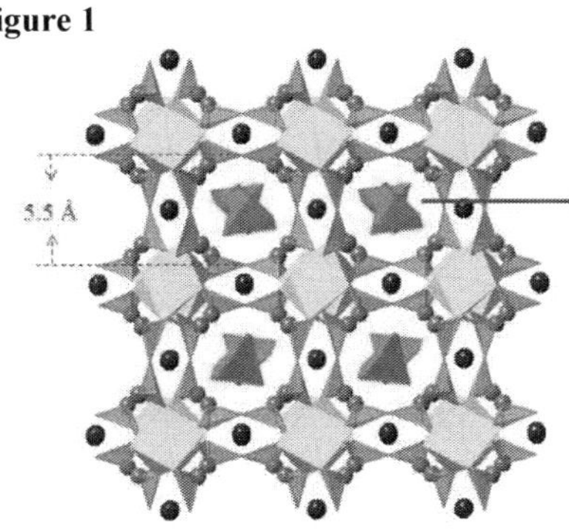

$Na_2Th(MO_4)(Te_2O_5)_2$ (M = S, Mo, W)

References
[1] E.V. Alekseev, O. Felbinger, S. Wu, T. Malcherek, W. Depmeier, G. Modolo, Th.M. Gesing, S.V. Krivovichev, E.V. Suleimanov, T.A. Gavrilova, L.D. Pokrovsky, A.M. Pugachev, N.V. Surovtsev, V.V. Atuchin, J. Solid State Chem. 204 (2013) 59.
[2] P. Zhao, M.M. Murshed, E.V. Alekseev, V.V. Atuchin, A.M. Pugachev, Th.M. Gesing, Mater Res Bull. 60 (2014) e258
[3] P. Zhao, M.M. Murshed, A. Huq, H.K. Grossmann, L. Mädler, E.V. Alekseev, Th.M. Gesing, submitted.
[4] A. Hussain, Chem Scripta. 11 (1977) 224.
[5] M.M. Murshed, C.B. Mendive, M. Curti, G. Nénert, P.E. Kalita, K. Lipinska, A.L. Cornelius, A. Huq, T.M. Gesing, Mater Res Bull. 59 (2014) 170.

MSC-P06

Thermal expansion behaviors of three alkali metal asenotungstates

P. Zhao[1], M. M. Murshed[1], A. Huq[2], E. V. Alekseev[3,4], T. M. Gesing[1]
[1]*University of Bremen, Chemische Kristallographie fester Stoffe, Institut für Anorganische Chemie, Bremen, Germany*
[2]*Oak Ridge National Laboratory, Chemical and Engineering Materials Division, Oak Ridge, Tennessee, United States*
[3]*Forschungszentrum Jülich, Institute of Energy and Climate Research (IEK-6), Jülich, Germany*
[4]*RWTH Aachen, Institut für Kristallographie, Aachen, Germany*

Materials with anisotropic thermal expansion, in particular with axial negative thermal expansion are of considerable interest to tune thermal expansion coefficients to be adjustable to electrical, optical and precision devices. Three alkali metal arsenotungstates $KAsW_2O_9$ [1], $NaAsW_2O_9$ [2] and $Li_3AsW_7O_{25}$ [3] were produced using solid state reactions. Both $KAsW_2O_9$ and $NaAsW_2O_9$ crystallize in the non-centrosymmetric space group $P2_12_12_1$, and $Li_3AsW_7O_{25}$ in the centrosymmetric space group *Pbca*. The crystal structure of lithium arsenotungstate possesses unique nanoscale building blocks similar to that of intergrowth tungsten bronzes [4]. Lattice thermal expansions of these compounds have been studied using temperature-dependent powder X-ray and neutron diffractions between 10 K and 973 K. Although $NaAsW_2O_9$ and $KAsW_2O_9$ are isotypic their thermal expansion behavior is significantly different. $KAsW_2O_9$ displayed anisotropic thermal expansion with contraction along the **b**- and expansion along the **a**- and **c**-axes, leading to an overall thermal expansion of the cell-volume. On the other hand, $NaAsW_2O_9$ displayed expansion anomaly with a broad trough in the temperature-dependent thermal expansion coefficient (TEC) of the cell volume. $Li_3AsW_7O_{25}$ showed positive coefficients of thermal expansion for respective metric parameter for the investigated temperature range. The thermal expansion of the metric parameters have been modeled using first-order Grüneisen approximation to the zero-pressure equation of state, where the change of the vibrational energy was calculated with a Debye-Einstein-Anharmonicity (DEA) model [5]. The model includes harmonic, quasi-harmonic and intrinsic anharmonic contributions to the internal energy as a function of temperature. The Raman and the infrared spectra were analyzed based on the density functional theory (DFT) calculation of the phonon modes, which help set the Debye and/or Einstein characteristic temperatures. The lack of definite saturation of TEC at high temperature and the expansion anomaly justified the anharmonicity term to be taken into account in the additive frame of the DEA model.

MSC-P07

Structural investigations of $CH_3NH_3PbI_3$

A. Franz[1], D. Toebbens[1], S. Schorr[1]
[1]*Helmholtz-Zentrum für Materialien und Energie, Kristallographie, Berlin, Germany*

Perovskites with ABX_3 - structure show a huge possibility on element substitutions on A-,B- and C-site which leads to a broad variety of physical properties. In recent years the interest become focused on hybrid perovskites as a future photovoltaic material. Our field of interest lays in lead methylammonium triodide in which A is the organic unit $[CH_3NH_3]^+$, $B=Pb^{2+}$ and $X=I_3^-$.

The perovskite crystal structures can be classified by their octahedral coordinated B-cation. The aristotype-structure (P m-3m) is symmetry lowered due to tilting, distortion of $[BX_6]$ - octahedra or displacement of B-cation from center of octahedron. These types of displacements are shown in the 'Bärnighausen-Stammbaum' [1] a tree diagram which was extended in 2002 by Bock & Müller [2] who presented a tree diagram where all group-subgroup relations of octahedral tilted space groups are taken into account, based on the work of Bärnighausen and 'Glazer's tilt systems' [3].

First structure analysis and Rietveld refinements confirmed the $MAPbI_3$ - perovskite belongs to the space group I 4/mcm. Further investigations showed that $MAPbI_3$ shows a highly disordered structure with large displacement factors of the anions. The analyses furthermore revealed a distortion and statistical distributed tilting of the octahedra. This in turn takes influence on the organic unit.

The synthesis was done by two different routes: first, single crystals as well as polycrystalline material was prepared by precipitation from hydroiodic solution, described by Poglitsch and Weber [4]. Second, polycrystalline perovskite samples were synthesized from an equimolar mixture of methylammonium iodide [5].

For structural investigations Synchrotron and Neutron diffraction experiments were performed at the Helmholtz-Centre Berlin for Materials and Energy. The data treatment was performed by F.O.X. and FullProf Suite software.

The presentation will give an overview of the results of this structural study with a certain focus on the statistic disorder of the methylammonium molecule, and anion displacement. Moreover, structural trends will be discussed.

1.Bärnighausen, H., *Group-subgroup relations between space groups: A useful tool in crystal chemistry.* MATCH Communications in Mathematical and in Computer Chemistry, 1980. 9: p. 139-175.

2.Bock, O. and U. Müller, *Symmetrieverwandtschaften bei Varianten des ReO3-Typs.* Zeitschrift für anorganische und allgemeine Chemie, 2002. 628(5): p. 987-992.
3.Glazer, A.M., *The classification of tilted octahedra in perovskites* Acta Crystallographica B, 1972. 28: p. 3384-3392.
4.Poglitsch, A. and D. Weber, *Dynamic disorder in methylammoniumtrihalogenoplumbates (II) observed by millimeter-wave spectroscopy.* The Journal of Chemical Physics, 1987. 87(11): p. 6373-6378.
5.Im, J.H., et al., *6.5% efficient perovskite quantum-dot-sensitized solar cell.* Nanoscale, 2011. 3(10): p. 4088-93.

MSC-P08

From square-planar to octahedral: Crystal Structures of Cycloplatinated Primary Amines in Octahedral Coordination

W. Raven[1], I. Kalf[1], U. Englert[1]
[1]*RWTH Aachen University, Institute for inorganic chemistry, Aachen, Germany*

The cycloplatination of primary amines became possible in 2008[1] and many Pt(II) complexes could be synthesized and crystallized, yet the analogous reaction using a Pt(IV)-precursor never yielded a reasonable product. Hence, we let square-planar platinum(II) compounds react with elemental iodine to undergo oxidative addition to obtain the corresponding octahedral platinum(IV) complexes. On the poster we will discuss this coordination sphere and its crystallographic properties on one example (see Figure 1)

Figure 1: Molecular structure of the organoplatinum-triiodido-triphenylphosphine-complex in the crystal; displacement ellipsoids are drawn at 50% probability, hydrogen atoms have been omitted for clarity.

[1] B. Calmuschi-Cula and U. Englert, *Organometallics* 27, 3124-3130, 2008

Figure 1

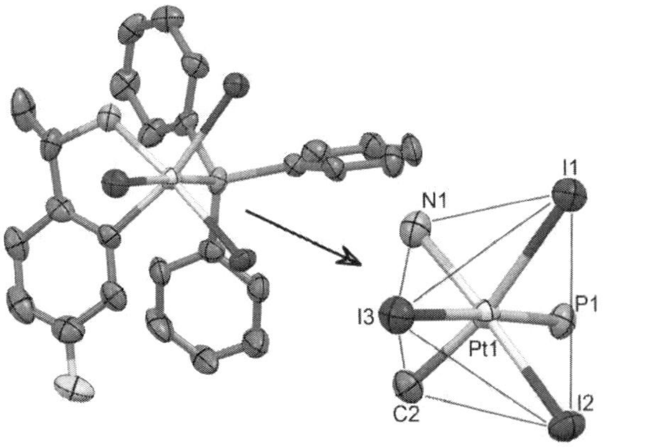

MSC-P09

Thermal expansion anomaly in lithium tungsten bronzes

M. S. Rahman[1], M. M. Murshed[1], M. Fischer[2], A. Huq[3], T. M. Gesing[1]
[1]*Universität Bremen, Chemische Kristallographie fester Stoffe, Institut für Anorganische Chemie, Bremen, Germany*
[2]*Universität Bremen, Kristallographie, FB Geowissenschaften, Bremen, Germany*
[3]*Oak Ridge National Laboratory, Oak Ridge, Chemical and Engineering Materials Division, Tennessee, United States*

Lithium tungsten bronzes (Li_xWO_3) have been known for years [1], exhibiting composition (x) dependent interesting electrical and optical properties [2, 3]. Of members of the alkali metal tungsten bronzes, Li_xWO_3 bronzes show peculiarities in terms of structural pitfalls and phase transformations [4]. Polycrystalline Li_xWO_3 (x = 0.1 and 0.4) bronzes were synthesized by solid state reactions using low pressure of 10^{-7} MPa and high temperature of 973 K for 7 days. Products were characterized by X-ray and neutron powder diffractions as well as Raman spectroscopy. The room-temperature diffraction data Rietveld refinements showed that $Li_{0.1}WO_3$ crystallized in the *Pcnb* space group and $Li_{0.4}WO_3$ in the $Im\bar{3}$ space group. Temperature-dependent *in-situ* neutron powder diffraction (NPD) data revealed nonlinear expansion of the metric parameters for both orthorhombic and cubic bronzes. Temperature-dependent anisotropy factor of the cell parameter of $Li_{0.1}WO_3$ showed a sigmoid behavior with a slowing down saturation at about 686(5) K. Moreover, the intensity of at least one Bragg reflection gradually decreased and finally disappeared at 680 K, which is a clear hint of high-temperature phase transition. Analysis of the low-temperature in-house powder X-ray diffraction data (PXRD) and Raman spectra also demonstrated anomaly in thermal expansion coefficient of the metric parameters as well as in Raman phonon modes between 200 K and 250 K. The NPD data Rietveld refinements allowed to unambiguously locate lithium atoms distributed on both *2a* and *6b* sites in $Li_{0.4}WO_3$ bronze. With increasing temperature the loosely bonded lithium in the *6b* site hopped to *2a* sites. These observations support the ionic conductivity of some lithium bronzes at high-temperatures [1]. Thermal expansion behavior of the metric parameters of $Li_{0.4}WO_3$ bronze showed a kink at low-temperature region. Phonon dispersion of fully ordered $Li_{0.75}WO_3$ (fully occupied *6b* position) calculated by density functional theory (DFT) showed an imaginary branch. This Γ-point instability seems to be a precursor to either phase transition, or occurrence of metastability due to local rotations of some WO_6-octahedra blocks leading to eventually a soft phonon for electron-phonon interaction as seen in superconducting bronzes [5].

References

[1] M. E. Straumanis, S. S. Hsu, J. Am. Chem. Soc. 72 (1950) 4027.
[2] M. J. Sienko, T. B. N. Troung, J. Am. Chem. Soc. 83 (1961) 3939.
[3] M. Green, Z. Hussain, J. Appl. Phys. 74 (1993) 3451.
[4] M. S. Rahman, M. M. Murshed, Th. M. Gesing, Z. Kristallogr. 229 (2014) 797.
[5] K. L. Ngai, T. L. Reinecke, J. Phys. F: Metal Phys. 8 (1978) 151.

MSC-P10

Tow Modifications of Tin(II) Bromide

P. Eckold[1], W. Hügel[1], R. Niewa[2]
[1]*Robert Bosch GmbH, Automotive Electronics, Stuttgart, Germany*
[2]*University of Stuttgart, Institute of Inorganic Chemistry, Stuttgart, Germany*

In 2000, Abrahams *et al.*[1] reported the crystal structure of tin(II) bromide, $SnBr_2$, which has been refined by Rietveld analysis of X-ray powder diffraction data and compared to other dihalides of Sn and Pb.[2-4] The structure was described as a packing of polymeric chains $^1\infty[SnBrBr_{2/2}]$ parallel to the *c*-axis vector.

We present the synthesis of single crystalline $SnBr_2$. Different temperature conditions in crystal growth of $SnBr_2$ lead to the formation of two modifications of the title compound. Applying a temperature gradient from 650 °C to room temperature, $SnBr_2$ (**1**) crystallizes in the orthorhombic space group *Pnma* (No. 62), with a = 839.08(4) pm, b = 423.54(2) pm, c = 1052.03(4) pm and Z = 4. Changing the reaction conditions towards a decreased temperature gradient from 700 °C to 400 °C, $SnBr_2$ (**2**) crystallizes in the hexagonal space group $P6_3/m$ (No. 176), with a = 1228.24(5) pm, c = 435.68(2) pm and Z = 6. The structural data of (**1**) is in agreement with the refined crystal structure data reported by Abrahams *et al.*[1] The trigonal coordination of Sn with three Br atoms and the resulting one-dimensional chains of corner sharing units $^1\infty[SnBrBr_{2/2}]$ can be assigned as common structural motifs of both modifications. In (**1**) pairs of these chains exhibit the motif of a hexagonal rod packing parallel to the crystallographic b-axis. In (**2**) the chains are rearranged resulting in a lower packing density and in a change in topology from 6^3 to $3^4.6$ (Fig. 1). Utilizing differential scanning calorimetry (DSC) on (**1**) endothermal processes at 224 °C and 226 °C were assigned as possible phase transition temperatures prior to the melting point at 227 °C. Based on the data, it can be assumed, that (**2**) is a high-temperature modification of (**1**).

References

[1] I. Abrahams, D. Z. Demetriou, *J. Solid State Chem.* 2000, *149*, 28-32.
[2] R. L. Sass, E. B. Brackett, T. E. Brackett, *J. Phys. Chem.-US* 1963, *67*, 2863-2864.
[3] J. M. van den Berg, *Acta Crystallogr.* 1961, *14*, 1002-1003.
[4] W. Nieuwenkamp, J. M. Bijvoet, *Z. Kristallogr.* 1933, *84*, 49-61.

Fig. 1: Projection of the crystal structure (50 % probability ellipsoids) of (**1**) along the crystallographic b-axis (left) and of (**2**) along the crystallographic c-axis (right).

Figure 1

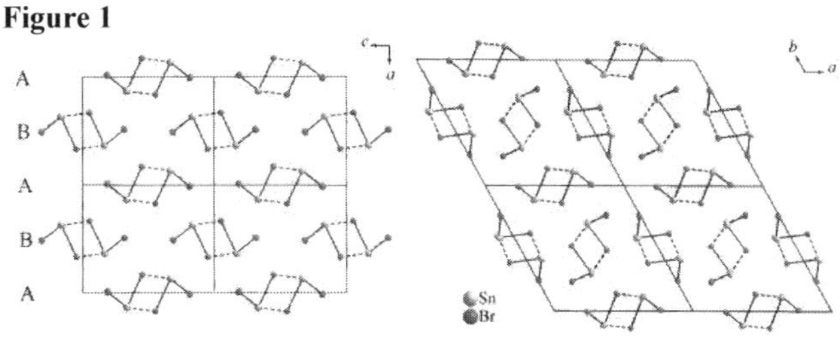

MSC-P11

The imidosulfates $Na_2(NHSO_3)$, $K_2(NHSO_3)$ and $BaNa_2(NHSO_3)_2$

V. Zimmermann[1], M. Wickleder[1]
[1]*Carl-von-Ossietzky-Universität Oldenburg, Institut für Chemie, Oldenburg, Germany*

First attempts aiming at the complete deprotonation of amidosulfonic acid, NH_2SO_3H, can be traced back to 1966,[1] although first structure elucidations were only reported in the 1980s by Popitsch et al. on the example of $Ag_3(SO_3N) \cdot 3\ NH_3 \cdot 2\ H_2O$ and $Ag_3(SO_3N) \cdot H_2O$.[2,3] Recently we started to investigate in more detail the anions $(NHSO_3)^{2-}$ and $(NSO_3)^{3-}$ and here we present the crystal structures of $Na_2(NHSO_3)$[4], $K_2(NHSO_3)$ and $BaNa_2(NHSO_3)_2$.

Structurally it seems interesting if $(NHSO_3)^{2-}$ and $(NSO_3)^{3-}$ display the same characteristics as the isoelectronic anions $(HPO_4)^{2-}$ and $(PO_4)^{3-}$. Furthermore it should be possible to take advantage of the presence of both oxygen and nitrogen donor atoms in these anions for the synthesis of compounds with oxo- and azophilic metal ions.

$Na_2(NHSO_3)$ (CSD 427714) (orthorhombic, *Pccn*, a = 811.83(2) pm, b = 1667.42(5) pm, c = 581.10(2) pm, V = 786.65(4) · 10^6 pm³, Z = 8) was obtained via the reaction of amidosulfonic acid in highly concentrated NaOH solution. Channels are formed along the c axis of the unit cell with the hydrogen atoms of the [NH] moieties pointing into these channels(Fig 1, left). A kind of undulated layers can be suggested via the a axis, where Na atoms connect the $NHSO_3$ tetrahedra.

The crystal structure of $BaNa_2(NHSO_3)_2$ (CSD 428728) (monoclinic, $C\ 2/c$, a = 936.04(4) pm, b = 588.33(2) pm, c = 13.8549(6) pm, β = 102.287(2)°, V = 745.51(5) · 10^6 pm³, Z = 4) shows a layer-type structure parallel to the ac plane. The remaining hydrogen atoms of the amidosulfonic acid are pointing into the space between the layers (Fig 1).

In contrast to these structures, the crystal structure of $K_2(NHSO_3)$ (CSD 428768) (orthorhombic, *Pnma*, a = 766.03(2) pm, b = 569.82(2) pm, c = 1017.57(3) pm, V = 444.17(2) · 10^6 pm³, Z = 4) provides no channels or layers for the remaining hydrogen atoms (Fig 1). All structures have in common that they lack significant hydrogen bonds as can be seen from the shortest D-A distances which are clearly above 306 pm.

Figure 1: left: crystal structure of $Na_2(NHSO_3)$, middle: crystal structure of $BaNa_2(NHSO_3)_2$, right: crystal structure of $K_2(NHSO_3)$. The displacement ellipsoids are drawn at a 80% probability level.

Figure 1

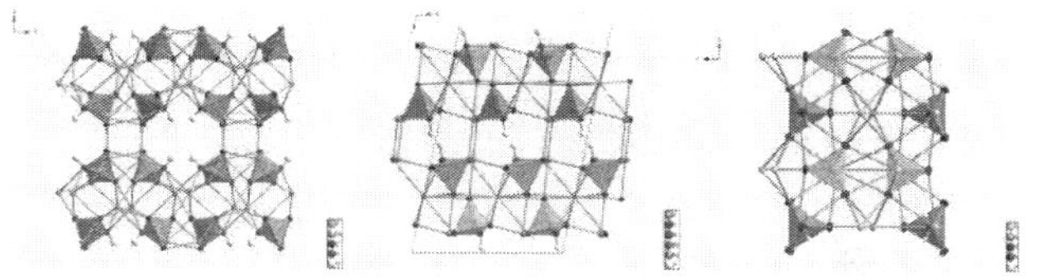

MSC-P12

Structural characterization of $Cu_2ZnSn(S_{1-x}Se_x)_4$: a comparative study

G. Gurieva[1], M. Dimitrievska[2], R. Gunder[1], H. Xie[2], S. Zander[1], V. Izquierdo-Roca[2], A. Pérez-Rodríguez[2], E. Saucedo[2], S. Schorr[1,3]
[1]*Helmholtz-Zentrum Berlin, Berlin, Germany*
[2]*Catalonia Institute for Energy Research (IREC), Barcelona, Spain*
[3]*Freie Universität Berlin, Berlin, Germany*

Quaternary $Cu_2ZnSn(S_{1-x}Se_x)_4$ compounds are promising semiconductor materials for absorber layer in thin film solar cells due to direct band gap in the range 1- 1.5 eV and high absorption coefficient (> 10^4cm^{-1}) [1,2]. All constituents of these films are abundant, low cost and non-toxic. The highest conversion efficiency of $Cu_2ZnSn(S, Se)_4$ solar cell is till now 12.6% [3].

Three types of $Cu_2ZnSn(S_{1-x}Se_x)_4$ were investigated: (1) Single phase powder of $Cu_2ZnSn(S_{1-x}Se_x)_4$ (with x=0.48, 0.58, 0.7, 0.79, 0.9, 1) grown by solid state reaction (2) $Cu_2ZnSn(S_{1-x}Se_x)_4$ thin films synthesized by deposition of Cu/Sn/Cu/Zn metallic multi-stacks onto Mo coated soda lime glass by DC magnetron sputtering and annealing under S+Se+Sn atmosphere and (3) $Cu_2ZnSn(S_{1-x}Se_x)_4$ monograins (with x=0.18, 0.19).

The structural characterization of the powder and monograin samples was carried out by neutron diffraction using the fine resolution neutron powder diffractometer E9 at BER II ($\lambda = 1.7986$ Å, RT). The polycrystalline thin films were investigated by grazing incidence X-ray diffraction. The Rietveld analysis of the diffraction data was performed using the FullProf software [4]. The structure model of kesterite (space group) was used as a starting point for the refinement procedure because it was shown that both of the end members of $Cu_2ZnSn(S_{1-x}Se_x)_4$ solid solution adopt kesterite type structure [5].

The CZTSSe samples studied show stoichiometric (powder) as well as off-stoichiometric (powder, monograins, thin films) chemical compositions, enabling a comparative study of the structural parameters of stoichiometric and off-stoichiometric CZTSSe. In general, for the lattice parameter a and c Vergards law is fulfilled for all samples, but the slope of the dependencies (lattice parameters in dependence on S/Se ratio) is different for stoichiometric and off-stoichiometric samples. Interestingly a lattice parameter ratio c/2a>1 was obtained for stoichiometric CZTSSe, whereas c/2a<1 was derived for off-stoichiometric CZTSSe bulk samples. The latter where found to belong to the A-type (Cu-poor/Zn-rich) and C-type (Cu-rich/Zn-poor) off-stoichiometry class. Off-stoichiometric CZTSSe thin films always show c/2a>1.

Applying the average neutron scattering length analysis it can be concluded, that the C-type CZTSSe shows Cu_{Zn} and Sn_{Zn} point defects additional to the Cu_{Zn}-Zn_{Cu} anti site defects in the planes at z=1/4 and ¾.

Acknowledgments: Financial supports from IRSES PVICOKEST 269167 and KESTCELLS 316488, FP7-PEOPLE-2012 ITN, Multi-ITN is highly appreciated.

[1] J.M. Raulot, et al., *J. Phys. Chem. Solids* 66 (2005) 2019.
[2] P.A. Fernandeset al., *Phys. Status Solidi C 7* (2010) 901.
[3] Wang, et al., *Adv. Energy Mater.* (2014), *4*, 1301465.
[4] Juan Rodriguez-Carvajal and Thierry Roisnel,www.ill.eu/sites/fullprof/
[5] Siebentritt, S. and Schorr, *S. Prog. Photovolt: Res. Appl.*, (2012), 20, 512.

MSC-P13
Microstructural Analysis of $Cu_2ZnSnSe_4$ Thin Films by Grazing Incidence X-Ray Diffraction

R. Gunder[1], S. Schorr[1], S. Mariño[2], M. Rodríguez[2], E. Saucedo[2]
[1] *Helmholtz Centre Berlin for Materials and Energy, Crystallography, Berlin, Germany*
[2] *Catalonia Institute for Energy Research, Barcelona, Spain*

The research on sustainable solar energy conversion technologies has further branched to another promising alternative, based on $Cu_2ZnSn(S,Se)_4$ (CZTSSe). In comparison to other compound semiconductor materials the elements contained in the CZTSSe system are abundant as well as less or even not toxic. These facts mediated increasing efforts in order to attain competitive, i.e. efficient and durable solar cells. The currently highest efficiency of a CZTSSe device was reported to be 12.6 % [1]. However, in the course of the incipient treatment as a semiconducting material, it turned out that the theoretical maximum, that is, the Shockley-Queisser limit, of attainable energy conversion efficiency of a single p-n junction device amounts $\eta_{max} = 32.2$ % [2]. Efforts dedicated to close the gap between actual and possible efficiency include the microstructural analysis of thin film solar cells since domain size and micro strain are affecting the behavior of charge carriers. By means of grazing incidence X-ray diffraction a depth-resolved characterization of the microstructure is performed for six thin film samples fabricated both by DC magnetron sputtering (PVD) as well as spray pyrolysis (CVD). Additionally, process conditions were varied with respect to composition, substrate material as well as number and temperature of annealing steps. The particular impact of different process conditions on microstructure is unraveled by appropriate comparisons. While the domain size of each absorber sample is beyond the resolution limit of laboratory XRD, the micro strain could be determined for any sample depth. In all samples the micro strain is emanating either from planar or point defects. At the junction between CZTSSe absorber and Mo back contact the micro strain is strongly correlated with unit cell parameters, thus pointing to distortional lattice mismatch. Segregation of secondary phases at the absorber surface rather introduces point defects as elements are removed selectively. From either source the micro strain propagates and thereby relieves towards the center of the absorber. The overall micro strain depends on the amount of secondary phase formed and the differences in unit cell parameters at the interface, which, in turn, is governed by the kind of process (i.e. PVD or CVD), temperature and composition. Generally, the micro strain is higher in case of, respectively, lower annealing temperature, austenitic steel substrate or CVD process. Conversely, absorbers fabricated on soda lime glass by PVD process promote the lowest micro strain.

References:

[1] Wang, W., et al., *Device Characteristics of CZTSSe Thin-Film Solar Cells with 12.6% Efficiency.* Advanced Energy Materials, 2014. 4(7).
[2] Guo, Q., H.W. Hillhouse, and R. Agrawal, *Synthesis of Cu_2ZnSnS_4 Nanocrystal Ink and Its Use for Solar Cells.* Journal of the American Chemical Society, 2009. 131(33): p. 11672-11673.

MSC-P14
Synthesis and structural characterization of off-stoichiometric kesterite type selenide compound semiconductor $Cu_2ZnSnSe_4$ (CZTSe)

L. E. Valle Rios[1], G. Gurieva[2], D. Többens[2], S. Schorr[1,2]
[1] *Freie Universität Berlin, Geowissenschaften, Berlin, Germany*
[2] *Helmholtz Zentrum Berlin, Crystallography , Berlin, Germany*

Quaternary $Cu_2ZnSnSe_4$ is a promising alternative absorber material for solar cells. A record efficiency of 12.6% was reported for a CZTSSe thin film solar cell [1]. The polycrystalline absorber layer exhibits an off-stoichiometric composition which causes intrinsic point defects (vacancies, anti-sites, interstitials). These defects determine the electronic properties of the material significantly.

This work focuses on the synthesis and characterization of off-stoichiometric CZTSe. In literature [2] off-stoichiometric kesterite types have been suggested, amongst them are A-type Cu-poor/Zn-rich, $Cu_{2-2x}Zn_{1+x}SnSe_4$; B-type Cu-poor/Sn-poor, $Cu_{2-2y}Zn_{1+3y}Sn_{1-y}Se_4$ and C-type Cu-rich/Sn-rich, $Cu_{2+2z}Zn_{1-3z}Sn_{1+z}Se_4$ materials. In our study we have synthesized powder samples of these types by solid state reaction from pure elements in sealed evacuated silica tubes in a one zone furnace. The first reaction took place at 750°C with several temperature steps (250°C, 450°C, 600°C) in between. After reaction all samples were ground, pressed in pellets and annealed again at 750°C.

In the ternary phase diagram Cu_2Se-ZnSe-$SnSe_2$ the single phase kesterite region is indicated to be very narrow [3], therefore the formation of secondary phases is highly expected. To determine phase content and chemical composition of the obtained samples, an electron microprobe system equipped with wavelength dispersive X-ray analysis was used. Measurements proved the presence of CZTSe as main phase within all the different off-stoichiometric type synthesized samples. The lattice parameters *a* and *c* of the CZTSe main phase were determined by Rietveld analysis from XRD collected data. Refinements were performed using FullProf software [4] with the kesterite structure model, because stoichiometric CZTSe crystallizes in the kesterite type structure [5].

However, isoelectronic cations Cu^{+} and Zn^{2+} cannot be distinguished by XRD. Therefore, anomalous X-ray diffraction near to the Cu-K and Zn-K edges have been performed at the diffraction station KMC-2 beamline of HZB-BESSY II. Anomalous scattering coefficients are highly wavelength-dependent close to the absorption edges. The usage of multiple wavelengths above, below and between the absorption edges of Cu and Zn, results on intensity variation from the reflections which contain information on cation distributions. Collected data was refined and the atomic positions within the kesterite type structure have been determined. The presentation will give a trend of the cation distribution in the crystal structure of the CZTSe phase concerning the different off-stoichiometry types.

[1] Wang, et al., Adv.Energy materials (2013).
[2] Lafond, et al., ZAAC 638, (2012) 2571-2577.
[3] Oleksyuk, et al., J. Alloys Comp. 368 (2004) 135-143.
[4] Carvajal, et al.,www.ill.eu/sites/fullprof/.
[5] Schorr, Sol. Energ. Mat. Sol. Cells 95 (2011) 1482-1488.

MSC-P15
Phase content and structural analysis by WDX and X-ray diffraction of off-stoichiometric Cu2ZnSnS4 (CZTS)

K. Neldner[1], S. Schorr[1], D. Többens[1], G. Gurieva[1]
[1]*HZB, Crystallography, Berlin, Germany*

Quaternary chalcogenides have seen rapid development in recent years leading to a world record efficiency for thin film solar cells based on $Cu_2ZnSn(S,Se)_4$ (CZTSSe) of 12.6% [1].

CZTS belongs to the $A^{I}_2B^{II}C^{IV}X^{IV}_4$ compound family and crystallizes in the tetragonal kesterite type structure, the certain ordering of the Cu and Zn cation layers results in the space group [2]. According to literature [3] the stability field of single phase CZTS is assumed to be quite narrow. Typical secondary phases coexisting with a quaternary kesterite type phase are ZnS, CuS, Cu_2S, SnS, Sn_2S_3, SnS_2 and Cu_2SnS_3. Nevertheless there is still a lack of knowledge about existing phases and phase relations in the vicinity of the intersection point of CZTS. In literature [4] four off-stoichiometric CZTS compounds, named A-, B-, C- and D-type, have been proposed.

The number of publications on intrinsic point defects in kesterites is very limited. Further studies on deviation from stoichiometry, distribution of the cations and formation of intrinsic point defects are of great importance to understand solar cell performance. Therefore, our experiments focus on the synthesis of off-stoichiometric CZTS reference powder samples with cation ratios Zn/Sn and Cu/(Zn+Sn) $\neq$ 1. All powder samples were synthesized by solid state reaction from pure elements in sealed evacuated silica tubes in a one zone furnace. The obtained samples have been well characterized regarding chemistry, phase composition and trends in lattice parameters using wavelength-dispersive X-ray spectroscopy (WDX) on an electron microprobe system and X-ray diffraction (XRD). Furthermore anomalous X-ray diffraction at the KMC-2 beam line was used to obtain the site occupation of the Cu and Zn sites in the kesterite structure. The phase content analysis presented here considers the samples before (intermediate samples) as well as after the annealing step.

The results of the XRD and WDX analysis show that CZTS is the main phase in all samples synthesized. The CZTS grains in the intermediate samples exhibit a slight variation of the chemical composition. The presence of secondary phases like CuS, ZnS and Cu_2SnS_3 was noticed for all samples. Moreover a limited solubility of Cu and Sn in ZnS, Zn and Sn in ZnS as well as Zn in Cu_2SnS_3 was determined. These results have been assistant to understand the growth process of CZTS and to improve the solid state synthesis procedure. After the annealing step the chemical composition of the off-stoichiometric samples is between C- and D-type. With increasing off-stoichiometry the lattice constants and also the a/2c ratios are decreasing. This leads to smaller tetragonal distortion with increasing off-stoichiometry.

[1] Wang, et al., Adv. Energy Mater., 2014. 4(7).
[2] Schorr, Sol. Energ. Mat. Sol. Cells, 2011. 95(6): 1482-1488.
[3] Olekseyuk, et al., J. All. Com., 2004. 368(12): 135-143.
[4] Lafond, et al., ZAAC, 2012. 638(15): 2571-2577.

MSC-P16
Further insight into U/Th metaphosphate crystal chemistry and effect of Nd^{3+} incorporation into U^{4+} metaphosphate

N. Yu[1], E. Alekseev[1]
[1]*Forschungszentrum Jülich GmbH, IEK-6, Juelich, Germany*

Single crystals of a tetragonal modification of uranium polymetaphosphate $U(PO_3)_4$, of a mixed U^{4+}/Nd^{3+} polymetaphosphate "$(U_{0.62}Nd_{0.38})(PO_3)_4$", and of two new tetraphosphates, $Th(P_4O_{12})$ and $U(P_4O_{12})$, were synthesized. The structures of obtained materials were characterized by X-ray diffraction and Raman spectroscopy. The presence of Nd in "$(U_{0.62}Nd_{0.38})(PO_3)_4$" was proven by EDX and its measured degree of substitution for U agrees well with the results of X-ray crystallography, however the mechanism of the necessary charge compensation could not be identified yet. The cation arrangement in the crystal structures of Th/U polymetaphosphates has been studied in terms of Voronoi-Dirichlet polyhedra. Different conformations of the tetraphosphate poly-anions were observed in the crystal structures of $Th(P_4O_{12})$ and $U(P_4O_{12})$. Variations in the U atoms arrangement in the considered crystal structures lead to different $(PO_3)_n^{n-}$ chain configurations. (Fig. 1) The Raman spectrum of $Th(P_4O_{12})$ single crystal was recorded and bands assigned.

Figure 1. Polyphosphate chains in crystal structures of the tetragonal (a) $U(PO_3)_4$, (b) $(U_{0.62}Nd_{0.38})(PO_3)_4$, and (c) β-$U(PO_3)_4$.

Figure 1

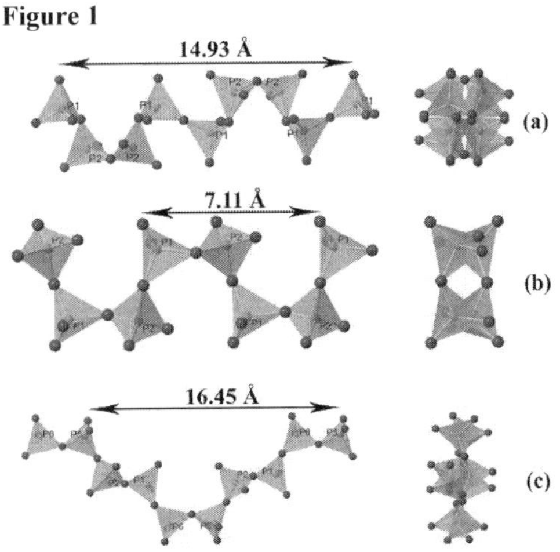

MSC-P17

Tunable k_2 phase transition in chain polymers $[Pd(acacCN)_2Ag]X$ ($X = BF_4$, ClO_4)

Q. Guo[1], U. Englert[1]
[1]RWTH-Aachen, Inorganic Chemistry, Aachen, Germany

The chain polymers $[Pd(acacCN)_2Ag]X$ ($X = BF_4$, ClO_4) have been investigated by temperature-dependent diffraction. Both compounds are isomorphous and undergo structural phase transitions at 198 K (BF_4) and 230 K (ClO_4) which can be monitored in single crystals. The high temperature form in space group *C2/c* undergoes a fully reversible k2 transition by de-centering of the conventional unit cell to the low temperature phase in subgroup $P2_1/c$ [1, 2]. A compound with statistical disorder for the counter anion may be synthesized and analyzed by diffraction and ^{19}F NMR; it features a phase transition temperature of ca. 216 K, intermediate to those of the parent compounds.

Fig. 1: Projections of the *C*-centered β phase (a) and the pseudo-centered α phase (b) of $[Pd(acacCN)_2Ag]BF_4$; conventional unit cells are indicated by solid, the centered pseudo-cell by dashed lines.

References :

[1] H. Bärnighausen, *MATCH*, 1980, 9, 139-175.
[2] U. Müller, *Symmetriebeziehungen zwischen verwandten Kristallstrukturen*, Vieweg + Teubner, 2011.

Figure 1

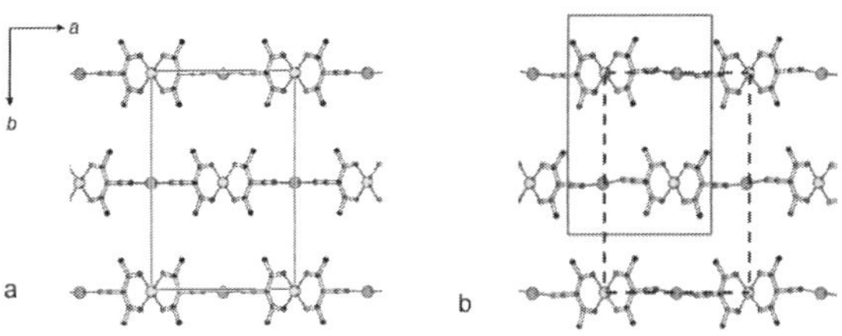

MSC-P18

Investigation of the phase transition of $CsGaSe_2$ by high-temperature *in situ* X-ray powder diffraction

D. Friedrich[1], M. Schlosser[1], A. Pfitzner[1]
[1]Universität Regensburg, Institut für Anorganische Chemie I, 93040 Regensburg, Germany

During explorative investigations of alkali metal group 13 chalkogenometallates two polymorphs of $CsGaSe_2$ were discovered.[1, 2, 3] The low temperature polymorph $CsGaSe_2$-II crystallizes in the $KInS_2$ structure type with a = 11.046(1), b = 11.051(1), c = 16.827(1), β = 99.402(9), V = 2026.5(3), and Z = 16 (space group *C2/c*). The anionic substructure of $CsGaSe_2$-II features anionic layers $^2_\infty[Ga_4Se_8{}^{4-}]$ (Figure 1) composed of corner-sharing Ga_4Se_8 supertetrahedra. The high temperature polymorph $CsGaSe_2$-I crystallizes in the $KFeS_2$ structure type with a = 7.651(1), b = 12.555(2), c = 6.179(1), β = 113.53(2), V = 544.2(3), and Z = 4 (space group *C2/c*). The anionic substructure of $CsGaSe_2$-I features anionic chains $^1_\infty[GaSe_2{}^-]$ (Figure 1) composed of edge-sharing $GaSe_4$ tetrahedra.

The reversible phase transition of $CsGaSe_2$ was studied *in situ* by high-temperature X-ray powder diffraction using a STOE Stadi P diffractometer equipped with a high temperature capillary furnace (monochromatized Mo-$K_{\alpha 1}$ radiation). Several experiments upon cooling/heating and at constant temperatures were performed to study the kinetics of this phase transition (Figure 2).

Figure 1: The different anionic substructures of both $CsGaSe_2$ polymorphs (left), featuring anionic chains $^1_\infty[GaSe_2{}^-]$ in $CsGaSe_2$-I (HT polymorph) and anionic layers $^2_\infty[Ga_4Se_8{}^{4-}]$ in $CsGaSe_2$-II (LT polymorph).

Figure 2: Evolution of the phase fractions of the $CsGaSe_2$ polymorphs during the phase transition from $CsGaSe_2$-II to $CsGaSe_2$-I.

[1] H. J. Deiseroth, *Z. Kristallogr.* 1984, *166*, 283.
[2] A. Kyas, Dissertation, TH Aachen 1981.
[3] D. Friedrich, M. Schlosser, A. Pfitzner, *Z. Anorg. Allg. Chem.* 2014, *640*, 2356.

Figure 1

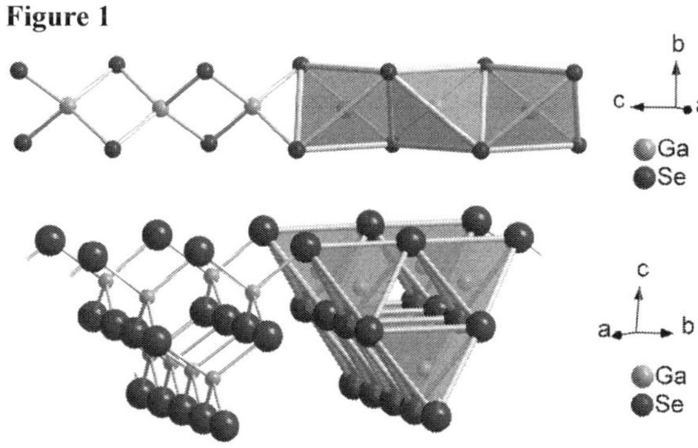

Figure 2

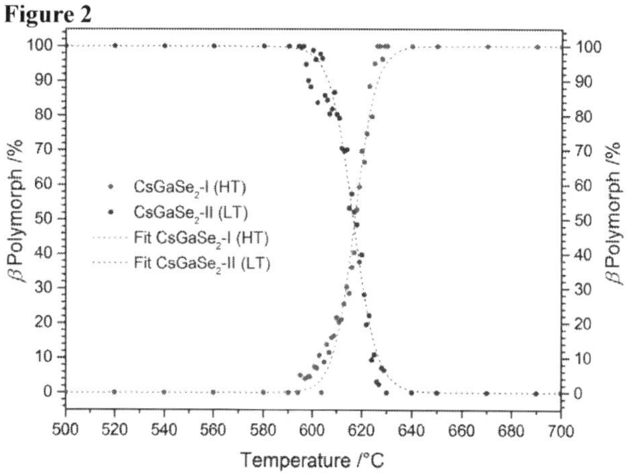

MSC-P19

$(La,Pr)PO_4$ - A model system for ceramic radioactive waste matrices

A. Hirsch[1], A. Neumann[1], C. Schausten[2], A. Thust[3], L. Peters[1], G. Roth[1]
[1]RWTH Aachen, Institute of Crystallography, Aachen, Germany
[2]RWTH Aachen, Institute of Mineral Engineering, Aachen, Germany
[3]Goethe University Frankfurt am Main, Department of Geosciences, Frankfurt am Mai, Germany

Potential host matrices for the storage of highly radioactive waste must fulfil several requirements such as high chemical durability, high resistance to radiation damages and long term stability. In natural monazites, all of these properties can be found, as well as incorporation of up to 30 weight% ThO_2 and UO_2 without amorphisation [1]. In this contribution, synthetic monazites are investigated, using La and Pr as surrogates for the radioactive elements Np, Am, and Cm.

(La,Pr)PO_4 powders were synthesised at 1000 °C in air [2]. Electron microprobe analyses (*EMPA*) yielded single phase monazite. The grains were highly porous due to the loss of water and ammonia during synthesis. The maximum deviation (4 mol%) from the expected composition probably covers both, inhomogeneity of the samples and experimental error.

X-Ray powder diffraction (*XRPD*) measurements were carried out for qualitative and structural phase analysis. The expected decrease of the lattice parameters and the average *Ln*-O bond distance due to the incorporation of the smaller Pr^{3+} for La^{3+} was confirmed. *In-situ* high temperature *XRPD* measurements will be discussed and the thermal expansion behaviour of the samples in this solid solution will be described.

Infrared and Raman spectroscopic analyses were performed to characterise the local environment of the phosphorous atom in the PO_4 tetrahedron in comparison with the results from *XRPD*. While in IR data only one asymmetric P-O stretching mode showed a significant linear shift to higher wave numbers with increasing Pr-content, in Raman data a linear shift of more modes can be seen. Within experimental resolution, all other modes remained at constant wave numbers. In both, Raman and IR, the observed shifts can be explained by stronger interactions between PO_4 tetrahedra and *Ln* for Pr-rich compounds [3]. Similar behaviour was reported for (La,Gd)PO_4 [4].

In addition to powder samples, single crystals were grown. Detailed structure analyses will be presented.

References:

[1] Ewing, R.; Weber, W. & Clinard, F., 1995. Radiation effects in nuclear waste forms for high-level radioactive waste. *Progress in nuclear energy, Elsevier,* 29, 63-127

[2] Bregiroux, D., Audubert, F., Charpentier, T., Sakellariou, D., & Bernache-Assollant, D, 2007. Solid-state synthesis of monazite-type compounds *Ln*PO 4 (*Ln* = La to Gd). *Solid State Sciences*, 9(5), 432-439.

[3] Chunhua, L., Jie, H., Zhongzie, X., Yaru, N., 2007. Preparation and IR spectra study of rare earth orthophosphates. *Journal of rare earth*, 25, 273-276.

[4] Heuser, J., Bukaemskiy, A. A., Neumeier, S., Neumann, A., & Bosbach, D., 2013. Raman and infrared spectroscopy of monazite-type ceramics used for nuclear waste conditioning. *Progress in nuclear energy, Elsevier,* **72**, 149-155.

MSC-P20

Flux growth of triphylite - lithiophilite, Li(Fe,Mn)PO_4 single crystals

P. Schmid-Beurmann[1], J. Sibbing[1], J. Thiäner[1], W. Lottermoser[2]

[1]*Universität Münster, Institut für Mineralogie, Münster, Germany*

[2]*Universität Salzburg, FB Materialforschung und Physik, Salzburg, Austria*

During the last decades there was a growing interest in the properties of the phospho-olivine triphylite, $LiFePO_4$ due to its importance in electrochemistry as well as in geoscience. In the field of geoscience the primary phosphates of the triphylite - lithiophilite, Li(Fe^{2+},Mn^{2+})PO4 solid-solution series are of importance as they are the educt of several reactions leading to secondary mineral associations which can be used to characterize the conditions of the evolution of pegmatites (Fransolet et al., 1985). One reaction is the oxidation of triphylite, Li(Fe^{2+},Mn^{2+})PO_4 to ferrisicklerite, $Li_{1-x}(Fe^{3+}{}_x,Mn^{2+}{}_{1-x})PO_4$ which was investigated using cuboids cut from a natural single crystal of triphylite (Schmid-Beurmann et al., 2013). Unfortunately natural phases suffer from the lack of pure end-member compositions, the presence of additional components as well as inclusions and cracks. Therefore it would be desirable to have synthetic single crystals of mm-size for experimental investigations. We synthesized crystals of the triphylite - lithiophilite solid-solution series in steps of 12.5 molar% using the flux growth method with LiCl as flux. The procedure was derived from the method of Zambonini & Malossi (1931) applied to lithiophilite, $LiMnPO_4$. In order to prevent the oxidation of Fe^{2+} we used welded Au-capsules (Ø 4mm, L = 40mm) with the addition of elemental Fe as oxygen getter. Homogenized mixtures of LiCl, $FePO_4$, $Mn_3(PO_4)_2$, Li_3PO_4 and Fe were heated to 900°C and then cooled down slowly to 600°C to allow crystallization. The growth resulted in mm-sized euhedral single crystals of gem-quality with colors ranging from brownish to pink with increasing Mn-content. Mapping by EPMA proved the homogeneity of the crystals with the exception of a rim of 20 to 30 µm. ^{57}Mössbauer-spectroscopy revealed an $Fe^{3+}/(Fe^{2+}+Fe^{3+})$ ratio of 5 to 7 %. As a result crystals of the triphylite - lithiophilite solid-solution series of good quality and homogeneity can be prepared by the flux growth method.

References

Fransolet, A.M., Keller, P., Fontan, F. (1985): The phosphate mineral associations the Tsaobismund pegmatite. Contrib. Mineral. Petrol., 92, 502 - 517.

Schmid-Beurmann, P., Ottolini, L., Hatert, F., Geisler, T., Huyskens, M., Kahlenberg, V. (2013). Topotactic formation of ferrisicklerite from natural triphylite under hydrothermal conditions. Min. Petrol., 107(4), 501-515.

Zambonini, F., Malossi, L. (1931): Riproduzione artificiale della litiofilite. Z. Krist., 80, 442.

MSC-P21

Synthesis and structural investigation of the layered sulfido metalates $A_{2x}[M_{2-x}S_2]$ (A=K, Rb, Cs; M=Fe, Co; x=0.4, 0.5)

M. Schwarz[1], P. Stüble[1], J. Kägi[1], C. Röhr[1]

[1]*Albert-Ludwigs-Universität Freiburg, Institut für Anorganische und Analytische Chemie, Freiburg, Germany*

Anionic layers of edge-sharing MS_4-tetrahedra are known for several transition metal elements like Fe, Co and Mn [1-3] with the heavy alkali metal cations K, Rb and Cs.

We have synthesized compounds of the series with the general formula $A_{2x}[M^{II}{}_{2-x}S_2]$ (A=K, Rb, Cs; M=Fe, Co; x=0.4, 0.5) in flux melting reactions starting from the pure elements (A=K) and the alkali sulfides (A=Rb, Cs), respectively. All sulfido metalates were investigated using X-ray powder and for the first time single crystal diffraction data.

For compounds $A_{0.8}[M^{II}{}_{1.6}S_2]$ (x=0.4) the single crystal diffraction patterns exhibit reflections fitting the $ThCr_2Si_2$-aristotype model (*I4/mmm*) with a=377.44(2)/379.27(1)/392.66(6) and c=1350.61(8)/1406.96(5)/1430.9(3) pm (A=K/Rb/Cs, M=Fe) and a=376.0(1) and c=12.89(7) pm, (A=K, M=Co). The data sets can be refined with a statistical occupation of both A and M cations. For $A_{0.8}[Fe^{II}{}_{1.6}S_2]$ (A=K, Rb) superstructure reflections are observed fitting two twinned domains of the tetragonal model (*I4/m*) with $a'=\sqrt{5}a$ and $c'=c$ already described for $K_{0.88}Fe_{1.6}S_2$ on the basis of PXRD data [1].

In addition, reflections are observed, which can be indexed using two twinned domains of the $Cs_2Mn_3S_4$-type structure model (*Ibam*

[3]) with $a''=a\sqrt{2}$, $b''=a2\sqrt{2}$ and $c''=c$ according to the respective selenide $K_{2x}(Fe,Co)_{2-x}Se_2$, which has been investigated applying electron diffraction and HAADF-STEM [4]. For the Cs ferrate, a different tetragonal superstructure with $a'''=3a$ and $c'''=c$ has been observed indicating a not yet finalised model, evidentially related to the recently published structure of $Cs_{11}[Fe_5S_8]_2O$ [5].

For the respective cobaltates $A_{2x}[Co_{2-x}S_2]$, the difference in the molar volumes clearly indicate the existence of two structurally and chemically different compounds for x=0.4 (=$A_{0.8}[Co_{1.6}S_2]$ [2]) and 0.5 (=$A[Co_{1.5}S_2]$=$A_2[Co_3S_4]$, A=Rb, Cs: [3]; A=K: this work).

The whole family of layered compounds with different metal vacancy ordering patterns is discussed applying crystallographic group-subgroup relations in a Bärnighausen tree.

[1] H. Lei, M. Abeykoon, E. S. Bozin, C. Petrovic, *Phys. Rev. B* 2011, 83, 180503.
[2] G. Huan, M. Greenblatt, *Eur. J. Sol. State Inor.* 1989, 26, 193.
[3] W. Bronger, P. Böttcher, *Z. Anorg. Allg. Chem.* 1972, 390, 1.
[4] S. M. Kazakov et al., *Chem. Mater.* 2011, 23, 4311.
[5] M. Schwarz, C. Röhr, *Z. Anorg. Allg. Chem.* 2014, 640, 2792.

Figure 1

Fig. 1: Vacancy patterns in the phases $A_{2x}[M_{2-x}S_2]$ (A=K, Rb, Cs; M=Fe, Co) for

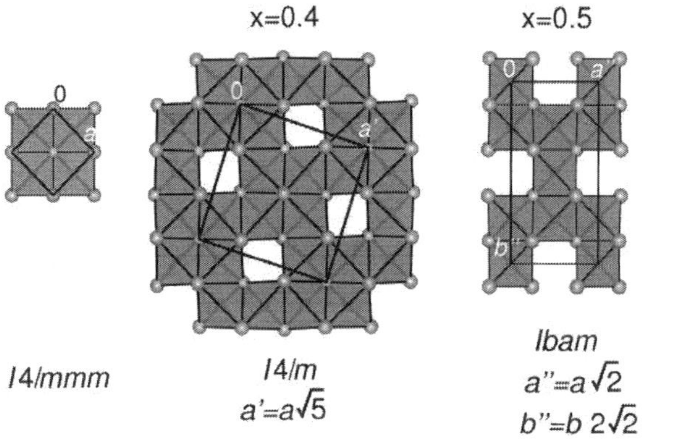

MSC-P22

Demixing of multinary metallic liquids during electrocrystallisation - synthesis and crystal structure of $CsIn_{12}$

C. Hoch[1], F. Tambornino[1], S. Hübner[1]
[1]*Ludwig-Maximilians-Universität München, Chemie, München, Germany*

The Cs-In phase diagram has recently been reviewed [1]. Two phases with compositions $CsIn_3$ [2] and Cs_2In_3 [3] were reported. In this work, we present the new indium-rich binary main group compound $CsIn_{12}$.

Single crystals of the new indium-rich binary main group compound $CsIn_{12}$ were prepared by isothermal electrolysis of a solution of CsI in N,N'-dimethyl formamide (DMF). The reactive cathode consisted of a liquid solution of indium in mercury (1:1) in a glass spoon [4]. Other than elemental Indium, the metallic solution is liquid at room temperature and enables the synthesis of indides at low temperatures. Mercury does not participate in the electrochemical reaction at the employed conditions. Another example for the demixing of a multinary metallic solution was observed for the ternary metallic eutectic "Galinstan" (Ga : In : Sn = 62.5 : 21.5 : 16). Electrolysis on a reactive Galinstan electrode with LiI in DMF leads to the formation of Li_3Ga_{14} , LiGa, LiIn and metallic indium. Significant amounts of tin were incorporated in Li_3Ga_{14} on one of the crystallographic Ga sites.

Single crystal X-ray diffraction patterns of $CsIn_{12}$ show a cubic face-centred subcell (a = 14.108(7) Å) with satellite reflections in all three directions, indicating a modulated structure. $NaZn_{13}$ is the aristotype to this new binary compound. Via distorting the snub-cubes in the $NaZn_{13}$ type to rhombicuboctahedra, the central icosahedron transforms into a cuboctahedron. In contrast to the $NaZn_{13}$ structure, the latter is uncentred, leading to the different sum formula. The electronic structures of both compounds were calculated by FP-LAPW-DFT methods implemented in the WIEN2k code [5]. The differences in the respective electronic structures are discussed.

References:

[1] Okamoto, H., *J. Phase Equilibra* 34, *3*, 251 (2013).
[2] Chuntonov, K. A.; Yatsenko, S. P.; Grin, Yu.; Yarmolyuk, Ya. P.; Orlov, A. N., *J. Less Common Met.* 99, *15* (1984).
[3] Yatsenko, S. P.; Chuntonov, K. A.; Orlov, A. N.; Yarmolyuk, Ya. P.; Grin, Yu., *J. Less Common Met.* 108, *339* (1985).
[4] Massalski, T. B. (ed.), *Binary Phase Diagrams* (2nd ed.), ASM International, Materials Park, Ohio, USA (1990).

[5] Blaha, P.; Schwarz, K.; Madsen, G. K. H; Kvasnicka, D. Luitz, J., *Wien2k - An Augmented Plane Wave and Local Orbital Program for Calculation of Crystal Properties*, TU Wien, Austria (2006).

Figure captions:
Figure 1: Structure of $CsIn_{12}$ with In atoms drawn as ellipsoids at probability level 90%. Cs atoms centre the rhombicuboctahedra. The centre of the cuboctahedron (dark grey) is empty. The analogy to the $NaZn_{13}$ structure type is clearly visible.

Figure 2: Result of a Rietveld refinement of the electrolysis product of a solution of LiI in DMF on a Galinstan cathode.

Figure 1

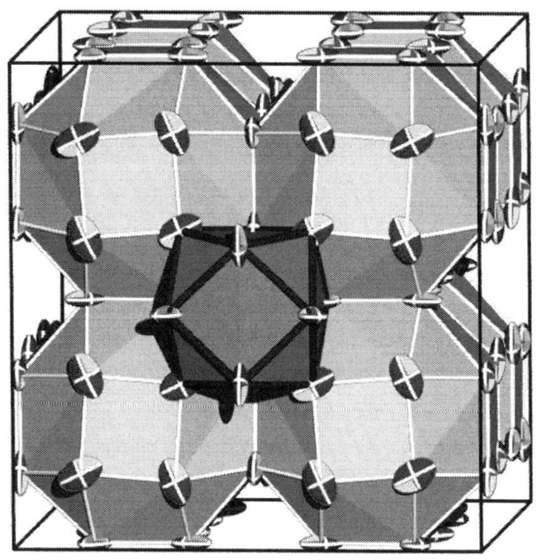

Figure 2

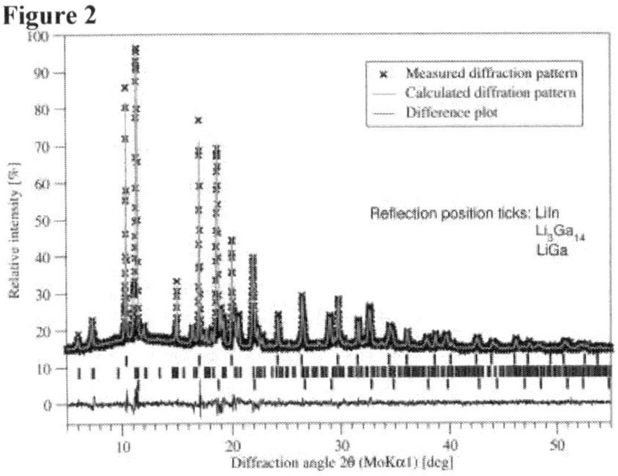

MSC-P23

The New Kappa-Diffractometer at Beamline P24 (Chemical Crystallography, PEX-E, Desy)

C. Paulmann[1], A. Berghäuser[1], D. Ropers[1], U. Bismayer[1]
[1]*Mineralogisch-Petrographisches Institut, Hamburg, Greenland*

The shutdown of DORIS III discontinued some successful beamlines which served as main stations for a wide field of crystallographic applications ranging from diffuse scattering studies, charge density analysis, phase transitions, disordered and modulated structures all at ambient and non-ambient conditions. As result of a joint researchBMBF project (coordination: University Hamburg) to build up a new beamline at PetraIII.14 dedicated to all fields of crystallographic research, layout and specifications for a new diffractometer were defined and a european call for tenders was placed in spring 2012. The diffractometer was delivered in December 2013 and preliminary installed in the former DORIS III experimental hall. Currently, final testing and commissioning is in progress and operation with a Mo microsource is expected to start in January 2015.

The four-circle diffractometer (ca. 3.5 tons), designed in Kappa-geometry, offers two independent detector circles able to take loads of 30 kg each equipped with motorized counter-weights to balance a detector travel of 500 mm. The sphere of confusion (SOC) of the main circles (incl. Omega) for multi axis movements is below 10 µm and repeatability below $2x10^{-4}$ degrees. The inner circles (Kappa, Phi) provide a SOC below 25 µm for loads up to 5 kg with a maximum load of 10 kg. The Phi circle includes a motorized xyz-stage offering an accuracy in the micrometer range. The available space for sample environments is 190 mm, extendable to 250 mm by removing the stage.In combination with a modular mounting system, an easy exchange of different detectors or sample environments is possible.A high-precision secondary optics system including a millisecond shutter system is under development.

The diffractometer will be installed at a standard PETRA III undulator (2m, U29) using a CEMO-type water-cooled DCM (Si111/Si311) with preceding water-cooled mirrors for higher harmonic rejection (Rh/Pt-coated, 1.5 - 3.0 mrad). Due to heat-load restrictions of the CEMO-DCM, additional Cu-coated diamond windows will reduce the low-energy range providing an optimised continous range between 15keV - 44keV and a smaller range around 8keV. The calculated horizontal and vertical beam profiles at 80m source distance and 17.7keV are 1.5 x 0.7 mm^2 and 0.6 x 0.3 mm^2 FWHM and FW90%M, respectively. Additional Be-CRLs are foreseen for further beam-focusing down to 1µm.

MSC-P24

Thermal Analysis and Crystal Chemistry of important AFM - phase solid solutions in CAC cement stones.

S. Stöber[1], H. Pöllmann[1]
[1]*Martin-Luther-Universität Halle, Mineralogie/Geochemie, Halle, Germany*

Hydration products of calcium aluminate cements consists of so called AFM - phases which crystallize if the clinker phases CA or C_4AF react with H_2O in excess. Those AFM - phases with the general formula $[Ca_2Al(OH)_6]^+$ $[X{\cdot}nH_2O]^-$ belong to the large family of layered double hydroxides (LDHs). Their crystal structures with mainly trigonal or hexagonal symmetries are composed of sequences $[Ca_2Al(OH)_6]^+$ and $[X{\cdot}nH_2O]^-$ perpendicular [001]. In the main layer $Ca_2Al(OH)_6]^+$ Ca^{2+} - and Al^{3+} ions are coordinated by 6 oxygen ions from hydroxide ions and Ca^{2+} by one further oxygen of a water molecule of the inter layer. For the charge balance the interlayer contains generally anions like X- = Cl^- & OH^- or Y = SO_4^{2-} & CO_3^{2-}. Additional water molecules, which are not necessary for the stability of the crystal structure can be reversibly added or removed from the interlayer, depending on temperature and relative humidity of the environment.

For our studies solid solutions with the chemical compositions $Ca_2Al(OH)_6]^+$ $[0.5Cl\ 0.25SO_4{\cdot}nH_2O]$- and $Ca_2Al(OH)_6]^+$ $[0.5OH\ 0.25SO_4{\cdot}nH_2O]^-$ were synthesized by paste reactions with a reaction time of 3 months at 25°C. The phases were preconditioned at 35% r. H. under CO_2 - exclusion. In order to determine the lattice parameters of further hydration levels X - ray data of the samples were collected applying a diffractometer from Panalytical plus a HTK1200 chamber in the temperature range of 25°...350°C. Water contents were determined quantitatively by thermal analysis in combination with Karl - Fischer titration. Onsett-temperatures for different hydration reactions were determined by DSC.

MSC-P25

Structural, spectroscopic and compαutational studies on $Li_2Ca_2Si_2O_7$

V. Kahlenberg[1], E. Brunello[1], C. Hejny[1], H. Krüger[1], D. Schmidmair[1], M. Tribus[1], D. M. Többens[2]
[1]*University of Innsbruck, Institute of Mineralogy & Petrography, Innsbruck, Austria*
[2]*Helmholtz-Zentrum Berlin für Materialien und Energie GmbH, Department of Crystallography, Berlin, Germany*

Synthesis experiments in the system Li_2O-CaO-SiO_2 resulted in the formation of single-crystals of $Li_2Ca_2Si_2O_7$. Structural investigations were based on single-crystal diffraction. At ambient conditions the compound has the following basic crystallographic data: hexagonal symmetry, space group $P\ 6_1\ 2\ 2$, a = 5.0961(2) Å, c = 41.264(2) Å, V = 928.07(6) $Å^3$, Z = 6. Structure solution was performed using direct methods. The final least-squares refinement calculations converged at a residual of R(|F|) = 0.0245. From a structural point the lithium calcium silicate belongs to the group of pyrosilicates containing $[Si_2O_7]$-groups. Additional lithium and calcium cations are incorporated between the silicate dimers and are coordinated by four and six nearest oxygen neighbours, respectively. Each $[LiO_4]$-tetrahedron shares two common corners with directly neighboring tetrahedra forming *zweier* single-chains which are running parallel to <100> in z-levels defined by the presence of the $6_1^{[001]}$-screw axes. From the corner-sharing $[LiO_4]$- and $[SiO_4]$-moieties a three dimensional framework can be constructed (see Figure 1). An interesting feature of this framework is the presence of an $O^{[3]}$-type bridging oxygen linking *three* tetrahedra (one $[LiO_4]$- and two $[SiO_4]$-units). Structural similarities with other silicates are discussed in detail. The high-temperature behavior of the Si-O, Ca-O and Li-O bond distances in $Li_2Ca_2Si_2O_7$ was investigated by *in-situ* single-crystal X-ray diffraction in the range between 65 and 700 °C. From the evolution of the lattice parameters, the thermal expansion tensor a_{ij} has been determined. The structural characterization has been supplemented by micro-Raman spectroscopy. Interpretation of the spectroscopic data including the allocation of the bands to certain vibrational species has been aided by DFT-calculations.

Figure 1

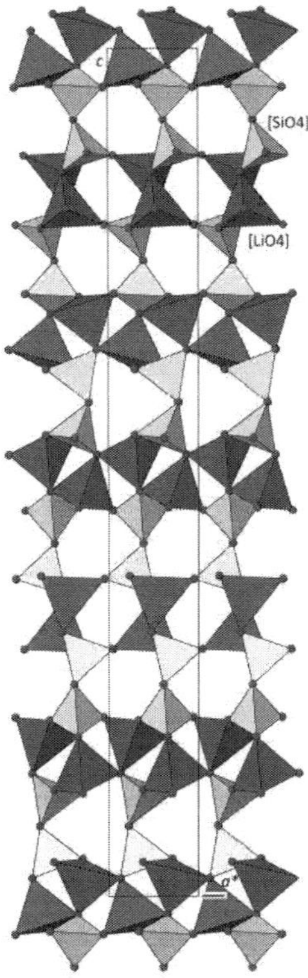

MSC-P26

Impedance spectroscopy of mullite type $Bi_2Fe_4O_9$ and $Bi_2Al_4O_9$ and their interface reactions at 800°C

F. Kiesel[1,2], R. Claus[1,2]
[1]*LUH, Mineralogy, Hannover, Germany*
[2]*University, Mineralogy, Hannover, Germany*

Mullite type compounds $Bi_2M_4O_9$ (M = Al, Fe, Ga) have recently attracted considerable attention concerning a suggested high oxygen conductivity at temperatures between 500 and 1000°C. However, it could be shown in ^{18}O tracer diffusion experiments that oxygen diffusivity remains in the range 10^{-13}-10^{-15} cm^2/s, i.e 6-7 orders of magnitude below that of good oxygen conductor, e.g. Y_2O_3 stabilized ZrO_2 [1-3]. Reinvestigations of the conductivity using impedance spectroscopy on single crystals [4] show the specific values of mullite and mullite type compounds quite well in the same range possessing similar activation energies of about 1-1.3 eV, too (Fig. 1). Whereas for mullite the conductivity was related to impurity conduction (e.g. Na+Mg-defects) for mullite type compounds a conduction mechanism via intrinsic Frenkel type defects remain a possibility. However, this possibility can also be ruled, too, as implied by the results of new experiments in the system $Bi_2Fe_4O_9/Bi_2Al_4O_9$ presented here: In a first step as prepared powders of $Bi_2Fe_4O_9$ and $Bi_2Al_4O_9$ were mixed, pressed to pellets and heated for 8 h and 24 h at 800°C. IR spectra (KBr-method) show that a new phase of composition $Bi_2(Fe_{1-x}Al_x)_4O_9$ of $x \approx 0.3$ was formed beside mainly unreacted $Bi_2Fe_4O_9$ and $Bi_2Al_4O_9$. The content of the new phase increases slightly with increasing heating time. In a second step a larger $Bi_2Fe_4O_9$ crystal were pressed into $Bi_2Al_4O_9$ polycrystalline material and vice versa and heated at 800°C for 11 and 14 days. $Bi_2Fe_4O_9/Bi_2Al_4O_9$ and $Bi_2Al_4O_9/Bi_2Fe_4O_9$ interfaces could be investigated in cross section on a micro-meter scale by means of micro-Raman spectroscopy. The results show the formation of a mixed phase of composition $Bi_2(Fe_{1-x}Al_x)_4O_9$ of $x \approx 0.3$ at certain interface zones while other remain unreacted. Therefore these observations imply that interface reaction probably occur via partial melt and recrystallisation enforced by local enrichment of impurities (alkali-metals). Any significant diffusion controlled exchange mechanism of cations Fe/Al can be ruled out. Therefore, we conclude that the dc conductivity as extracted by impedance spectroscopy should be realized by a random walk of impurity ions in the mullite structure.

[1] T. Debnath, C. H. Rüscher, P. Fielitz, S. Ohmann, G. Borchardt, J. Sol. St. Chem. 183 (2010) 2582-2588.
[2] P. Fielitz, G. Borchardt, M. Burianik, J. Ottinger, M. Mühlberg, Th. M. Gesing, R. X. Fischer, H. Schneider, Sol. St. Ionics 221 (2012) 40-42.
[3] M. Kilo, C. Argirius, G. Borchardt, R. A. Jackson, Phys. Chem. Chem. Phys. 5 (2003) 2219-2224.
[4] C. H. Rüscher, F. Kiesel, in Advances in Solid Oxide Fuel Cells IX, eds. N. P. Bansal et al. (2013) 103-114.

Fig. 1 DC-conductivity by impedance spectroscopy of single crystals of mullite and mullite type compounds ($Bi_2Fe_4O_9$, $Bi_2Al_4O_9$) (modified from Ref. [4]).

Figure 1

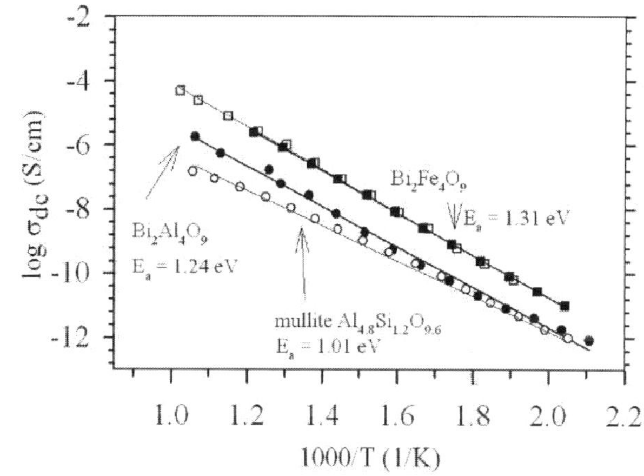

MSC-P27

Microstructural analysis of $CuInSe_2$ thin films by grazing incidence X-ray diffraction

J. Marquardt[1], S. Schorr[1,2], S. Brunken[2]
[1]*Freie Universität Berlin, Berlin, Germany*
[2]*Helmholtz-Zentrum Berlin, Kristallographie, Berlin, Germany*

By now, the progress in manufacturing $Cu(Ga,In)Se_2$ absorbers used for thin film solar cells has led to conversion efficiencies of more than 20% [Ref.a]. In general, compound semiconductors own the advantage of adjusting the band gap by changing the composition of the solid solution, as in $CuIn_{1-x}Ga_xSe_2$. However in this study we focused only on the $CuInSe_2$ end member. We fabricated six similar samples which differ only in substrate temperature during the 2nd and 3rd stage of the 3-stage-co-evaporation process [Ref.b]. With this series we determine the influence of the substrate temperature on microstructure. In addition we fabricated six identical samples with the same temperatures but with a NaF precursor, to identify these effects. The substrate temperatures, during the 2nd and 3rd stage of the 3-stage-co-evaporation process, vary in the range from 330°C to 525°C. The thin films have been characterized with respect to both, the microstructure as well as the phase content for various thin film depths, by means of grazing incidence X-ray diffraction (GIXRD). Bragg-Brentano geometry was used to ascertain a potential texturing. The microstructural analysis refers to the separation of size and strain induced peak broadening using the Thompson-Cox-Hastings-pseudo-Voigt profile function [Ref.c]. Through measuring a standard reference material (LaB6, NIST-SRM 660b) the finite resolution of the XRD instrument was taken into consideration to solely obtain the microstructural information of the XRD pattern. It will be shown that the micro strain decreases with increasing temperature and varies with depth. The NaF precursor increases the micro strain considerately and leads to the formation of planar faults.

Ref.a: Press Release 12/2014, ZSW, http://www.zsw-bw.de/uploads/media/pi18-2013-ZSW- WorldrecordCIGS.pdf

Ref.b: Walter T. et al Solar Energy Materials and Solar Cells 41/42, 355-373, 1996

Ref.c: Thompson P. et al Journal of Applied Crystallography 20, 79-83, 1987

MSC-P28

Spin and orbital disordering in the hole-doped system Pr1−xCaxVO3

M. Reehuis[1], C. Ulrich[2], P. M. Abdala[3], P. Pattison[3], J. Fujioka[4], S. Miyasaka[4], Y. Tokura[4], B. Keimer[5]

[1]*Helmholtz-Zentrum Berlin, EM-AQM, Berlin, Germany*
[2]*University of New South Wales, School of Physics, Sydney, Australia*
[3]*Swiss-Norvegian Beamlines at ESRF, Grenoble, France*
[4]*University of Tokyo, Department of Applied Physics, Tokyo, Japan*
[5]*Max-Planck-Institut für Festkörperforschung, Stuttgart, Germany*

Changes of the crystal structure and magnetic ordering in the system $Pr_{1-x}Ca_xVO_3$ was investigated by high-resolution powder x-ray diffraction and single-crystal neutron diffraction up to x(Ca) = 0.30, well above the metal-insulator transition (MIT) observed at x(Ca) = 0.23. The Mott-Hubbard insulator $PrVO_3$ shows a structural phase transition at T_S = 180 K from an orthorhombic (*Pbnm*) to a monoclinic ($P2_1/b$) structure, which is associated with the onset of orbital ordering and strong Jahn-Teller distortions of the VO_6-octahedra. A magnetic ordering transition driven by exchange interactions between vanadium moments is observed at T_N = 140 K. At lower temperature the C-type ordered vanadium moments oriented in the ab-plane induce a progressive magnetic polarization of the praseodymium sublattice resulting in a ferrimagnetic structure with coexisting modes (C_x, F_y) and (F_x, C_y). Up to x(Ca) = 0.30 the cell volume is continuously shrinking by 2.80 % (at 295 K) and 3.12 % (at 15 K) indicating a growing influence of a covalent bond character. A strong reduction of the transition temperatures (T_S = 108 K, T_N = 95 K), as well as a reduction of the ordered vanadium moments from 1.36 to 0.85 μ_B has been observed in the insulating state up to x(Ca) = 0.20, while the Jahn-Teller distortions are slightly weakened. At the MIT transition Jahn-Teller distortions and long-range orbital ordering are completely quenched. The crystal structure of metallic $Pr_{0.70}Ca_{0.30}VO_3$ remains orthorhombic down to low temperature, while a weakly ordered moment of μ_{exp} = 0.14(3) μ_B could be observed on the vanadium site.

MSC-P29

Experimental studies in boundary systems: low calcian magnesites

U.- N. Berninger[1,2], G. Jordan[1], M. Lindner[1], E. H. Oelkers[2], A. Reul[1], J. Schott[2]

[1]*LMU, München, Germany*
[2]*CNRS-UPS-IRD-CNES, Géosciences Environnement Toulouse, Toulouse, France*

Mg, after Ca, is the second most abundant cation of sedimentary carbonates (Lippmann, 1973). The influence of Mg on Ca-carbonate growth therefore has been studied extensively. However nothing is known about the effect of Ca on the growth of Mg-carbonates. The most common and stable Mg-carbonates are magnesite and dolomite.

Here we used hydrothermal atomic force microscopy (HAFM) and hydrothermal mixed-flow reactors (HMFR) to investigate the effect of aqueous Ca on magnesite growth. Experiments were performed at 100 °C, pH ~7.7 and at 0.1 molal ionic strength. HAFM measurements of obtuse step velocities at Ω~100 showed that only experiments with Ca^{2+} concentrations above 100 µmol/L reduce these step advancements. Such high concentrations of aqueous Ca are unfeasible for HMFR experiments due to Ca-carbonate nucleation during long residence times. Nevertheless, steady-state magnesite growth rates determined by HMFR experiments confirmed that there is no significant kinematic influence of aqueous Ca below 100 µmol/L.

Steady-state solution concentrations of HMFR experiments revealed Ca incorporation of up to 8 mol% into the growing magnesite. Raman spectra of these ~7 µm thick (104) growth layers were collected by vertical line mapping down to 30 µm in depth using a confocal Raman spectrometer with an Olympus BX51 microscope. A 100x objective with a numerical aperture of 0.8 was used for all measurements. Samples were excited using a 532nm LASER and a 2400 spectral grating per mm. Raman spectra were collected in a range of 50 cm^{-1} to 2000 cm^{-1} with an acquisition time of 2x60s.

An impact of Ca^{2+} on carbonate vibration modes was confirmed for translational T and librational L bands of magnesite. While the frequencies of the T and L bands were not affected by the presence of Ca^{2+}, pronounced peak asymmetries at the low vibrational frequency sides of these bands were detected. Subtraction of reference spectrums indicates the presence of secondary bands (T* and L*) at frequencies which are approx. 8 cm^{-1} lower than the T and L bands and contradicts a linear interpolation of frequency shifts between magnesite and dolomite as proposed by Bischoff et al. (1985). Assuming that the T* and L* bands are caused by carbonate groups with a Mg-Ca-coordination ratio of 5, the observations indicate no agglomeration of Ca^{2+} within the cation layers and points to a statistical Ca-Mg distribution within the cation layers of low calcian magnesite.

Bischoff, W.D., Sharma, S.K. & Mackenzie F.T. (1985): Carbonate ion disorder in synthetic and biogenic calcites: a Raman spectral study. Amer. Mineral. 70, 581-589.

Lippmann, F. (1973): *Sedimentary carbonate minerals*. Springer (New York).

MSC-P30

Crystallisation of 11 Å-tobermorite from industrial waste materials

D. Fregin[1], A. Hartmann[1], J.- C. Buhl[1]
[1]*Institut für Mineralogie, Leibniz Universität Hannover, Kristallographie, Hannover, Germany*

11 Å- tobermorite is the main phase in steam hardened building materials, formed from quartz, lime and water. Investigations of raw materials substitution by industrial waste residuals are important for waste recycling processes. Thus a study of crystallization of the CSH Phase 11 Å -tobermorite ($Ca_5Si_6(O, OH)_{18}·5H_2O$) from paper sludge combustion ashes and residues of industrial wastewater cleaning is presented.

Lime was substituted by two calcium-rich ashes with 48.61 M.-% (PA-DS) and 61.19 M.-% (PA-NL). Whereas gehlenite and calcite were found to be the main crystalline phases in PA-DS, PA-NL contains high amounts of calcite. The SiO_2 rich residues from waste water cleaning of the silica and zeolite production (FR1: 80.5 M-% SiO_2; X-ray amorphous) or from silane production (FR2: 90 M-% SiO_2; amorphous silica + minor parts of quartz) were inserted as substitute of quartz. All the waste materials also contain other elements, mainly Al, Fe, Mg, Na, K, S (also 4,84 M-% sulphate was analysed in FR1) and further trace elements. Especially the content of heavy metals must be carefully checked to keep the limiting values for a secure handling and environmental protection.

Synthesis series with 50% substitution of quartz and lime as well as total substitution of both by the waste materials were carried out. In a further series only FR1/2 and burnt lime were reacted to test if 11 Å-tobermorite forms generally from FR1/2. In addition a modified series was developed, using alkaline treated ashes to transformation calcite into $Ca(OH)_2$ by leaching of PA's in 16 M NaOH for 4 h. All experiments were realized in the CaO/SiO_2 interval (C/S-ratio) 0.53 - 0.68 under hydrothermal conditions (200°C, autogenous pressure, 40.5 h). Each product was characterized by XRD, SEM and qualitative EDX-analysis.

Results show the formation of CSH from wastes but not only 11 Å-tobermorite but also CSH (I), xonotlite, plombierite and gyrolite. Furthermore calcite was observed in all products. 11 Å-tobermorite of sufficient quality was analyzed using PA-DS compared with crystals from the calcite rich PA-NL. For substitution of quartz highest amounts of tobermorite were found with FR 1, but also ettringite formation was observed due to the high sulphate concentration in FR1 (Fig. 1). This result have to keep in mind during insertion of industrial residues [1-2].

The optimal 11 Å-tobermorite crystals could be obtained in the modified crystallization series, using ashes, leached in 16 M NaOH for 4 hours, before synthesis together with FR 1 (see XRD patterns in Fig. 2) [3].

Fig. 1: Ettringite with CSH-phases from of the basic materials FR 1 and caustic lime

Fig. 2: XRD of products of the modified series (sludges leached in 16 M NaOH for 4 h before insertion in synthesis).

Acknowledgements: We thank Dr. K. Rübner, BAM-Berlin, for the assignment of the residues.

References: [1] Fregin, D.: Masterarbeit. Institut für Mineralogie, Leibniz Universität Hannover, Hannover 2014. [2]Pöllmann, H.: Compunds with ettringite structure, Stuttgart 1989. [3] Fregin,D.: Projektarbeit. Institut für Mineralogie, Leibniz Universität Hannover, Hannover 2014.

Figure 1

Figure 2

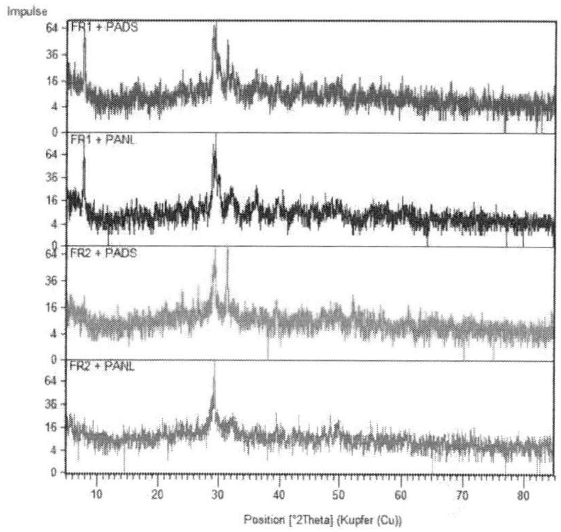

MSC-P31

Characterization of Lithium Niobate Sol-Gel Thin Films with embedded Crystalline Particles

D. Eger Passos[1], D. Karsch[1], H. Stöcker[1], E. Mehner[1], U. Bergmann[2], D. Spitzner[2], D. C. Meyer[1]
[1]*TU Bergakademie Freiberg, Freiberg, Germany*
[2]*TU Dresden, Dresden, Germany*

Lithium niobate is a widely applied material owing to its pyroelectic properties. Most novell applications for pyroelectrics, like anti-icing and disinfection, require thin films on substrates. The wet chemical sol-gel synthesis is an environmentally friendly alternative for coating. Nevertheless, the parallel crystallization of lithium niobate and removal of by-products during the annealing requires long high temperature treatments. This process may be shortened by addition of crystallization nuclei. Hence, in this contribution investigations on the crystallization processes at the nano-scale of wet chemical coatings of silicon and stainless steel substrates are presented.

Samples were produced by dip-coating the different substrates with a sol, followed by a drying procedure of about 1 hour at 100 °C and an annealing procedure of about 4 hours at 600 °C or 800 °C. The crystallization nuclei in form of lithium niobate single crystal powder was either directly added to the sol or air brushed in solution immediately after dip-coating. The samples were analysed by microscopic methods and energy dispersive X-ray spectroscopy. With XRD the high amount of lithium niobate is found. But REM shows, that there are different crystalline structures, and so that the lithium niobate is not homogenously distributed over the surface. EDX provide a relation between the structures and the niob density so that the positions of lithium niobate can be found (fig. 1).

In general, the crystallization starts at the substrate. Some particles of the powder indeed work as additional crystallization nuclei, but some are only covered with the sol (fig. 2). Further research is aiming at a better understanding of lithium niobate crystal growth in sol-gel thin films.

Figure 1

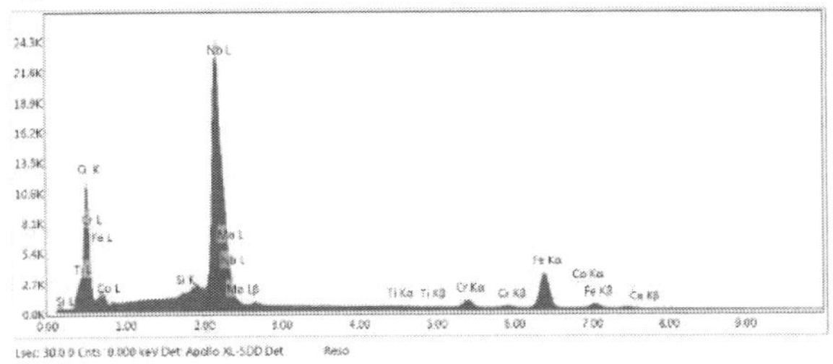

Figure 2

MSC-P32

Experimental determination of transformation textures in various semiconductor materials

N. Pukallus[1], H. Klein[1], H. Sowa[1]
[1]*Geowissenschaftliches Zentrum der Universität Göttingen, Abteilung Kristallographie, Göttingen, Germany*

Under non-ambient conditions the crystal structure of a compound may become energetically unfavourable and a phase transformation occurs which can be driven by different mechanisms. High-pressure experiments were performed in diamond-anvil cells (DAC). It is assumed that the orientation relations of the unit cells of both phases give clues about the actual transformation path. The investigated phase transformations are considered to be reconstructive. During the phase transformation the single crystal turns into a polycrystalline material with a highly preferred orientation. The investigated materials include indium arsenide, copper indium selenide, lithium indium selenide, cadmium sulfide, and cadmium selenide. All of these materials adopt a NaCl-type structure in their high-pressure phase.

The experiments were carried out at the high energy synchrotron beamlines BW5 (λ~0.12 Å) and P02.1 (λ~0.2 Å) at DESY, Hamburg. Before and after the phase transition pole figures were measured in a range $-42° \leq \omega \leq 42°$ of the orientation angle ω according to the opening angle of the DAC. The data were recorded with either a mar345 or a Perkin Elmer 1621 area detector.

For the determination of the experimental pole figures the detector images are cut into concentrical rings which only contain the texture information of a single peak in the diffraction pattern. From the detected intensity distribution along these rings the experimental pole figures were determined from which the orientation distribution function (ODF) was calculated. The ODFs of the starting and resulting phases of the considered material can be directly used to find out the orientation relationships. One common feature of all substances is the recovery of the initial orientation of the single crystal in the post-experiment low-pressure phase, even if the orientation of the high-pressure phase is complex.

MSC-P33

The different water contents of LDH-type structure - Monosulfate

S. Stöber[1], H. Pöllmann[1]
[1]*Martin-Luther-Universität Halle, Mineralogie/Geochemie, Halle, Germany*

Monosulfate (Kuzelite) is a typical LDH structure type mineral composed of a positively charged main layer $Ca_2[Al(OH)_6]$ and a negatively charged interlayer $[SO_4.nH_2O]$. The amount of water in the interlayer is variable due to relative humidity, interlayer anion and temperature. Monosulfate can play a very important role as a typical hydration product of sulfate containing cementitious materials. The following different hydration stages can be described. Due to lower amounts of sulfate a partial replacement of sulfate by hydroxide can occur forming Hemisulfate $Ca_4[Al_2(OH)_{12}]$ $[1/2SO_4.OH.12H_2O]$.

Composition	Temperature	Lattice parameter	Water molecules
$Ca_2[Al(OH)_6]$ $[1/2SO_4.10H_2O]$.	> r.h.	30.67	16
$Ca_2[Al(OH)_6]$ $[1/2SO_4.8H_2O]$.	> r.h.	28.69	14

$Ca_2[Al(OH)_6]$ $[1/2SO_4.6H_2O]$.	25°C	26.8	12
$Ca_2[Al(OH)_6]$ $[1/2SO_4.5H_2O]$.	45°C	26.37	11
$Ca_2[Al(OH)_6]$ $[1/2SO_4.4H_2O]$.	55°C	24.3	9.5-10
$Ca_2[Al(OH)_6]$ $[1/2SO_4.3H_2O]$.	80°C	23.73	9
$Ca_2[Al(OH)_6]$ $[1/2SO_4.0H_2O]$.	>170°C		6 (=OH of main layer)

Literature

PÖLLMANN, H.: AFm-phase in cementitious materials. Proceeding of the 33rd international conference on cement microscopy, San Francisco California, USA, (ISBN 1-930787-07-3), 94 pages, April (2011)

MSC-P34

Joining forces: Combination of IR spectroscopy and DFT calculations to study the adsorption of carbon monoxide in Na-exchanged chabazites

M. Rodríguez Delgado[1], C. Otero Areán[1], C. Oliver Duran[1], M. Fischer[2]
[1]*University of the Balearic Islands, Department of Chemistry, Palma, Spain*
[2]*University of Bremen, Crystallography/Geosciences, Bremen, Germany*

Carbon monoxide is frequently used as an IR probe molecule to study the cation sites in zeolites. Variable-temperature infrared (VTIR) spectroscopy does not only provide insights into the interaction strength (enthalpy of CO adsorption), but also permits to distinguish adsorption sites constituted by a single cation from those in which the adsorbed CO molecule interacts simultaneously with two or more cations. On the computational side, it has been shown previously that density-functional theory (DFT) calculations are a valuable complementary tool for such studies, as they can deliver additional insights into the structure and bonding of the adsorbed CO complexes [1].

In this contribution, we compare the CO adsorption behaviour of two synthetic Na-exchanged chabazites having different Si:Al ratios: A silicon-rich system, termed Na-SSZ-13 (Si:Al ratio of 20:1), and an aluminium-rich system, termed Na-CHA (Si:Al ratio of 2.6:1). The VTIR spectroscopic experiments for CO-loaded samples show qualitative differences in the shift of the C-O stretching frequency. The single (and well-defined) band observed for Na-SSZ-13 exhibits a rather large frequency shift towards higher frequency with respect to the free CO molecule (2143 cm^{-1}), which is indicative of interaction with a single-cation site. On the other hand, the observation of a broader IR absorption band having a smaller frequency shift in Na-CHA points to a significant fraction of CO molecules interacting simultaneously with two or several (nearby) cations.

In order to study the local environment of the adsorbed carbon monoxide molecules, periodic DFT calculations using the dispersion-corrected PBE-D2 functional were performed with the CASTEP code [2]. In the silicon-rich model, which has an Si:Al ratio of 11:1, the adsorbed CO molecules can only interact with single cations. A comparison of the two most probable cation sites shows that the interaction strength varies moderately depending on the cation location. In contrast to this, a variety of local minima were observed in the aluminium-rich model (Si:Al ratio of 2:1). Here, both single-site and dual-site interactions occur. The observed changes of the C-O bond length upon adsorption are found to be in qualitative agreement with the experimentally measured shifts of the stretching frequency, with dual-site interactions exhibiting longer C-O distances and smaller frequency shifts. While the DFT adsorption enthalpies are somewhat higher than the values obtained from VTIR measurements, the differences in interaction strength between Na-SSZ-13 and Na-CHA are reproduced rather well.

[1] C. Otero Areán, M. Rodríguez Delgado, C. López Bauçà, L. Vrbka, P. Nachtigall, Phys. Chem. Chem. Phys. 9 (2007) 4657.

[2] S.J. Clark, M.D. Segall, C.J. Pickard, P.J. Hasnip, M.I.J. Probert, K. Refson, M.C. Payne, Z. Kristallogr. 220 (2005) 567.

MSC-P35

Influence of electric field and domain structure on the low-temperature Raman spectra of $SrTiO_3$

H. Gibhardt[1], J. Leist[1], G. Eckold[1]
[1]*University of Goettingen, Institute for Physical Chemistry, Goettingen, Germany*

The perovskite $SrTiO_3$ is an interesting example of a material with multiferroic properties. It exhibits a phase transition into an antiferrodistortive phase at 105 K via a softmode mechanism and a field induced ferroelectric phase at lower temperatures. Although this is known since more than 20 years, this material is still in the focus of research due to its unusual properties. A particular interesting feature is the formation of ferroic domains.

The domain structure of $SrTiO_3$ below its cubic to tetragonal phase transition depends not only on the stress-state of a particular sample but can also be modified very efficiently by the application of electric fields [1, 2]. Hence, the interpretation of field induced phenomena needs a careful consideration of the domain distribution which is often neglected in existing studies.

In this work, we present polarized Raman spectra of a $SrTiO_3$ single crystal with and without applied electric fields (up to 6.2 kV/cm) for temperatures between 120 K and 20 K. The results clearly show that the strong variation of polarized Raman spectra under external fields observed within both, the paraelectric tetragonal phase and the field induced ferroelectric phase, is a direct consequence of the redistribution of domains. Raman spectroscopy can therefore be used to characterize the domain structure of a particular sample [3].

[1] Leist J, Sidoruk J, Gibhardt H, Hradil K, Meven M and Eckold G 2010
Diffusion Fundamentals 12 69.
[2] Sidoruk J, Leist J, Gibhardt H, Meven M, Hradil K and Eckold G 2010
J. Phys.: Condens. Matter 22 235903.
[3] Gibhardt, H.; Leist, J.; Eckold, G.; Mater. Res. Express, in press.

MSC-P36
The Effect of additive V_2O_5 ON sinter mechanism and roperties of $NiFe_2O_4$ spinel

Y. Liu[1], G. Song[1], J. Zhang[1]
[1]*Northeastern University, China, Materials & Metallurgy, Shenyang, China*

The unique crystal structure hold by Natural magnesium aluminate spinel ($MgAl_2O_4$) is called the spinel type structure. The structure belongs cubic crystal, and possesses face centered cubic lattice. The spinel structure can be considered as the oxygen ions to form cubic closest packing, and then occupied 1/8 interstice of 64 tetrahedrons by X ions, namely 8A sites; and Y ions occupied 1/2 gap of 32 of octahedron, namely 16B sites. Thus we obtained the formula of spinel unit cell is $X_8Y_{16}O_{32}$, and XY_2O_4 for contract expression[1~5].

Ferrite, as its name implies, is a iron oxides originating from magnetite in which the Fe_3O_4 is as main component. And it is the earliest applied nonmetallic magnetic materials in the world. Each single component ferrite possesses typical spinel structure and its typical magnetic properties. $NiFe_2O_4$ is, in itself, a good spin magnetic material and can be used for X band microwave ferrite devices.

In order to improve the properties of $NiFe_2O_4$ spinel, a few amount of additive V_2O_5 was added to the raw materials during the course of synthesizing. The powders of NiO、Fe_2O_3 and slight amount of V_2O_5 were mixed, then moulded and sintered at 1200°Cfor 6h. The sinter mechanism of powders of NiO and Fe_2O_3 with some additive V_2O_5 was researched. The effect of V_2O_5 on density, electric conductivity and corrosion resistance of $NiFe_2O_4$ spinel was studied at the same time. The results show that the sinter mechanism of powders of NiO and Fe_2O_3 with some additive V_2O_5 is liquid-phase sintering; Additive V_2O_5 can increase the density of the samples; especially it improves the corrosion resistance of the samples remarkably. When the amount of V_2O_5 is 1.5%, the sample's corrosion rate is 1/80 of that of sample without V_2O_5. But the electric conductivity of the samples with V_2O_5 is lower than that of the sample without V_2O_5.

Reference

[1] Evans R C. A Introduction to Crystal Chemistry[M]. Cambridge University Press, 1976.
[2] West A R. Solid State Chemistry and Its Application[M]. John Wiley & Sons, 1984.
[3] Kingery W D, et al. Introduction to Ceramics[M]. John Wiley & Sons, 1976.
[4] Zhou Zhigang, et al. The Magnetic Ferrite Materials[M]. Beijing: Science Publication, 1981.
[5] Li Mengyuan, Li Guodong. The Ferrite Physics[M]. Beijing: Science Publication, 1978.

Figure 1

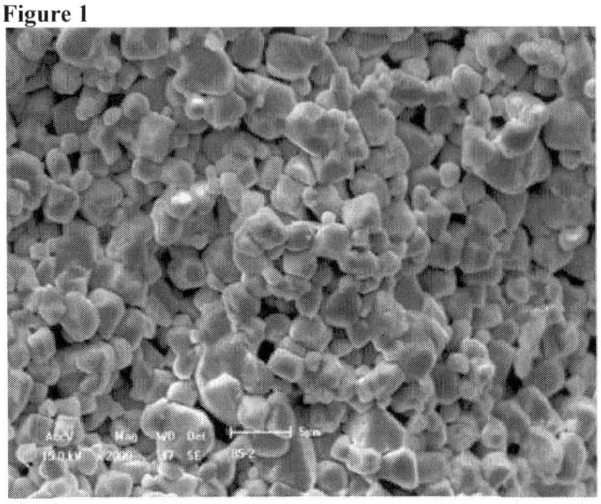

SEM photograph of sample without additive

Figure 2

SEM photograph of sample with 1.5% V_2O_5

MSC-P37
Investigations of the solid state of protected aminoacids in presence of dicyclohexylamine

K. Pappert[1], C. Schauerte*[2], U. Schmitz*[3], K. Merz*[1]
[1]*Ruhr-Universität Bochum, Department of Inorganic chemistry I, Bochum, Germany*
[2]*solid-chem GmbH, Bochum, Germany*
[3]*chemcube, Bochum, Germany*

Protected amino acids precipitated with dicyclohexylamine (DCHA) are commercially available and used for polypeptide synthesis as well as some are precursors of active pharmaceutical ingredients (*APIs*). However, their chemical properties in the solid state have hardly been characterized [1]. So it is totally unclear if those compounds are salts, co-crystals or physically mixtures.

Protecting groups like carbobenzyloxy (Cbz) and *tert*-butyloxycarbonyl (Boc) have functional groups, which are theoretically capable to form molecular synthons and motifs in the solid state. Due to the aromatic rings of Cbz and the building of carboxylate esters using Cbz and Boc, compounds with protecting groups are be able to build up molecular interactions like π-π stacking motifs and hydrogen bonds during crystallization. DCHA possesses an amine as a functional group. Secondary amines are likely to form hydrogen bond motifs like N-H···N. If DCHA is added to a protected amino acid it can be considered that a molecular synthon of the type N-H···O might be formed. However, it can also be considered that the amino acid and the amine will perform a salt formation by transferring the hydrogen cation from the hydroxyl group to the lone pair of the amine. To clarify the nature of the solid state and the formation of polymorphic forms we have carried out several experiments using three different protected amino acids precipitated with dicyclohexylamine. The three used compounds are based on one naturally occurring amino acid (*tryptophan*), one synthetic amino acid (*2,4-diaminobutyl acid*) and a derivate of the aspartic acid (*aspartic acid-beta-t-butyl ester*) protected by Cbz- and Boc- protecting groups. The used compounds are shown in Fig. 1.

Fig. 1: Used compounds and their trivial names.

[1] Pathik Sahoo et al. Gel Sculpture: Moldable, Load-Bearing and Self-Healing Non Polymeric Supramolecular Gel Derived from a Simple Organic Salt, Chem. Eur. J. 18, 8057 - 8063, 2012.

Figure 1

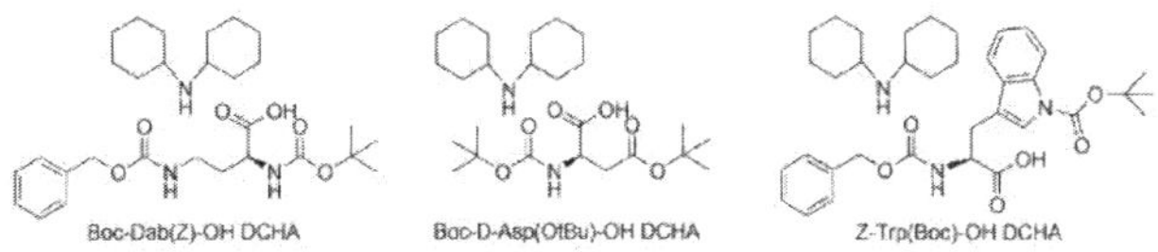

MSC-P38

Unusual reduction route to a novel Mn^{III}-polymer with a tetraazaadamantane-like ligand as linker

D. Premužić[1], M. Hołyńska[1]
[1]*Philipps-Universität Marburg, FB Chemie, Marburg, Germany*

Symmetry plays an important role in the field of SMMs, where axial symmetry of the system enhances the magnetic anisotropy.[1] Tripodal ligands can be used to impose a threefold symmtry on the resulting metal complex. Tris(1-propan-2-onyl oxime)amine (Ox_3H_3) is an example of a tripodal oxime ligand with different known metal complexes, mainly containing Ni.[2] This ligand undergoes an unusual intramolecular cyclization leading to a 4,6,10-trihydroxy-1,4,6,10-tetraazaadamantane product.[3] In combination with Mn, Ox_3H_3 transforms *in situ* and yields different Mn^{IV}-dimers depending on the used additives. In order to improve the magnetic properties and obtain Mn^{III}-complexes with much higher intrinsic magnetic anisotropy, these complexes were reduced with KC_8. In the case of the azide-brigded complex a polymeric compound was synthesized (Fig. **1**). Herein we present the synthesis, structure and first properties of this polymer, incorporating the tetraazaadamantane-like ligand showing two different coordination modes and Mn atoms with two completely different coordination spheres, one distorted octahedrally and the other trigonal bipyramidal.

Fig. **1**: Molecular structure of polymeric compound (hydrogen omitted for clarity).

[1] T. Glaser, *Chem. Comm.*, 2011, *47*, 116-130. [2] A. N. Semakin, A. Yu. Sukhorukov, A. V. Lexiv, S. L. Ioffe, K. A. Lyssenko, Y. V. Nelyubina, V. A. Tartakovsky, *Org. Lett.*, 2009, *11*, 4072-4075. [3] M. J. Goldcamp, S. E. Robison, J. A. Krause Bauer, M. J. Baldwin, *Inorg. Chem.*, 2002, *41*, 2307-2309.

Figure 1

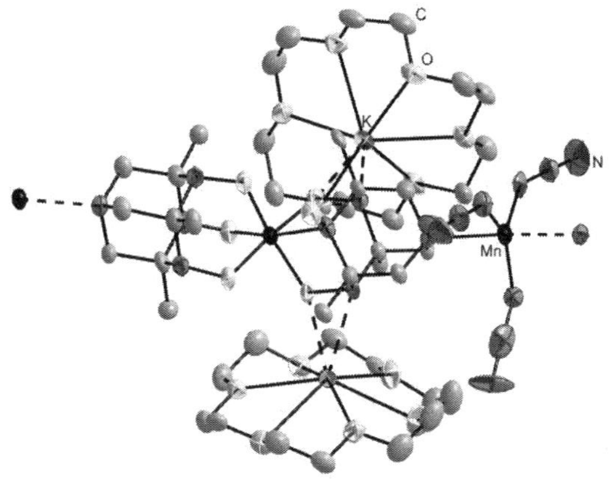

MSC-P39

Influence of H/D exchange on the crystallization process of oxalyl dihydrazide

V. Grasmik[1], R. Zabatyuk[2], K. Merz*[1]
[1]*Ruhr-Universität Bochum, Department of Inorganic Chemistry I, Bochum, Germany*
[2]*National Academy of Science of Ukraine, Division of Functional Materials Chemistry, Kharkiv, Ukraine*

Polymorphism occurs in the solid phase when one type of molecule is able to form different crystal structures although the chemical composition stays the same. Since the crystal structures differ upon intermolecular interactions (like hydrogen bonds) in polymorphs, some properties may differ as well. The investigation of polymorphism is important for pharmaceuticals, organic semiconductors and other applications [1]. Experiments on acridine and deuterated acridine as well as pyridine-N-oxide showed that the smallest change of the molecule - the hydrogen/deuterium exchange - results in a change of the polymorph [2-6].

Oxalyl dihydrazide (ODH) is an interesting molecule since it exhibits five polymorphs which can be obtained through recrystallization in different solvents using different temperatures [1]. Molecules with four and more experimentally obtained polymorphs are rare [1]. Due to a low energy barrier, rotation of the terminal NH_2 group around the NH-NH_2 bond occurs easily and leads to distinct conformations. This results in diverse energies of intermolecular interactions for each polymorph (figure 1). The question is whether H/D exchange has an influence on the polymorphism of oxalyl dihydrazide.

The second part of the investigations deals with the influence of H/D exchange on the formation of co-crystals involving oxalyl dihydrazide. Co-crystals play a crucial role in the field of crystal engineering which is used to synthesize crystals with special properties [7]. Since oxalyl dihydrazide is build up of two CO-NH-NH_2, suitable co-formers are hydrazides, amides or carboxylic acids. An example for a suitable co-former is cyanuric acid (CA) as shown in figure 2.

Figure 1: oxalyl dihydrazide dimers of the α polymorph (top) and γ polymorph (bottom)

Figure 2: arrangement of oxalyl dihydrazide, cyanuric acid and water in the co-crystal [ODH/CA/2H_2O]

[1] S. Ahn, F. Guo, B.M. Kariuki, K.D.M. Harris, *J. Am. Chem. Soc.*, 2006, *128*, 8441.
[2] A. Kupka, V. Vasylyeva, D.W.M. Hofmann, K. Merz, *Cryst. Growth Des.*, 2012, *12*, 5966.
[3] O.V. Shishkin, S.V. Shishkina, A.V. Maleev, R. Zabatyuk, V. Vasylyeva, K. Merz, *ChemPhysChem*, 2013, *14*, 847.
[4] V. Vasylyeva, T. Kedziorski, N. Metzler-Nolte, C. Schauerte, K. Merz, *Cryst. Growth Des.*, 2010, *10*, 4224.
[5] K. Merz, V. Vasylyeva, *CrystEngCom.*, 2010, *12*, 3989.
[6] K. Merz, A. Kupka, *Cryst. Growth Des.*, 2014, in revision.
[7] K.D.M. Harris, N.M. Stainton, A.M. Callan, R.A. Howie, *J. Mater. Chem.*, 1993, *3*, 947.

Figure 1

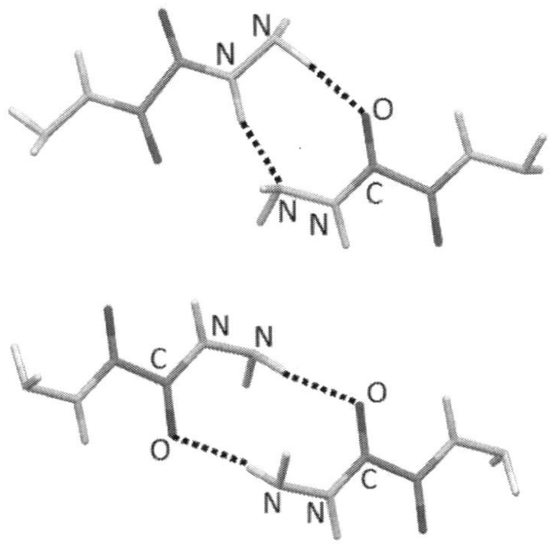

Figure 2

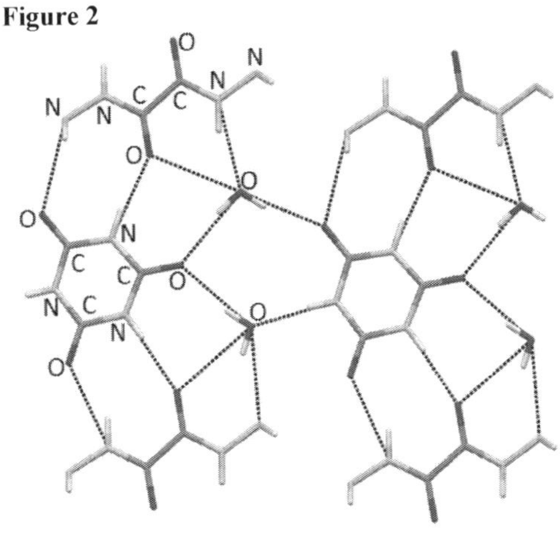

MSC-P40

Unusual Raman signals in multiferroic $Bi_2Mn_4O_{10}$

M. M. Murshed[1], F. Ziegler[2], H. Gibhardt[2], G. Eckold[2], T. M. Gesing[1]
[1]*Universität Bremen, Chemische Kristallographie fester Stoffe, Institut für Anorganische Chemie, Bremen, Germany*
[2]*Georg-August Universität Göttingen, Institut für Physikalische Chemie, Göttingen, Germany*

Mullite-type $Bi_2Mn_4O_{10}$ [1] was reported as a multiferroic although its crystal structure [2] is not typical for ferroelectric since the dipoles of the constituent $Mn^{4+}O_6$ (along *c*-axis, bond valence sum (BVS) = 3.90(2) v.u.) and $Mn^{3+}O_5$ (along *b*-axis, BVS = 3.04(2) v.u.) polyhedra are oriented in different directions [3]. Both its nuclear and magnetic structural features differ from those of other rare-earth members of the $R_2Mn_4O_{10}$ family. For instance, the $Mn^{4+}O_6$, $Mn^{3+}O_5$ and $Bi^{3+}O_8$ (BVS = 3.18(2) v.u.) polyhedra show distinctive distortions due to slight but significant lone electron pair activity of the Bi^{3+} cation. Whereas each member of $R_2Mn_4O_{10}$ family possesses incommensurate magnetic character [4], $Bi_2Mn_4O_{10}$ shows a commensurate [5] magnetic structure. We report possible electromagnon, the spin waves excited by the electric component of the electromagnetic wave, observed in $Bi_2Mn_4O_{10}$ at temperatures far away from the Néel temperature of 39 K. The crystal structure (*Pbam*) has been refined from both single crystal and powder X-ray diffraction data. Temperature-dependent Raman spectra showed tens of intense bands at low frequency regions where the optical phonons are not expected according to lattice dynamics calculations [6]. Selective polarization either in *Z(XX)Z* or *Z(XY)Z* configuration produced similar number of bands at identical positions, however with different intensity. As much as two sets of peaks were identified, which are equally spaced (frequency span of 7.90(3) cm^{-1} and 11.40(1) cm^{-1}), and each set followed an arithmetical sequence. The frequency and line-width of these bands are temperature independent, which unambiguously decline the possibility to originate from magnons. Since powder samples showed identical spectra at ambient temperature, the modes cannot be explained as an anomalous surface-induced phenomenon. Based on experimental works [7], the repeated observations of these inelastic light scattering features could possibly be related to electromagnons which are also reported on some isostructural materials [4, 8]. The symbiotic control of electric and magnetic signals associated with the unique property of an electromagnon bears an attractive implication for advanced technologies.

References

[1] H. Schneider, R.X. Fischer, Th.M. Gesing, et al. Int. J. Mat. Res. 103 (2012) 422. [2] M. Burianek, Th.F. Krenzel, M. Schmittner et al. Int. J. Mat. Res. 103 (2012) 449. [3] L.M. Volkova, D.V. Marinin, J. Phys.: Condens. Matter 21 (2009) 015903. [4] S. Petit, V. Balédent, C. Doubrovsky et al. Phys. Rev. B 87 (2013) 140301. [5] C. Vecchini, L.C. Chapon, P.J. Brown et al. Phys. Rev. B 77 (2008) 134434. [6] F.M. Silva Júnior, C.W.A. Paschoal, R.M. Almeida et al. Vibrat. Spec. 66 (2013) 43. [7] M. Cazayous, Y. Gallais, A. Sacuto et al. Phys. Rev. Lett. 101 (2008) 037601. [8] A.B. Sushkov, R.V. Aguilar, S. Park et al. Phys. Rev. Lett. 98 (2007) 027202.

MSC-P41

Fast XRD studies with pink beam at the Extreme Conditions Beamline P02.2 at PETRA III

W. Morgenroth[1], H.- P. Liermann[2], B. Winkler[1]
[1]*Goethe-Universität, Kristallographie, Frankfurt, Germany*
[2]*Deutsches Elektronen-Synchrotron DESY, Photon Science, Hamburg, Germany*

The Extreme Conditions Beamline (ECB) P02.2 at PETRA III [1] is a beamline for micro X-ray diffraction studies of matter at simultaneous high-pressure and high/low-temperature in the diamond anvil cell (DAC). In 2015 two important upgrades will be commissioned to strengthen time-resolved X-ray diffraction experiments:

a) pink beam with a bandwidth of 20 % centered around 42 keV and
b) two fast GaAs based LAMBDA detectors.

Pink beam at the ECB offers a broad energy band and high flux and will be used to study rapid reactions (*in situ* or *operando* research), small-molecule structure determination in the laser heated DAC without sample rotation and, additionally, μ-fluorescence spectroscopy experiments which benefit from the high flux.

Two fast LAMBDA detectors, a 3 module GaAs bonded area detector from *X-Spectrum GmbH*, will be available at the beamline to conduct time-resolved powder and single-crystal diffraction experiments in the kHz to MHz time regime. Two detectors will be used to collect diffraction data in forward and perpendicular direction.

Within this presentation we will describe the current and upcoming capabilities of the Extreme Conditions Beamline to conduct time-resolved X-ray diffraction experiments.

The authors gratefully acknowledge financial support from the BMBF (project 05K13RF1).

[1] H.-P. Liermann et al., "The Extreme Conditions Beamline P02.2 and The Extreme Conditions Science Infrastructure at PETRA III", submitted to JSR (2014)

MSC-P42

In situ Grazing Incidence Diffraction in electrochemical catalysis - from theory to experiment (Part II)

M. Scherzer[1], F. Girgsdies[1], M. Friedrich[1], E. Frei[1], R. Schlögl[1]
[1]*Fritz Haber Institut der Max Planck Gesellschaft, Inorganic chemistry, Berlin, Germany*

Surface sensitive X-ray diffraction techniques like grazing incidence diffraction (GIXRD) or X-ray reflectivity (XRR) give a great possibility to characterize thin layers and their structural properties[1].

Over the last decades, as well as in recent heterogeneous catalysis research, the water splitting reaction on noble metal surfaces is of particular interest. The reaction itself is well known but the mechanism not well understood. Spectroscopic and structural proofs are largely missing. Recent works show first hints and possible mechanism pathways for certain potential regimes[2]. In those works the role of oxides as intermediates and catalyst within the oxygen evolution reaction (OER) is mentioned, but they are structurally rarely characterized.

In the present work we present first results of the structural investigation in these oxide containing surface layers. Ex-situ anodic oxidized (OER) noble metal foils of iridium (at ~ 1.5V) and platinum (at ~ 2.5 V) were measured by GIXRD and XRR. To extend the characterization, additional synchrotron XPS spectra of an oxidized Pt foil was taken.

The metal foils exhibit a thin oxide layer after the oxidation step. The platinumoxide layer is not long term stable and decomposes after a few days. Fig.1 shows a 20 nm thick platinum particle layer (TEM image) after the decomposition of the oxide. XRR patterns of freshly oxidized foils of and iridium (at different time scales) show a clear dependence and shift of the critical angle. Thicker oxide layers (~ 2 h oxidation) show a smaller critical angle than thinner ones (~ 30 min oxidation). Fig.2 shows the XRR trend for three different thicknesses against pure platinum as reference. XPS spectra of a thin layer sample was recorded at BESSY II (Berlin). The measurements showed that the approximately 2 nm thin layer consists of Pt^{2+} oxide and at very surface near regions of Pt^{4+} oxide. The high valent oxide is not stable under the beam radiation and decomposes.

GIXRD measurements do not exhibit any characteristic phase peaks. The observations are referred to a shift of the platinum peak itself, which originates from the refraction effect of the measurements close to the critical angle.

The results of GIXRD, XRR and XPS give a first view on the expectations for the upcoming planned in situ measurements.

Fig. 1: The metastable material formed on top (Pt - oxo - hydroxo-layer) of a polycrystalline Pt-foil decomposes (is reduced) to form Pt nanoparticles under open circuit conditions.

Fig. 2: XRR measurements of oxidized platinum foils (30min (blue) - 1h (black) - 2 h (red)) and pure platinum as reference (green). The critical angle shifts with the thickness of the oxide layer towards lower values.

[1] W.Marra, P.Eisenberger, A.Cho; J.Appl.Phys. 50, 6827 (1979)

[2] O. Diaz-Morales et al; ChemSci 4, 2334-2343 (2013),

Mail to the corresponding author: scherzer@fhi-berlin.mpg.de

Figure 1

Fig. 1

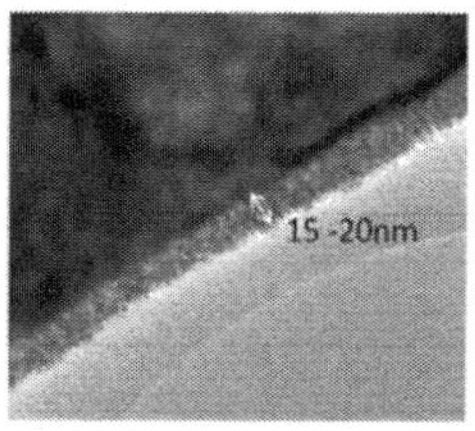

Fig. 2

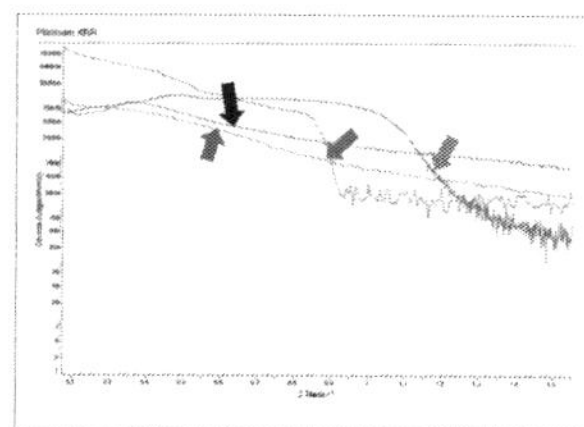

MSC-P43

Chemical Modification of x and Exchange of M in $M_xV_2O_5$ and M_xFePO_4

J. Bauer[1], G. Kickelbick[1], R. Haberkorn[1]
[1]*Universität des Saarlandes, Anorganische Festkörperchemie, Saarbrücken, Germany*

Energy storage is an increasing demand nowadays that can be achieved *via* rechargable batteries for stationary as well as mobile applications. Cathode materials of lithium-ion batteries, e.g. $LiFePO_4$ and $LiCoO_2$, are versatilely researched in and are already established as technical storage materials. However, research in the field of sodium-ion batteries achieved a boost in the last few years, due to several reasons. Sodium is less toxic, less expensive and naturally more abundant in the Earth's crust than lithium.

Our research focuses on vanadates and phosphates, especially $Na_xV_2O_5$, $Na_xV_3O_8$, Na_xFePO_4, and the corresponding lithium compounds. Electrochemical sodiation/lithiation or desodiation/delithiation is often described in the literature [1-4]. Especially in the case of sodium the corresponding chemical reactions are scarcely reported.

By chemical sodiation of V_2O_5 using Na_2S as reducing and sodiation agent $Na_xV_2O_5$ with x up to 1 was synthesized [5] after optimization of the reaction conditions and of the surface area of V_2O_5. Using NaI instead of Na_2S only negligible sodiation was observed. Notably all these products belong to the α/α'-phase while solid state reactions lead to different phases (α, β, τ, α') depending on the value of *x*.

Chemical sodiation of $Na_{1.2}V_3O_8$ by Na_2S (NaI) yielded $Na_{1.2+x}V_3O_8$ with *x* up to about 1 (1/3). Again, NaI proved to be a less suitable sodiation agent, nevertheless a higher amount of sodiation was observed compared with V_2O_5.

Because the preparation of V_2O_5 showed great influence on the reactivity during sodiation [5], delithiation/desodiation was taken into account for preparation of an activated educt. For example, delithiation of $LiFePO_4$ by an aqueous solution of Br_2 or $K_2S_2O_8$ provided $FePO_4$ for sodiation experiments. Using Na_2S Na_xFePO_4 with *x* up to 1 could be achieved. Both, delithiation and sodiation, are topotactical reactions. The product $NaFePO_4$ is therefore isotypic to $LiFePO_4$ (triphylite type) while 'normal' $NaFePO_4$ (maricite type) shows a different structure type. Both types may be derived from the olivine type as already reported in literature [6].

[1] A. Hammou, A. Hammouche, *Electrochim. Acta* 1988, *33*, 1719.
[2] J. Farcy, R. Messina, J. Perichon, *J. Electrochem. Soc.* 1990, *137*, 1337.
[3] H. R. Bak, J. H. Lee, B. K. Kim, W. Y. Yoon, *Electron. Mater. Lett.* 2013, *9*, 195.
[4] K. West, B. Zachau-Christiansen, T. Jacobsen, S.Skaarup, *Solid State Ionics* 1988, *28-30*, 1128.
[5] R. Haberkorn, J. Bauer, G. Kickelbick, *ZAAC* (in press), DOI: 10.1002/zaac.201400381.
[6] M. Avdeev, Z. Mohamed, C. D. Ling, J. Lu, M. Tamaru, A. Yamada, P. Barpanda, *Inorg. Chem.* 2013, *52*, 8685.

MSC-P44

KMC-2, a versatile diffraction and XAS beamline at BESSY II

D. Többens[1], S. Zander[1], S. Schorr[1]
[1]*Helmholtz-Zentrum Berlin, Abteilung für Kristallographie, Berlin, Germany*

The KMC-2 beamline operates two permanent endstations, "XANES" and "Diffraction". A graded SiGe monochromator provides radiation in the energy range of 4 - 15 keV with energy resolution of 1/4000 and beam intensity stabilized to an accuracy of 0.3 %. XANES uses this high-quality radiation EXAFS, XANES, μEXAFS, μXANES, and X-ray fluorescence measurements, in which Si-PIN photodiodes, ionization chambers and an energy-dispersive Röntec X-Flash detector can be mounted for transmission and fluorescence. DIFFRACTION is a six-circle Ψ-geometry goniometer, equipped with either a Vantec 2000 area detector or a scintillation detector with motorized detector slits, for maximum resolution. It is suitable for a wide range of scattering and diffraction measurements, including grazing incidence diffraction (GID), reciprocal space mapping, diffuse scattering, anomalous diffraction, reflectometry, and powder diffraction.

Sample environments to control temperature and atmosphere are available on site, and user-provided sample environments can be adapted.

The time needed to switch between XANES and DIFFRACTION is only two hours. This makes it possible to combine both stations in the same experiment, making KMC-2 a very versatile beamline especially suited for a wide range of non-standard experiments.

MSC-P45

Influence of substitution and nanostructures on the thermoelectric properties of disordered tellurides

S. Maier[1], F. Fahrnbauer[2], T. Schröder[1], N. Giesbrecht[1], M. Nentwig[1,2], G. J. Snyder[3], O. Oeckler[2]
[1]*University of Munich (LMU), Chemistry, Munich, Germany*
[2]*Leipzig University, IMKM, Leipzig, Germany*
[3]*California Institute of Technology, Materials Science, Pasadena, United States*

Multinary thermoelectric materials, especially when quenched from melts, are often inhomogeneous, which may, however, lead to favorable nanoscopic heterostructures.[1,2] In addition to a high Seebeck coefficient (S) and electrical conductivity (σ), both combined in the so-called power factor ($S^2\sigma$), a low thermal conductivity is the key property for a high thermoelectric figure of merit (ZT). While the electronic contribution to the thermal conductivity (κ_e) is intrinsically tied to the electrical properties via the Wiedemann-Franz law, the phononic controbution (κ_{ph}) can be lowered by heterostructuring, substitution or short-range vacancy ordering as a consequence of phase transitions as shown e.g. for $(GeTe)_nSb_2Te_3$. For n ≥ 3, a rocksalt-type high-temperature (HT) phase is accessible above ~ 450 °C, in which the cation position is randomly occupied by Ge, Sb and vacancies. Quenching this HT phase leads to a pseudocubic metastable phase with a nanodomain structure that contains finite defect layers perpendicular to the cubic <111> directions. Cation diffusion at T ~ 300 °C leads to vacancy layers associated with a transition from the metastable phase to a trigonal layered one. In quenched $(GeTe)_nSb_2Te_3$, κ_{ph} is especially low due to a parquet-like nanostructure.[3]

The thermal conductivity can be further reduced by heterostructuring, e.g. in $[CoSb_2(GeTe)_{0.5}]_x(GeTe)_{10.5}Sb_2Te_3$ materials which consist of a disordered $(GeTe)_{10.5}Sb_2Te_3$ matrix and a varying fraction x of $CoSb_2(GeTe)_{0.5}$ skutterudite type precipitates. The thermoelectric properties of such composite materials were determined as a function of the precipitate content x. The measurement strategy has a crucial influence on the results. Thus, the thermoelectric performance depends on the thermal treatment, especially in the existence range of the pseudocubic metastable phase of the matrix. The low thermal conductivity found for samples cooled from the melt, annealed at the existence range of the rocksalt-type HT phase and subsequently quenched in air is compensated with a low electrical conductivity and results in a ZT similar to the one found for homogeneous materials. However, quenching the melt in water also yields a low thermal conductivity, which in combination with a high power factor improves ZT noteably compared to the one of the pure matrix material.

In related germanium silver antimony telluride (TAGS) materials, the introduction of cation vacancies and additional cation disorder due to substitution of Sb with In led to a low κ_{ph}. The thermoelectric properties of $Ge_{0.53}Ag_{0.13}Sb_{0.27}\square_{0.07}Te_1$, $Ge_{0.61}Ag_{0.11}Sb_{0.22}\square_{0.06}Te_1$ and $(GeTe)_{5.5}AgIn_{0.5}Sb_{0.5}Te_2$ as a member of the solid solution series $(GeTe)_xAgIn_ySb_{1-y}Te_2$ are improved compared to conventional TAGS materials.

[1] Y. Zhang, G. D. Stucky, *Chem. Mater.* 2014, *26*, 837.
[2] W. Liu, X. Yan, G. Chen, Z. Ren, *Nano Energy* 2012, *1*, 42.
[3] T. Rosenthal. M. N. Schneider, C. Stiewe, M. Döblinger, O. Oeckler, *Chem. Mater.* 2011, *23*, 4349.

MSC-P46

A new low-cost method to take and store Laue patterns

W. Lottermoser[1]
[1]*University of Salzburg, Materials Engineering and Physics, Salzburg, Austria*

The Laue method, which was derived at the beginning of the 20th century by Max von Laue and coworkers, was the first procedure to show reciprocal lattice planes by their X-ray reflections on a fluorescent screen or film. Insofar it was the birth of modern crystal structure determination. However, in the following decades this method received a loss of interest by the uprise of powerful diffractometers where not only the special Laue symmetry and orientation of a crystal could be determined but also the space group, positions and thermal motions of ions in the unit cell.

With the development of high-performance computers, however, the Laue method experienced a splendid renaissance [1]. In particular for comparably big unit cells of organic structures and with the wide use of synchrotron radiation Laue patterns were recorded with amazing details and resolution. As storage materials films, image plates, "electronic" films or counter tubes have been applied with their disadvantages being well-known: The first are chemically hazardous, the second and third very expensive and the last rather time-consuming. A fifth possibiility, the fluorescent screen has been widely neglected because the common weakness of reflections render storage almost impossible.

However, in collaboration with PHYWE Goettingen [2], their slightly modified XR 4.0 X-ray expert unit, a fluorescent screen and a digital camera have been combined in order to give the crystallographer a low-cost, rather quick and powerful method at hand which moreover allows digitally storing. This will be demonstrated at the comparably "easy" example LiF - at present optimization of the components, especially the use of a high-tech fluorescent foil from a Swiss manufacturer, is being performed.

[1] Steiner, K., Computerunterstützte Orientierung von Einkristallen mit Hilfe des LAUE-Verfahrens, diploma work, Salzburg, unpublished (1995)
[2] PHYWE Systems GmbH und Co. KG, TESS expert Laboratory experiments (X-ray), manual order-No. 01200-02,TEP 5.4.16-01, 4th edition, pp. 1-6 (2012)

MSC-P47

Crystallography and crystal chemistry of alkaline earth chloroacetates and their hydrates

R. Kaden[1], H. Pöllmann[1]
[1]*Martin Luther University, Halle (Saale), Mineralogy / Geochemistry, Halle (Saale), Germany*

Although chloroacetic acids are used in different fields of pharmacy, cosmetics, production of dyes, pesticides and microporous polymers, the alkaline and alkaline earth salts of these acids are scarcely reported. Crystal structural data are available for NH_4-, Li-, Na-, Ca-, Ag-, Ni-monochloroacetates/-hydrates [1-7]. Thus the present study focuses on the crystal chemistry, structure and thermal stability of alkaline earth salts of chloroacetic acids and their hydrates.

The compounds were synthesised by reaction of chloroacetic acids with the carbonates in aqueous solution and crystallised by slow evaporation of the solvent water. Owing to their high alkalinity, for all the alkaline compounds stored in mother liquor, a hydrolysis of monochloroacetate into glycolic acid and Cl^- results [8]:

$$ClCH_2COOA + H_2O \leftrightarrow ClCH_2COOH + A^+ + OH^- \rightarrow HOCH_2COOH + A^+ + Cl^-$$

Only for A = Li this conversion reaction is not complete and an addition compound $Li(ClCH_2COO)(HOCH_2COOH)$ with a chain like crystal structure crystallised. Also for Mg a hydrolysis can be observed leading to $Mg(HOCH_2COO)(ClCH_2COO)$. The Ca, Sr and Ba mono-, di- and trichloroacetates (MCA, DCA and TCA) were found to be stable also when stored in the mother liquor and to form layered structures. Characteristic for all of these compounds the chloromethyl groups are roughly directed in stacking direction on both outsides of the layers and the structural bonded water is always part of the cationic coordination polyhedron. However, significant differences occur in the design of the layers. In case of $Ca(TCA)_2{\cdot}4H_2O$ the layers consist of coordination polymeric chains with only interlinking hydrogen bonds (Fig. 1). The polymeric chains of $Ca(MCA)_2{\cdot}H_2O$, $Ca(DCA)_2{\cdot}H_2O$ and $Sr(DCA)_2{\cdot}2H_2O$ are interlinked by the carboxylic groups of the chloroacetate ligands. Whereas, $Sr(MCA)_2$, $Ba(MCA)_2{\cdot}H_2O$, $Ba(DCA)_2$ and $Ba(TCA)_2{\cdot}H_2O$ form 2-dimensional coordination polymeric layer types.

The inter layer spacing d_{il} primarily depends on the cationic radius and the number of Cl of the chloromethyl groups. For the monochloroacetates at 295 K d_{il} increases from 10.5 Å for $Ca(MCA)_2{\cdot}H_2O$ to 10.7 Å for $Sr(MCAc)_2$ and 10.9 Å for $Ba(MCAc)_2{\cdot}H_2O$. Whereas, in the same order of cations d_{il} decreases from 11.95 Å → 11.7 Å → 11.3 Å for the dichloroacetates and from 11.9 Å → 11.4 Å → 11.0 Å for the trichloroacetates.

Fig. 1. Crystal structure of $Ca(TCA)_2{\cdot}4H_2O$ projected || [001], the solid rectangle marks the unit cell. Right: single chain of $[Ca(TCA)_2{\cdot}4H_2O]_\infty$, length = 4*c*.

[1] M. Ichikawa (1972) Acta Cryst. B28, 755
[2] H. Ehrenberger *et al.* (1999) Acta Cryst. B55, 517
[3] L. Elizabé *et al.* (1997) J. Phys. Chem. 101, 8827
[4] A. Karipides, K. Peiffer (1988) Inorg. Chem. 27, 3255
[5] R. Kaden, H. Pöllmann (2011) ICDD Grant In Aid, pdf 00-062-1696
[6] M. Epple, H. Kirschnick (1997) Chem. Ber. 130, 291
[7] X. Solans, C. Miravitlles (1981) Acta Cryst. B37, 1407
[8] W. A. Drushel, G. S. Simpson (1917) J. Am. Chem. Soc. 39, 2453

Figure 1

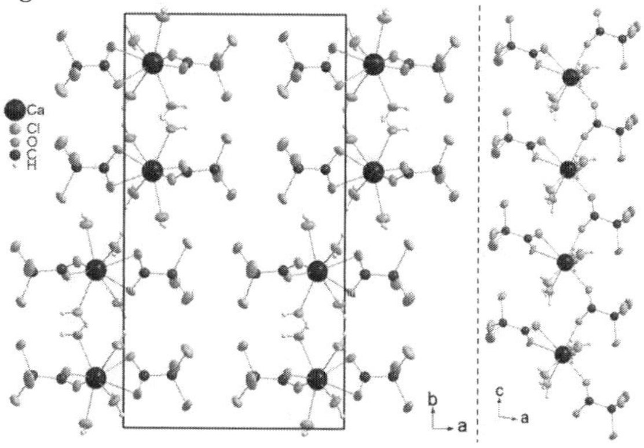

MSC-P48

Crystallographic investigations of ball-milled gold supported on α-Fe_2O_3

S. Ortatatli[1], S. Immohr[1], C. Weidenthaler[1]
[1] *MPI für Kohlenforschung, Heterogeneous Catalysis, Mülheim an der Ruhr, Germany*

Au supported on α-Fe_2O_3 is tested as catalyst for fuel cell applications and is known to be an active catalyst for CO oxidation [1]. Mechanochemical treatment of catalysts under reactive gas atmospheres in high energy ball-mills has been shown to significantly improve catalytic activity [2]. Aim of the work was to study the influence of milling time and reactive gas atmospheres on the crystalline catalyst and the support. Quantification of the crystalline phases and their microstructure properties were obtained from X-ray powder diffraction data. Starting from α-Fe_2O_3, Fe_3O_4 is formed under milling conditions. Increase of milling time leads to higher amounts of magnetite and larger oxide particles (Figure 1).

Samples (1 g each) were milled with varied milling time in air and under catalytic conditions with a continuous gas flow of $CO/O_2/He$. XRD measurements were performed at room temperature on a STOE (θ-θ) Bragg-Brentano diffractometer with Cu $K\alpha_{1/2}$ radiation and a secondary monochromator. Quantitative Rietveld refinements were performed using TOPAS-Academic (version 4.2) [3] software.

Figure 1: Effect of the milling time on the diffraction patterns of 10 wt% Au supported on α-Fe_2O_3 milled in air. In the figure, star (*) indicates the peaks from Fe_3O_4 while empty circle (o) indicates Au peaks. Unmarked ones belong to α-Fe_2O_3.

References
[1] Haruta, M., Yamada, N., Kobayashi, T., Iijima, S., J. Catal. 115 (1989) 301.
[2] Immohr, S., Felderhoff, M., Weidenthaler, C., Schüth, F., Angew. Chem. Int. Ed. 125 (2013) 12920.
[3] Coelho, A. A. (2007). TOPAS-Academic, version 4.1 (Computer Software), Coelho Software, Brisbane.

Figure 1

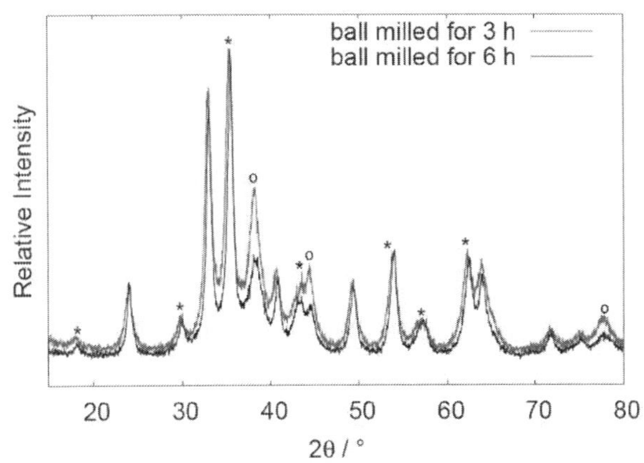

MSC-P49

Synthesis and characterization of monophase, polycrystalline erbium calcium oxoborate ($ErCa_4O(BO_3)_3$) for single-crystal growth by Czochralski

K. Mielke[1], J. Götze[1], M. Hengst[1]
[1] *TU Bergakademie Freiberg, Freiberg, Germany*

The group of rare-earth-oxoborates (RECOB) shows some unique mechanical, dielectrical, piezoelectrical und optical properties resulting in several applications in laser und sensor industries (Vivien et al. 2002). Based on previous investigations of selected calcium-rare-earth-oxoborates (e.g. GdCOB, Möckel 2012), the present work was focused on the synthesis of monophase, polycrystalline ErCOB by sintering, and the subsequent growth of single crystals using the Czochralski technique.

A first step was the determination of optimal parameters for the sintering process such as the stoichiometry of the starting products or the temperature regime for the synthesis. Different starting materials as well as sintering products of syntheses running under varying time and temperature conditions were analysed using X-ray powder diffraction and IR spectroscopy. Furthermore, the stoichiometric composition of the starting materials was studied with by a combination of thermo-analytical methods (STA, TG-DTA).

Single crystals of ErCOB were successfully grown by the Czochralski method using monophase, polycrystalline ErCOB from the sintering experiments (Fig. 1). The crystals were free of impurities proved by microscopy, X-ray powder diffraction and IR spectroscopy. Nevertheless the crystals cracked a short time after crystallization during the cooling process.

Figure 1: Single crystal of ErCOB synthesized by the Czochralski technique

The results of simultaneous thermo-analysis (STA) of single ErCOB crystals including several consecutive heating and cooling sequences indicate that the formation of cracks is caused by complex reactions and anisotropic thermal expansion of the material (Fig. 2).

Figure 2: DTA curves of an ErCOB crystal measured with consecutive heating and cooling sequences up to 1300°C

UV-VIS transmission spectroscopy of the grown ErCOB crystals showed a transparency for visible light until the near infrared and the absorption edge in the UV at 220 nm. Furthermore, it was possible to assign the absorption lines of Erbium to the diagram of Dieke (1968).

Structural investigations of the ErCOB material with X-ray powder diffraction revealed differences in the intensity ratios of the diffraction lines compared to data of Norrestam and Nygren (1992). Therefore, additional single crystal analyses were conducted on pieces of a grown single crystal which could improve the structural data.

References:
Dieke, G.H. (1968) Spectra and energy levels of rare earth ions in crystals. Interscience Publishers, New York.
Möckel, R. (2012) Growth and properties of $GdCa_4O(BO_3)_3$ single crystals. PhD thesis, TU Bergakademie Freiberg.
Norrestam, R. and Nygren, M. (1992) Structural Investigations of New Calcium-Rare Earth (R) Oxyborates with the Composition $Ca_4RO(BO_3)_3$. Chemical Materials, 4, 737-743.

Vivien, D., Aka, G., Kahn-Harari, A., Aron, A., Mougel, F., Benitez, J.-M., Ferrand, B., Klein, R., Kugel, G., Le Nain, N., Jacquet, M. (2002) Crystal growth and optical properties of rare earth calcium oxoborates. Journal of Crystal Growth, 237-239, 621-628.

Figure 1

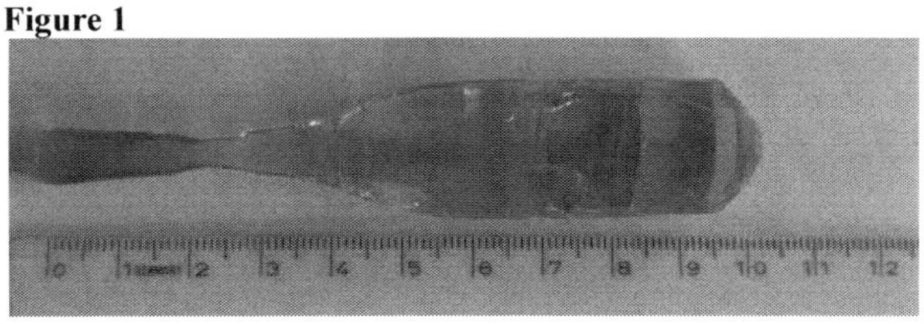

Figure 2

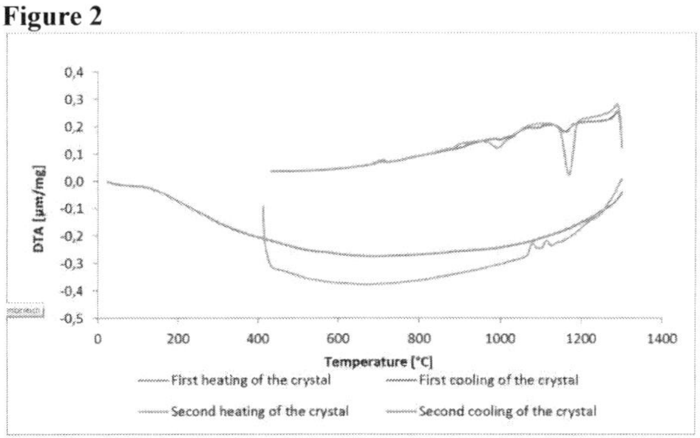

MSC-P50

Modulation, twinning and crystal structures of Λ-Co(sepulchrate) trinitrate at low temperatures

S. Dey[1], A. Schönleber[1], S. van Smaalen[1], F. K. Larsen[2]
[1] *University of Bayreuth, Laboratory of Crystallography, Bayreuth, Germany*
[2] *Aarhus University, Department of Chemistry, Aarhus, Denmark*

Λ-Co(sepulchrate) trinitrate, an organic salt with formula $C_{12}H_{18}N_8Co^{3+}\cdot 3(NO_3)^-$, crystallizes at room temperature in symmetry $P6_322$ [1,2]. The Co(sepulchrate) cation and two of the nitrate anions are all centred on threefold rotation axes and are linked to each other via N-H···O hydrogen bonds. The third nitrate anion close to the origin exhibits sixfold orientational disorder [2].

Two phase transitions reported at T_1 = 133 K and at T_2 = 107 K, respectively, have been observed by means of light microscopy and spectroscopic measurements [1,3]. As found by temperature dependent single crystal neutron diffraction satellite reflections are present in the diffraction pattern below T_1, varying their positions continuously as function of temperature. Also a third phase transition was observed at T_3 = 98 K [4].

We have performed single crystal X-ray diffraction at beamlines F1 and D3 of Hasylab (DESY, Hamburg) in all three low temperature phases (at temperatures 115 K, 104 K and 95 K). The main reflections are regularly surrounded by satellite reflections up to third order. All observed peaks can be indexed in an hexagonal unit cell with lattice parameters of $a = b \approx 8.45$ Å and $c \approx 15.75$ Å and two **q**-vectors (σ, σ, 0) and (-2σ, σ, 0) with $\sigma \approx 0.088$ (T = 115 K) in superspace group $P6_322(\sigma\sigma 0)000(-2\sigma\sigma 0)000$. However, it is also noticed that mixed hexagonal satellite reflections of second or third order are absent, indicating a possible lowering of symmetry at transition T_1. We propose therefore to describe the diffraction pattern at T = 115 K by an incommensurately modulated threefold orthorhombic twin in superspace group $C222_1(2\sigma 00)000$ with $a \approx$ 8.45 Å, $b \approx 14.63$ Å and $c \approx 15.75$ Å, reducing the modulation to one dimension and enabling a structure description in (3+1)-dimensional superspace. In our model the symmetry is further lowered at T_2, resulting in an incommensurately modulated sixfold monoclinic twin. At T_3 a lock-in phase transition takes place towards a sixfold superstructure (with $2\sigma = 0.1667$).

[1] L. Dubicki, J. Ferguson, B. Williamson, *J. Phys. Chem.* 88 (1984) 4254-4258.
[2] A. Schönleber, S. van Smaalen, F. K. Larsen, *Acta Crystallogr. C* 66 (2010) m107-m109.
[3] L. Dubicki, J. Ferguson, R. J. Geue, A. M. Sargeson, *Chem. Phys. Lett.* 74 (1980) 393-397.
[4] F. K. Larsen, P. Jørgensen, R. Grønbæk Hazell, B. Lebech, R. Thomas, R. J. Geue, A. M. Sargeson: Phase transitions in Λ-Co(sepulchrate)(NO_3)$_3$, Chapter 49 in *Molecular structure: Chemical reactivity and biological activity*, J. J. Stezowski *et al.* (*ed.*) IUCr, Oxford University Press (1988) 499-504.

MSC-P51

Crystal structures, magnetic properties and electrochemical properties as anode electrode in sodium-ion batteries of chromium (III)-based phosphates α- and β-$Na_2Ni_2Cr(PO_4)_3$

R. Essehli[1], M. Ait haddouch[1], I. Belharouak[1], S. benmoktar[1]
[1] *université hassan II mohammedia, chemistry, casablanca, Morocco*

The ternary phosphate $Na_2Ni_2Cr(PO_4)_3$ exists in two polymorphs form α and β with three-dimensional frameworks. This compound has been obtained by flux method at high temperature. The crystal structures of both polymorphs have been determined by single crystal X-ray diffraction analysis. The α modification crystallizes in an orthorhombic system (space group *Ibnm*, Z=6, a= 6.478930 (9) Å, c= 13.203676 (4) Å) and the β modification in a monoclinic system (space group *C2/c*, Z = 4, a= 11.729(7) Å, b = 12.433(5) Å, c= 6.431(2) Å, β= 113.66(4) °, R_1= 0.043, wR_2= 0.111). The crystal structure of β belongs to the alluaudite type, characterized by the $X(2)X(1)M(1)M(2)_2(PO_4)_3$ general formula. The open framework results from [Ni_2O_{10}] units of edge-sharing [NiO_6] octahedra, which alternate with [CrO_6] octahedra that form infinite chains. These chains are linked together through the common corners of [PO_4] tetrahedra yielding two distinct tunnels with sodium cation occupation. The magnetization data of the β form reveal antiferromagnetic (AFM) interactions at low temperature by the onset of deviations from a Curie-Weiss behaviour. The corresponding temperature dependence of the reciprocal susceptibility (c^{-1} = H/M) follows a typical Curie-Weiss behaviour for T > 105 K. A canted AFM state is proposed for β below 46 K with a field-induced magnetic transition at H » 19 kOe, revealed in the hysteresis loop measured at 5 K. This transition is most probably associated with a spin-flop transition. During the first discharge, α-$Na_2Ni_2Cr(PO_4)_3$ delivers a capacity of 750 mAh/g. This high capacity is equivalent to an intercalation of more than 5 sodium ions per formula unit. This is a strong indication of conversion-type behaviour with metallic Cr and Ni formation. As expected for such a conversion mechanism, not all Na-ions could be extracted again during the subsequent charge.

MSC-P52

Synthesis, structural and corrosion inhibition properties of Phenylhydrazine dihydrogenophosphate

A. Kheireddine[1], M. Tridane[1,2], S. Benmokhtar[1], Y. Abboud[3], S. Belaaouad[1]
[1]Laboratory chemistry and physics of materials - University Hassan II Casablanca, Casablanca, Morocco
[2]regional center of PROFESSIONS of education and training of grand casablanca, Casablanca, Morocco
[3]Laboratory physics and chemistry of applied materials - University of Hassan II Casablanca, Chemistry, Casablanca, Morocco

The new salt Phenylhydrazine dihydrogenophosphate $(C_6H_9N_2)_2H_2P_2O_7$ denoted PHDP is synthesized and characterized by X-Ray, IR and NMR spectral studies. Molecular structure of PHDP are determined by single crystal X-ray diffraction technique. This salt crystallizes in the monoclinic system, space group P21/c a=7.1991(9) Å b=31.207(3) Å c=8.0209(4) Å, β= 93.577° (7).The structures of PHDP is stabilized by inter-molecular H-bonding. The electrochemical study of these compounds and the corrosion inhibition study of the PHDP has been performed by using potentiodynamic polarization measurements. The compound PHDP show an appreciable corrosion inhibition property against corrosion of mild steel in some acid medium at 298 K. As the concentration of the PHDP increase, corrosion inhibition property of the compounds increases.

MSC-P53

The structure of the common limestone corrosion product thecotrichite

N. Wahlberg[1], R. E. Dinnebier[2], A. Fischer[3], G. Eggert[3]
[1]Aarhus University, Department of Chemistry, Aarhus, Denmark
[2]Max-Planck-Institute, Solid state Research, Stuttgart, Germany
[3]State academy of art and design Stuttgart, Stuttgart, Germany

The corrosion of calcareous specimens in museum environments has been known for more than a century (Byne, 1899). Still the structure of the efflorescent salts is in some cases unknown, since the corrosion product is of a polycrystalline quality inhibiting structure determination by single crystal diffraction. However, the quality of powder diffraction data and the structure solving algorithms have reached a level where structure solution from powder is feasible.

Thecotrichite is found on limestone objects, often Egyptian artifacts, stored in wooden cabinets. The high concentration of acetic acid emitted by the wood and fluctuations in the relative humidity is the main contributors to the corrosion (Gibson *et al.* 2005). Characterization with various techniques (IR, NMR, ion chromatography) let to the chemical formula $Ca_3(CH_3COO)_3Cl(NO_3)_2 \cdot 6.6H_2O$ (Gibson *et al.* 1997).

The structure was solved *ab initio* from high resolution laboratory powder diffraction by a combined use of charge flipping, simulated annealing, and visual investigation of the residual Fourier maps (*$P2_1/c$, a=23.5933(4)Å, b=13.8459(3) Å, c=6.8010(1) Å, β=95.195(2) Å, V=2212.57(7)$Å^3$*). Crystallographically the chemical formula was determined to $Ca_3(CH_3COO)_3Cl(NO_3)_2 \cdot 6H_2O$, where the acetate, nitrate and five water molecules are directly coordinating to the calcium cation. The result is a metal-organic framework structure (MOF) containing long chains of calcium ions linked by acetate, figure 1a. The chains are interlinked by nitrate ions in sets of three, forming broader strands. The strands are again interlinked by bridging nitrate units, figure 1b, forming a brickwork-structure, with open-ended channels parallel to the *c*-axis. The chloride ions and uncoordinated water molecules are located in these open-ended channels, suggesting that it may be possible to remove the water.

The new gained knowledge on the structure of the efflorescent salts on limestone will allow a fast and unambiguous identification of thecotrichite especially in the field of conservation of historical artifacts. The structure stresses the need to control and diminish the vapor pressure of acetic acid if this corrosion product is to be avoided.

References

Byne, L. S. G. (1899), The corrosion of shell in cabinets, J. Conchology 9, 172-178
Gibson, L. T., Cooksey, B. G., Littlejohn, D., Linnow, K., Steiger, M., Tennent, N. H. (2005), The mode of Formation of Thecotrichite, a Widespread Calcium Acetate Chloride Nitrate Efflorescence, Stud. Conserv. 50, 284-294
Gibson, L. T., Cooksey, B. G., Littlejohn, D., Tennent, N. H. (1997), Characterisation of an unusual crystalline efflorescence on an Egyptian limestone relief, Anal. Chim. Acta 337, 151-164

Figure 1

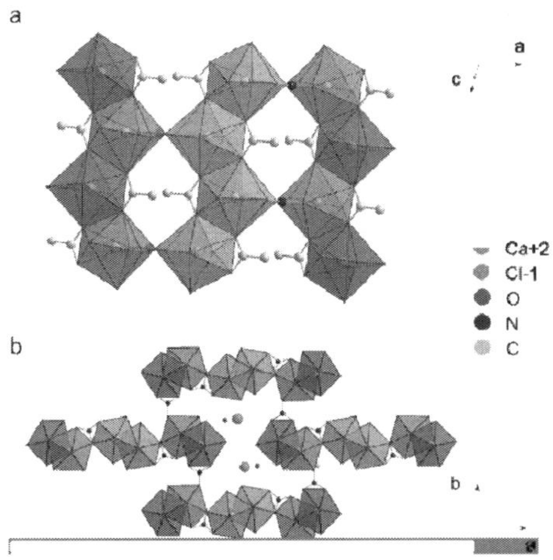

Figure 1. Extended packing motif of thecotrichite viewed (a) along the b-axis and (b) along the c-axis. All polygons have a central calcium ion, but are color coded to differentiate the three chemically different calcium ions.

MSC-P54

Light diffraction experiments with micron-sized models of 3D-crystals

J. Birkenstock[1], M. Bülters[2], R. X. Fischer[1]
[1]Universität Bremen, FB5-Geowissenschaften, Bremen, Germany
[2]BIAS - Bremer Institut für angewandte Strahltechnik GmbH, Bremen, Germany

The geometric aspects of X-ray diffraction experiments may be well described in terms of P.P. Ewald's famous construction and the math is fully developed to calculate the Bragg-reflected and scattered intensity. However, when facing students with diffraction theory for the first time it is usually helpful to demonstrate diffraction in analogous experiments. Visual electromagnetic radiation, i.e., the monochromatic light of a laser, is diffracted on transmitting thin semitransparent objects with one- or 2D-periodic patterns and typical distances at the micrometer-scale. It is fairly easy to provide such flat objects for scattering and diffraction, e.g. a traditional photographic slide displaying nonperiodic and up to 2D-periodic objects at the micrometer scale or gauze-nets with micrometer-sized interwoven threads.

For one- or 2D-periodic objects the diffraction condition is always fulfilled so that a pseudo-periodic diffraction pattern is always present. However, this is much different in case of 3D-periodic objects - such as single crystals which are the usual objects examined in X-ray diffraction experiments. Now the conditions for diffraction are only fulfilled in very specific, well-defined orientations and they are rarely fulfilled for many reflections at a time. Accordingly it is even more desirable to demonstrate this case with visual light. However, it is much harder and yet necessary to provide 3D-periodic objects with repeat distances of a few microns in all spatial directions.

State-of-the-art 3D printers may print micrometer-sized 3D structures with a resolution at the nanometer scale. Periodic structures at such scales are often designated "photonic crystals". We designed simple 3D-periodic crystals at sizes of a few hundred micrometers, e.g., an orthorhombic crystal with dimensions of 290 x 290 x 120 μm^3 and with lattice parameters of a = 3 µm, b = 2 µm and c = 4 µm. After printing the structure (nanoscribe Photonic Professional) the "micro-crystals" will be mounted on a standard goniometer head in a suitable goniometer, e.g., a Huber 2-circle reflection goniometer. There it can be exposed to laser light, expecting that we may mimic the process of single crystal diffraction with visible light. With a red laser (λ = 650 nm) typical diffraction angles for the above orthorhombic structure are for example $2\theta_{001} = 9.32°$, $2\theta_{100} = 12.43°$, $2\theta_{010} = 18.70°$ and $2\theta_{011} = 20.94°$.

The experimental setup is currently under construction and will primarily be used for teaching purposes in our X-ray diffraction courses. Public demonstrations are planned during the exhibition "Kristallographie. Kunst. Technik" (crystallography.art.technology) at the Haus der Wissenschaft in Bremen in July and August 2015.

Acknowledgement: We are grateful to Hermann Gies and Bernd Marler of the crystal-chemistry group at Universität Bochum for providing us with a 2-circle reflection goniometer.

MSC-P55
Ti-Substituted $BaFe_{12}O_{19}$ Single Crystal Growth and Characterization

S. Nemrava[1], D. A. Vinnik[2], D. A. Zherebtsov[2,3], L. S. Mashkovtseva[2], A. S. Semisalova[3], N. S. Perov[3,4], I. V. Krivtsov[2], L. I. Isaenko[5,6], G. G. Mikhailov[2], R. Niewa[1]
[1]University of Stuttgart, Institute of Inorganic Chemistry, Stuttgart, Germany
[2]South Ural State University, Chelyabinsk, Russian Federation
[3]Moscow State University, Moscow, Russian Federation
[4]Baltic Federal University, Kaliningrad, Russian Federation
[5]Russian Academy of Sciences, Novosibirsk, Russian Federation
[6]Novosibirsk State University, Novosibirsk, Russian Federation

Hexaferrites with the magnetoplumbite structure, such as $BaFe_{12}O_{19}$, are interesting materials for applications as magnetic materials or for microwave absorption due to large axial anisotropy, high resistivity and high permeability. To understand and modify these properties partial substitution of iron by different cations is a suitable tool. [1]

Here we present Ti-substituted barium hexaferrite crystals with the formula $BaFe_{12-x}Ti_xO_{19}$ with x up to 1.3. All crystals were grown from a flux composed of Na_2CO_3 and $BaCO_3$ with γ-Fe_2O_3 and TiO_2 (60 wt% rutile and 40 wt% anatase) as educts. The amount of TiO_2 in the initial mixture correlates with x in the crystal composition. After being ground in an agate mortar and placed in a platinum crucible, the mixture was heated for 3 h at 1260 °C. After cooling to room temperature, crystals with sizes up to 8 mm were separated from the flux by leaching in hot nitric acid (Fig. 1).

PXRD refinements show increasing axes and cell volumes for $BaFe_{12-x}Ti_xO_{19}$ for $x \leq 0.8$ compared to pure $BaFe_{12}O_{19}$; for $x > 0.8$ axes and volumes decrease. The first effect originates from a reduction of Fe^{3+} to Fe^{2+} due to charge balance when incorporating Ti^{4+} into the structure, the second one is supposed to derive from vacancies in the transition metal substructure. SCXRD confirmed the typical M-type structure with space group $P6_3/mmc$. For the distribution of Ti over the five crystallographic sites occupied by iron we obtain following results: The 4e, 4f_1 and 4f_2 sites are similarly substituted by Ti (≈ 13 % Ti, Fig. 2), while the 2a and 12k sites do not accumulate significant amounts of Ti.

Magnetic characterization of the samples leads to the typical hysteresis curves for M-type ferrites and shows decreasing saturation magnetizations and Curie temperatures with increasing amount of Ti, which, carrying no magnetic moment, disturbs the magnetic order in its surrounding. [2]

Figure 1: Single crystals obtained.

Figure 2: Transition metal assignments of $BaFe_{12-x}Ti_xO_{19}$ (M = Fe, Ti).

[1] B. Halbedel, D. Hülsenberg, S. Belau, U. Schadewald, M. Jakob, *Cfi/Ber. DGK* 2005, *82*, 182 - 188.
[2] D. A. Vinnik, D. A. Zherebtsov, L. S. Mashkovtseva, S. Nemrava, N. S. Perov, A. S. Semisalova, I. V. Krivtsov, L. I. Isaenko, G. G. Mikhailov, R. Niewa, *Cryst. Growth Des.* 2014, *14*, 5834 - 5839.

Figure 1

Figure 2

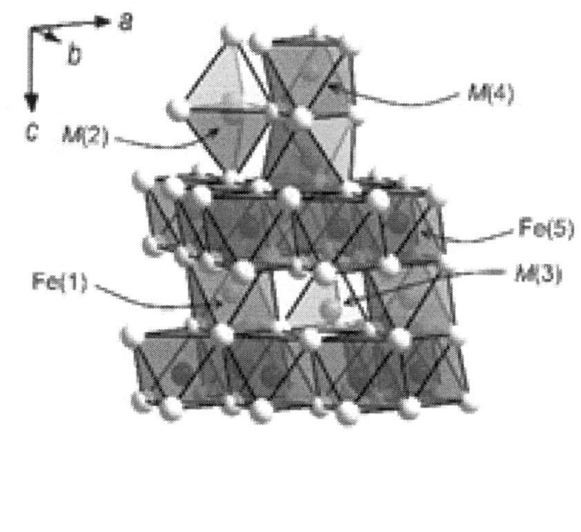

MSC-P56

Electrostatic potential and Nonlinear Optical Properties of Zwitterionic Compound : 6-methyl-2-oxo-3-[1-(ureidoiminio)ethyl]-2H-pyran-4-olate monohydrate

A. Chouaih[1,2], N. Benhalima[1], R. Rahmani[1], N. Boukabcha[1], F. Hamzaoui[1]
[1]*University of Mostaganem, Chemistry, Mostaganem, Algeria*
[2]*,, Algeria*

We report here a theoretical study on molecular structure and nonlinear optical properties of Zwitterionic 6-methyl-2-oxo-3-[1-(ureidoiminio)ethyl]-2H-pyran-4-olate monohydrate molecule. The molecular geometry in the ground state was investigated by ab initio and density functional method (B3LYP) with LANL2DZ, 6-31G(d) and 6-31+G(d,p) basis sets. The results show that the computed geometrical parameters are in better agreement with experimental values. An accurate electron density distribution around the molecule has been calculated based on a high-resolution X-ray diffraction study in order to understand the nature of inter- and intramolecular charge transfer and to obtain the experimental electrostatic potential distribution of the molecule. The electron density distribution was obtained from the multipolar model. The X-ray data were collected at low temperature (173 K). The molecular electrostatic potential and HOMO-LUMO of the title molecule have been also calculated using the theoretical methods. The calculated HOMO and LUMO energies show that charge transfer occurs within the molecule. To investigate nonlinear optical behavior, the electric dipole moment μ, the polarizability α and the hyperpolarizability β were computed. All β values that we report here are the magnitude of static hyperpolarizability. Theoretically predicted β values exhibit the high nonlinear optical activity.

MSC-P57

Molecular and solid-state structure of 5-(5-nitro furan-2-ylmethylen), 3-N-(2-methoxy phenyl),2-N'-(2-methoxyphenyl) imino thiazolidin-4-one

R. Rahmani[1], A. Chouaih[1], F. Hamzaoui [1]
[1]*University of Mostaganem, Chemistry, Mostaganem, Algeria*

The crystal structures of 5-(5-nitro furan-2-ylmethylen), 3-N-(2-methoxy phenyl),2-N'-(2-methoxyphenyl) imino thiazolidin-4-one have been solved by direct methods and by direct-space simulated annealing, respectively, from powder synchrotron X-ray diffraction data. The obtained structure is monoclinic, space group C2/c, with a = 7.38042 (4), b = 10.38220 (5), c = 9.75092 (5) A° and β = 95.359 (1)°.

The molecular geometrical parameters, of the title compound have been calculated by using DFT/B3LYP with 6-31G (d, p) level. From results, experimental and theoretical structural properties are in good agreement. In the solid state molecular structures and packing are discussed. To study the photovoltaic application of this compound, HOMO, LUMO and gap energies were determined.

MSC-P58

The anisotropy of the thermal expansion of Mg-, Fe-, and Co-olivines

P. Schmid-Beurmann[1], H. Kroll[1], A. Sell[1], A. Kirfel[2], P. Adelmann[3], H. Pentinghaus[3]
[1]*Universität Münster, Institut für Mineralogie, Münster, Germany*
[2]*Steinmann Institut, Universität Bonn, Bonn, Germany*
[3]*KIT-IFP, Eggenstein-Leopoldshafen, Germany*

The value of the thermal expansivity enters many thermoelastic properties and parameters that are required for modelling density, temperature and seismic wave propagation profiles in the upper mantle. Axial thermal expansivities, α_i, and axial isothermal compressibilities, $\beta_{T,i}$, are related via the thermodynamic identity $\alpha = \beta_T\,(\delta P/\delta T)_V$, i.e. "thermal expansion is a response to thermal pressure $(\delta P/\delta T)_V$ and is the greater the more compressible the material or direction" (White et al., 1985). Consequently, as a general rule crystalline compounds show sequences of axial expansivities that agree with those of the axial compressibilities. For olivines, isothermal compressibilities can be related to structural properties. Compression along *a* involves compression of Si tetrahedra which are rigid. M2 octahedral layers, which are free of Si sites and thus more easily to compress, extend normal to *b*. The M2 octahedra share O-O edges with Si tetrahedra being parallel *c*. Accordingly, for all olivines it is found that $\beta_{T,b} > \beta_{T,c} > \beta_{T,a}$ (SG *Pbnm*). Since Mg and Fe olivines are isomorphous, it is striking that the axial thermophysical properties of Mg olivine (Fo) are distinctly less anisotropic than those of Fe olivine (Fa). While in Fo, both the axial compressibilities and the axial expansivities follow the same sequences, the expectation is violated in Fa where $\alpha_c > \alpha_a > \alpha_b$ above 400 K. It is shown that the different anisotropies of the expansivities and hence of the α-dependent thermophysical properties can be related to differing structural responses to heating, in particular to the rotation and extension of M1 and M2 octahedral axes.

Fe and Co possess closely related electronic configurations. It is thus interesting to explore whether Co olivine follows the general rule of α sequences or deviates from it as does Fe olivine. Lattice parameters were determined between 300 K and 1300 K from X-ray powder diffraction experiments. Thermal expansivity was modelled using the same Equation-of-State related expression of Kumar (2003) as used for Fo and Fa. In order to avoid the correlations between fit parameters, Debye temperatures were fixed with the help of low temperature data of Sazonov (2009) and pressure derivatives of the axial compressibilities were taken from Kroll et al. (2014). The axial expansivities of Co olivine turned out to neither follow the scheme of Fo nor Fa, but show the sequence $\alpha_c > \alpha_b > \alpha_a$.Work is in progress to obtain from Rietveld refinements the structural evolution of Co olivine and compare it to Fo and Fa in order to structurally interpret the various thermal expansivities of olivines.

References

Kroll H, Kirfel A, Heinemann R (2014) Eur J Mineral 26: 607-621.
Kumar M (2003). Phys Chem Miner 30: 556-558.
Sazonov A (2009) Dissertation TU Aachen.
White GK, Roberts RB, Collins JG (1985) High Temp - High Press 17: 61-65.

MSC-P59

Change of modulation in relaxor ferroelectric $Ca_{0.28}Ba_{0.72}Nb_2O_6$ (CBN28) on quenching

H. Graetsch[1]
[1]*Ruhr Universität Bochum, Institut für Geologie, Mineralogie & Geophysik, Bochum, Germany*

Electric properties and Curie temperatures of SBN crystals with the tetragonal tungsten bronze type structure can be altered by quenching from temperatures higher than 600°C [1]. The changes are ascribed to increased disorder in the distribution of Ba^{2+} and Sr^{2+} ions over two incompletely filled cation sites [1], [2].

Tetragonal SBN and isotypic CBN mixed crystals are incommensurately modulated in two directions perpendicular to the c-axis [3]. The displacive modulations consist of a wave-like tilting of the NbO_6 octahedra which seem to be coupled to variations in the occupation of the large cation sites [4], [5]. CBN crystals quenched from 1,200 °C to room temperature merely show a one dimensional modulation which reduces the symmetry from tetragonal to orthorhombic [6]. The average structure remains unchanged.

CBN crystals quenched from 600 or 800°C show both systems of satellite reflections; however, with different intensities indicating either incomplete transformation or twinning of the orthorhombic phase with different volumes of the twin individuals (see figure below: projection of the diffraction patterns of untreated and quenched CBN28 crystals along the c* axis on to the reciprocal unit cell using Jana2006 [7]; the peak at the center is due to the $\lambda/2$ effect). Similar patterns were also found for CBN31 and SBN53.

Contrasting with SBN, the overall occupation of the large cation sites in quenched CBN28 does not differ from untreated crystals. However, due to the lower symmetry, two independent Me2 sites (Me2a and Me2b) can be distinguished for quenched orthorhombic CBN28. The Me2a and Me2b sites mainly differ by the Ca^{2+} incorporation. The tetragonal phase can be restored by annealing the quenched crystals above ca. 400°C and subsequent slow cooling. The overall tetragonal symmetry of CBN28 might be emulated by twinning of the local orthorhombic structure in addition to the usual inversion twinning of unpoled ferroelectric crystals.

[1] Guo, R. T., Bhalla, A. S., Burns, G., Dacol, F.H. (1989) Ferroelectrics 93, 397-405
[2] Trubelja, M.P., Ryba, E., Smith, D.K. (1996) J. Mat. Scie. 31, 1435-1443
[3] Woike, T., Petříček, V., Dušek, M., Hansen, N. K., Fertey, P., Lecomte, C., Arakcheeva, A., Chapuis, G, Imlau, M., Pankrath, R. (2003) Acta Cryst. B59, 28-35
[4] Schefer, J., Schaniel, D., Petříček, V., Woike, T., Cousson, A., Wöhlecke, M. (2008) Z. Kristallogr. 223, 399-407
[5] Graetsch, H. A., Pandey, C. S., Schreuer, J., Burianek, M., Mühlberg, M. (2012) Acta Cryst. B68, 101-106
[6] Graetsch, H. A., Pandey, C. S., Schreuer, J., Burianek, M., Mühlberg, M. (2014) Acta Cryst. B70, 743-749
[7] Petříček, V., Dušek, M., Palatinus, L. (2006) JANA2006, Academy of Sciences of the Czech Republic, Praha

Figure 1

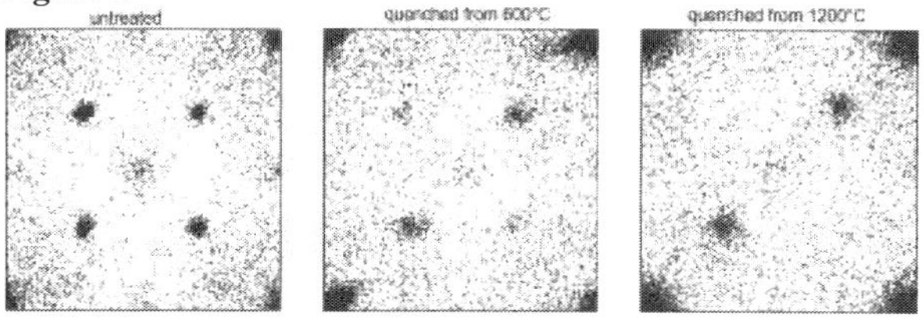

MSC-P60

Structural distortion of biogenic aragonite in Ranella Olerea mollusc shell layers

S. Ouhenia[1], D. Chateigner[2]
[1]*Université de Béjaia, Physique, Béjaia, Algeria*
[2]*Université de Caen, Laboratoire Crismat, Caen, France*

Abstract (Poster)

Mollusc shells, mainly made of calcite and aragonite crystalline polymorphs of calcium carbonate, are fascinating organic-mineral biocomposites with high mechanical performances, as they attain a fascinating increase in both strength and toughness compared to the geological mineral. The major part of organic materials is intercrystalline, and in a minor way intracrystalline (Pokroy et al., 2006).The organic represents less than 5% in volume, is a biopolymer dispersed in inorganic crystal of calcium carbonate (Barthelat and Espinosa, 2007). This organic part behaves as nanometer growth-control of the inorganic crystals and also plays an important role in stopping crack propagation in nacre (Cortie et al.,2006). This has stimulated chemists and materials scientists to design and synthesize high performance materials with a microstructure similar to that of nacre (Wang et al., 2013).

In the present work we made use of Combined Analysis to determine the structure and unit-cell distortions of constituting aragonite crystallites of the shell layers (figure 1) of the gastropod *Ranella olearea*. This approach was chosen because it allows working on real samples, without grinding operation (Ouhenia et al., 2008). SEM analyses show the presence of three distinct layers; an inner layer composed of Radial Lamellar, an intermediate comarginal crossed lamellar layer and an outer crossed lamellar layer. The refinement of X-ray diffraction diagrams, gives quantitatively the structure of the three layers and their respective aragonite unit-cell distortions. An anisotropic unit-cell distortion is quantified for the three layers which is attributed to the combined effects of inter- and intra-crystalline macromolecules.

References.

Pokroy, B., Fitch, A.N., Lee, P.L., Quintana, J.P., Caspi, E.N., Zolotoyabko, E., 2006. Anisotropic lattice distortions in mollusc-made aragonite: a widespread phenomenon. Journal of Structural Biology 153, 145-150.
Barthelat, F., Espinosa, H.D., 2007. An experimental investigation of deformation and fracture of nacre-mother of pearl. Exp. Mech. 47, 311-324.
Cortie, M.B., Mc Bean,E. M., Margaret M. Elcombe, M. M., 2006. Fracture mechanics of mollusc shell, Physica B 385-386, 545-547.
Ouhenia, S. D. Chateigner, M.A. Belkhir, E. Guilmeau. 2008. Microstructure and crystallographic texture of Charonia lampas lampas shell Journal of Structural Biology 163, 175-184.

Legend

Figure 1: Cross-section SEM image of the fractured shell at the location indicated in G, M and N indicate the Growth, Margin and Normal directions, respectively.

Figure 1

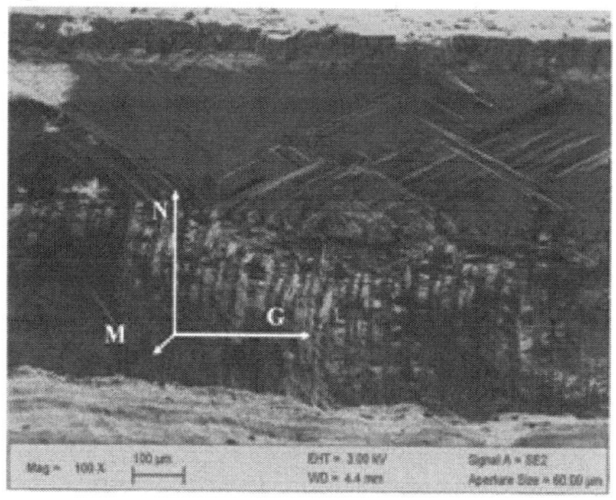

MSC-P61

The role of Ca in Al-, Li- and $^{[4]}$B-bearing tourmalines

A. Ertl[1,2], E. Tillmanns[2], G. Giester[2]
[1]*Naturhistorisches Museum, Mineralogisch-Petrographische Abteilung, Wien, Austria*
[2]*Universität Wien, Institut für Mineralogie und Kristallographie, Wien, Austria*

Single crystals of three Al-rich Li-bearing tourmalines, which contain also significant amounts of Mn, were recently refined. The refinements show that these samples are Na dominant at the 9-coordinated *X* site, but also contain significant amounts of Ca (~0.4-0.5 apfu). The tetrahedral <*T*-O> distances are in the range ~1.615-1.617 Å. Interestingly, the refined amount of B, which substitutes for Si, is in the range of ~0.2-0.3 *apfu*. Similar tourmaline samples, but with a significantly lower Ca content (0.02-0.14 *apfu*), studied by Ertl et al. (2013), have refined $^{[4]}$B values, which are lower (≤0.1 *apfu*). Ertl et al. (2006) have shown that ^{Y}Al (reflected by the octahedral <*Y*-O> distance) can be associated in short-range orders with $^{[4]}$B (reflected by the <*T*-O> distance). To learn more about a possible correlation between <*Y*-O> and <*T*-O> distances in Ca-rich and Li-bearing tourmalines, the bond-lengths of seven tourmalines with significant Ca contents in the range ~0.3-0.5 *apfu* were plotted, including also the data of Ertl et al. (2006) and Hughes et al. (2000). A pronounced positive correlation [<*Y*-O> = 8.392(<*T*-O>) - 11.546] between <*Y*-O> and <*T*-O> distances was found (r^2 = 0.92). We conclude that Ca in natural Al-rich Li-bearing tourmalines can be an important part of short-range orders associated with ^{Y}Al, ^{Z}Al and $^{[4]}$B. Such Ca-rich tourmalines can contain higher amounts of $^{[4]}$B than similar samples, which are poor in Ca. One explanation could be that Ca^{2+} with one more positive charge than Na^{1+} could help to produce a charge balanced formula when B^{3+} is substituting for Si^{4+}. Usually, Ca can be mobilized from surrounding Ca-rich rocks and eventually occupies the *X* site in Li-bearing tourmaline of a pegmatitic rock. Hence, especially Al-rich tourmalines from pegmatites in contact to a (calc-silicate) marble could be candidates for pronounced contents of $^{[4]}$B.

References

Ertl, A., Hughes, J.M., Prowatke, S., Ludwig, T., Prasad, P.S.R., Brandstätter, F., Körner, W., Schuster, R., Pertlik, F. & Marschall, H. (2006): Tetrahedrally-coordinated boron in tourmalines from the liddicoatite-elbaite series from Madagascar: Structure, chemistry, and infrared spectroscopic studies. Am. Mineral., 91, 1847-1856.

Ertl, A., Giester, G., Schüssler, U., Brätz, H., Okrusch, M., Tillmanns, E. & Bank, H. (2013): Cu- and Mn-bearing tourmalines from Brazil and Mozambique: Crystal structures, chemistry and Correlations. Mineral. Perol., 107, 265-279.

Hughes, J.M., Ertl, A., Dyar, M.D., Grew, E.S., Shearer, C.K., Yates, M.G. & Guidotti, C.V. (2000): Tetrahedrally coordinated boron in a tourmaline: boron-rich olenite from Stoffhütte, Koralpe, Austria. Can. Mineral., 38, 861-868.

MSC-P62

Growth, structural and magnetic characterization of Al-substituted barium hexaferrite single crystals

S. Nemrava[1], D. A. Vinnik[2], D. A. Zherebtsov[2], L. S. Mashkovtseva[2], M. Bischoff[1], N. S. Perov[3,4], A. S. Semisalova[3], I. V. Krivtsov[2], L. I. Isaenko[5], G. G. Mikhailov[2], R. Niewa[1]
[1]*University of Stuttgart, Institute of Inorganic Chemistry, Stuttgart, Germany*
[2]*South Ural State University, Chelyabinsk, Russian Federation*
[3]*Moscow State University, Moscow, Russian Federation*
[4]*Baltic Federal University, Kaliningrad, Russian Federation*
[5]*Russian Academy of Sciences, Novosibirsk, Russian Federation*

Many applications in modern industry, like radio technology or microelectronics, require materials with high Curie temperatures, chemical stability and magnetic anisotropy. Barium hexaferrites with the magnetoplumbite crystal structure comply with these requirements. The mentioned characteristics can be changed by partial substitution of iron by different cations. [1]

Al-substituted barium ferrite crystals with the chemical formula $BaFe_{12-x}Al_xO_{19}$ with x up to 1.5 were grown from a Na_2CO_3 flux. The starting materials γ-Fe_2O_3, $BaCO_3$ and γ-Al_2O_3, in which the amount of γ-Al_2O_3 in the starting batch correlates with x in the resulting products, were ground in an agate mortar and placed in a platinum crucible. After heating to 1260 °C and subsequent cooling to room temperature, crystals with sizes up to 8 mm were separated from the flux by leaching in hot nitric acid.

PXRD refinements reveal that the axes and cell volumes for Al-substituted barium hexaferrite decrease with higher substitution levels, as expected from the smaller ionic radius of Al^{3+} compared to Fe^{3+}. According to SCXRD Al is distributed over all iron sites except for $4f_1$, which is tetrahedrally coordinated. The highest amount of Al with 26 % can be found at the octahedrally coordinated $2a$ site. The remaining sites $4e$, $4f_2$ and $12k$ accumulate similar amounts of Al (Fig. 1).

The magnetic characterization shows the typical hysteresis curves for M-type ferrites. The substitution of Fe^{3+}, which carries a magnetic moment, by a non-magnetic ion like Al^{3+} leads to reduced saturation magnetizations and Curie temperatures in comparison with pure $BaFe_{12}O_{19}$ (Fig. 2). [2]

Figure 1: Section of the crystal structure (M = Fe, Al). **Figure 2:** Hysteresis curves at room temperature for different Al concentrations.

[1] B. Halbedel, D. Hülsenberg, S. Belau, U. Schadewald, M. Jakob, *Cfi/Ber. DGK* 2005, *82*, 182 - 188.
[2] D. A. Vinnik, D. A. Zherebtsov, L. S. Mashkovtseva, S. Nemrava, M. Bischoff, N. S. Perov, A. S. Semisalova, I. V. Krivtsov, L. I. Isaenko, G. G. Mikhailov, R. Niewa, *J. Alloy. Compd.* 2014, *615*, 1043 - 1046.

Figure 1

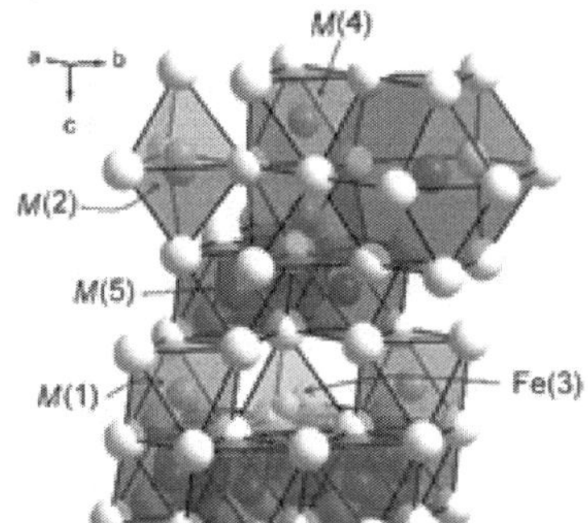

Figure 2

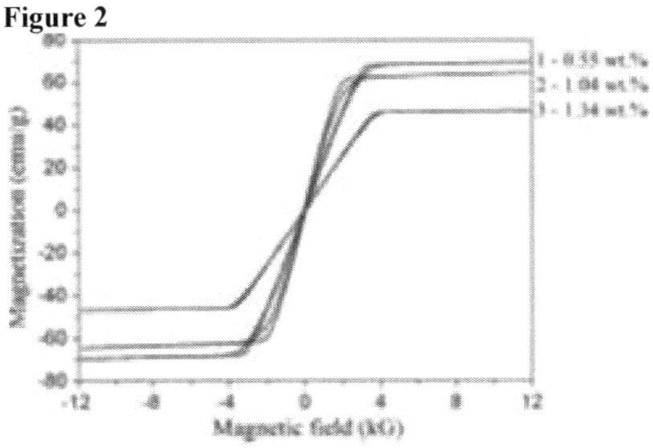

MSC-P63

Accurate determination of structural characteristics TiNI-based shape memory alloys

B. Kodess[1], P. Kodess[1], J. Chumlakov[1]
[1]ICS&E-VNIIMS, Crystals metrology, Moscow, Russian Federation

Samples of alloys near equiatomic composition (Ti- 50-51at.% Ni) has been analyzed by using high-accurate X-ray diffraction method. Single crystals of alloy being grow by the Bridgman method in the helium environment and homogenized at T = 1173 K 12 h, followed by quenching in water. Martensitic transformation temperature was determined by data of e, dependence of electric resistance on temperature. X-ray data have been collected in full Ewald sphere (7-12 thousand reflections) in Mo-K alpha radiation. Data collection was repeated by rotating the sample (analog azimuthally scanning). To separate the effects of coexisting phases used a special method of processing images obtained using CCD detector. This data allowed us to be unambiguously determined the structural characteristics of all three base and supplementary phases in this alloys. Determining the size of the unit cells was carried out according to the results of measurements in the high angle range. The value for the cubic (parent B2 austenitic phase) a=0.301 nm, in good agreement with reference data. Also, a good agreement was obtained for the hexagonal R-phase, the appearance of which is preceded by a complete martensitic transformation in B19`-phase. At the same time, the angle in the unit cell of the monoclinic phase of almost one degree lower than the values presented in the early structural work for TiNi. The distinction may be due that in these studies [1] the measurement carried out only for part of the equivalent reflections and therefore the values of R-factors were sufficiently high (17-25% for the isotropic approximation and 6-7% when used the anisotropic approximation). It is found that the hydrogen saturation leads to a noticeable increase in the size of the unit cell and the martensitic transformation temperature, decreases the ductility and as results in the appearance of superelasticity (up to 2.0–2.5% in this samples [2]). This is because are the influence of hydrogen on the change of phase composition of TiNi alloy under hydrogenation (particle formation TiNi3 and TiH). This stabilization is important in the use of these various compositions, especially in high cycle load, for example, in miniature aerospace actuators, energy sources and other applications, including medicine.

1] Kudoh Y. and Tokonami M., Miyazaki S., Otsuka K. Acta metall. Vol. 33, N11, p. 2049, 1985

[2] Kireeva I.V. Chumlakov, Yu.I., Platonova Yu. N. JTPh Letters 41,6, p.58 , 2015

MSC-P64

PDF and Rietveld Analysis of Disordered Rock-Salt-Type Li_2VO_2F as Li-ion cathode material

M. Yavuz[1,2,3], I. Peral[4], M. Knapp[1,3], R. Chen[3,5], H. Ehrenberg[1,2,3]
[1]Karlsruhe Institute of Technology (KIT), Institute for Applied Materials - Energy Storage Systems (IAM-ESS), Karlsruhe, Germany
[2]Technische Universität Darmstad, Materials Science, Darmstad, Germany
[3]Karlsruhe Institute of Technology (KIT), Helmholtz Institute Ulm for Electrochemical Energy Storage (HIU), Karlsruhe, Germany
[4]CELLS-ALBA, Barcelona, Spain
[5]Karlsruhe Institute of Technology (KIT), Institute of Nanotechnology (INT), Karlsruhe, Germany

Lithium-ion batteries (LIBs) having high energy density, high discharge power and long cycle life, are a promising battery technology for various technological applications, such as electric vehicles, power tools and portable electronics. Finding suitable electrode materials is key point for their applications. Li_2VO_2F was developed recently as cathode material for lithium- ion batteries. This Li-rich material was synthesized by a ball-milling route using the Li_2O, LiF and V_2O_3 precursors [1]. The material crystallizes in a disordered cubic rock-salt structure. Electrochemical performance was evaluated galvanostatically between 1.3 and 4.1 V versus lithium. Up to 1.8 Li^+ in Li_2VO_2F can be cycled, corresponding to a high capacity of 420 mAh g^{-1}. In comparison with state-of-the-art commercial cathode materials, the new material has the following merits: (i) Li-rich chemical composition, (ii) intercalation reaction mechanism with large amount of Li^+ storage capability, (iii) crystal host structure is stable for Li^+ (de)intercalation.

While the traditional Rietveld refinement only takes into account the Bragg scattering from the average structure, the PDF or total scattering method utilizes both, Bragg and diffuse scattering, and provides information about long- and short-range order in materials. However, traditional Rietveld refinements are not sensitive to a local structure distortion [2,3]. PDF analysis of pristine $Li_{2-x}VO_2F$ shows structural distortion of the octahedral units around the F-atoms. Local order and average structure was also investigated for this material at different charge/discharge states of $Li_{2-x}VO_2F$.

[1] R. Chen, S. Ren, S. Indris, M. Fichtner, H. Hahn, European patent application: EP 14160894.3, 2014.
[2] S. L. J. Billinge and M. F. Thorpe, *Local Structure from Diffraction*, Fundamental Materials Research, 1998.
[3] Egami T and S. L. J. Billinge, *Underneath The Bragg Peaks Structural Analysis of Complex Materials*, Pergamon, 2003.

MSC-P65

Hardening of calcium sulphate based binders and strength decrease in presence of moisture

C. Pritzel[1], R. Trettin[1]
[1]University of Siegen , Chemistry , Siegen, Germany

Calcium sulphate based binders were used in the building industries for different applications. For example calcium sulphate hemihydrate or anhydrate were used as binding materials, they react with water to dihydrate. When calcium sulphate hemihydrate is reacting with water a new structure of calcium sulphate dihydrate crystals is formed. This new structure constitutes very good compressive and bending tensile strength of the created gypsum stone, but the strengths are decreasing in the presence of moisture. In literature different theories about the mechanism of this hardening were described, but none of them is explaining all observed effects. The strength decrease in presence of moisture is connected to these theories about hardening and it is also not completely explained.

To investigate the processes of hardening different experiments were done. Gypsum stone samples of different gypsum materials were prepared. The main difference between the gypsum materials which were used was the crystal morphology of the dihydrate crystals which were created during the hydration process. These samples were stored in different environments with different moisture, in water or in saturated gypsum solution. The strengths of the stored samples were measured and compared depending on crystal morphology and the storage conditions.

The samples were also investigated using different other methods. The crystal morphology was measured by optical microscopy during the hydration process of the investigated hemihydrates. These results were checked by SEM in the samples which were used for the strength tests.

With these experiments it was possible to evolve a new theory about hardening of calcium sulphate based binders and the strength decrease in presence of moisture which explains all experimental results. Depending to this theory the main factor for hardening of gypsum stone is the friction between the single gypsum crystals. Smaller factors are van der Waals forces and felting of the crystals. Branches lead to felting which prevents gliding of gypsum crystals on each other. Small water layers could work as a glue for the crystals, which explains the lightly increase of strength at lower humidity. Larger water layers like in liquid water or at higher humidity could change the friction mechanism from a solid to solid friction to a solid to liquid friction, which leads to a change of the coefficient of friction of a factor round about 10. It could be assumed that there will be always a mixture of solid-solid and solid-liquid friction so that 80% of strength decreases for gypsum stone is realistic. This theory could explain all experimental results but it is still part of the research.

[1] Roland J.-M. Pellenq and Henri Van Damme; Why does concrete set?: The Nature of cohesion forces in hardend cement-based materials; MRS Bulletin; pp. 319-323

[2] Mineralien Magazin; Heft 9 September 1981; Kosmos-Verlag; Stuttgart (1981)

[3] Pritzel, C.; Trettin R.; Herstellung von Beta-Halbhydrat im Labor; 1. Weimarer Gipstagung (2011)

MSC-P66

Investigating the annealing behaviour of Eifel sanidine

J. Kähn[1], D. Többens[1], M. Reehuis[1], S. Schorr[1]

[1]*Helmholtz-Zentrum Berlin, Kristallographie, Berlin, Germany*

Sanidine is the monoclinic high-T modification of K-rich alkalifeldspars. Annealing at T > 900°C usually causes disordering of Al/Si distribution at the two non-equivalent tetrahedral sites, but it is supposedly possible to disorder samples of sanidine from Volkesfeld/Eifel at notably lower temperatures and shorter times [1]. To investigate this behavior and compare various approaches to obtain the Al/Si distribution, samples from different Eifel locations and Madagascar have been studied. Al/Si order was determined by direct and indirect methods.

Neutron powder diffraction experiments were executed at Neutron Powder Diffractometer HB-2A at Oak Ridge National Laboratories and at Fine Resolution Neutron Powder Diffractometer E9 at the Berlin Research Reactor BERII. Samples of Madagascar Sanidine were measured at the High-Resolution Powder Diffractometer HRPT at Paul-Scherrer Institute. Single crystal neutron diffraction was performed at the 4-circle neutron diffractometer E5 at BERII. Al/Si distribution was determined directly, refining site occupancies by applying Rietveld analysis to powder diffraction data and XTAL for single crystal data. This approach is inapplicable when using X-ray data, due to similar atomic form factors of Al^{3+} and Si^{4+}, thus indirect methods [2,3] were applied. X-ray powder diffraction was performed at Helmholtz Centre Berlin, single crystal X-ray diffraction at Ruhr-University Bochum. The obtained data was processed using Rietveld refinement and ShelXL software respectively.

It was possible to verify a dependency of decreasing Al/Si order on increasing annealing times and temperatures. Surprisingly we could not observe any difference in annealing behaviour of sanidine from Eifel and Madagascar, suggesting the Al/Si distribution is not the origin of the optical anomalies observed in sanidine from Volkesfeld. We observed different results from direct and indirect methods, regardless whether samples were untreated or annealed. A stronger change of Al/Si distribution during annealing was revealed applying direct determination.

[1] Zeipert&Wondratschek, An unusual annealing behavior of Eifel Sanidine, N.Jb.Miner.Mh., 1981

[2] Kroll&Ribbe, Lattice parameters, composition and Al,Si order in alkali feldspars. Miner.Soc.Am. Rev. in Miner., 1983

[3] Carpenter&Salje, Thermodynamics of nonconvergent cation ordering in minerals. 3. order-paramenter coupling in potassium-feldspar, Am.Miner., 1994

SPE - Spectroscopy

SPE-P01

Sarcosine and betaine crystals on cooling: structural motifs unstable at high pressure become stable at low temperatures

E. Kapustin[1], V. Minkov[1,2], E. Boldyreva[1,2]
[1]*Novosibirsk State University, Natural Sciences, Novosibirsk, Russian Federation*
[2]*Institute of Solid State Chemistry and Mechanochemistry, Novosibirsk, Russian Federation*

The crystal structures of N-methyl derivatives of the simplest amino acid glycine, namely sarcosine ($C_3H_7NO_2$) and betaine ($C_5H_{11}NO_2$), were studied on cooling by single-crystal X-ray diffraction and single-crystal polarized Raman spectroscopy. The effects of decreasing temperature and increasing hydrostatic pressure on the crystal structures were compared. In particular, we have studied the behavior on cooling of those structural motifs in the crystals, which are involved into structural rearrangement during pressure-induced phase transitions. In contrast to their high sensitivity to hydrostatic compression, the crystals of both sarcosine and betaine are stable to cooling down to 5 K. Similarly to most α-amino acids, the crystal structures of the two compounds are most rigid on cooling in the direction of the main structural motif, namely head-to-tail chains (linked *via* the strongest N-H...O hydrogen bonds and dipole-dipole interactions in case of sarcosine, or exclusively by dipole-dipole interactions in case of betaine). The anisotropy of linear strain in betaine does not differ much on cooling and on hydrostatic compression, whereas this is not the case for sarcosine. Although the interactions between certain structural motifs in sarcosine and betaine weaken as a result of phase transitions induced by pressure, the same interactions strengthen when volume reduction results from cooling.

SPE-P02

Nanoparticle precursor into polycrystalline $Bi_2Fe_4O_9$: temperature-dependent in-situ investigations by X-ray diffraction, Raman and Mössbauer spectroscopy

A. Kirsch[1], M. M. Murshed[1], A. Friedemann[1], P. Gaczynski[2], K.- D. Becker[2], T. M. Gesing[1]
[1]*Universität Bremen, Chemische Kristallographie fester Stoffe, Institut für Anorganische Chemie, Bremen, Germany*
[2]*TU Braunschweig, Institut für Physikalische und Theoretische Chemie, Braunschweig, Germany*

Mullite-type $Bi_2Fe_4O_9$ has attracted great attention due to its multiferroic [1] and photocatalytic properties [2, 3]. For nanoscale materials, there is an ongoing search to exploit the associated electronic, optic, catalytic and magnetic properties for advanced applications. In the present study, nano-crystalline (X-ray amorphous) material was produced using $Bi(NO_3)_3{\cdot}5H_2O$ and $Fe(NO_3)_3{\cdot}9H_2O$ dissolved in diethylenglycol and heated at 423 K for 2 hours under reflux. While heating, the amorphous powder transformed into a rhombohedral perovskite-type $BiFeO_3$ at 680 K followed by a second transformation to polycrystalline $Bi_2Fe_4O_9$ starting from 800 K on. This observation was confirmed by complementary in-situ X-ray diffraction and Raman spectroscopy. Moreover, temperature-dependent ^{57}Fe Mössbauer spectra provided evidence for structural transformation in the given temperature range. The fit of the room-temperature ^{57}Fe Mössbauer spectrum of the as-synthesized amorphous powder prepared by the polyol method revealed the existence of two Fe^{3+} species in two different octahedrally coordinated sites with parameters different from those of the bulk material either $BiFeO_3$ or $Bi_2Fe_4O_9$. The spectral majority component possesses identical isomer shift (IS) as that of the octahedrally coordinated Fe-site in $Bi_2Fe_4O_9$ [4], thus almost similar bond lengths can be expected. The minority component shows a slightly higher IS, indicating a lower electron charge density at the nucleus, pointing to larger bond lengths. The broad room-temperature line-widths refer to some sort of disorder around these two sites. The signal intensity suggests formation of a $(Bi_2Fe)Fe_3O_9$ perovskite-type - as a non-equilibrium phase, where the Bi-site is partially shared by Fe. Structure refinements on the high-temperature X-ray data of this perovskite-type phase clearly shows partial occupation of Fe on the Bi position and vice versa. The oxygen position could be refined slightly off the special position (½, ½, 0) with 8-fold disorder. This comprehensive investigation sheds light on some open questions on why the formation of pure nanoparticle $BiFeO_3$ is difficult, and how the diffusion of the constituents into the reaction zone is hindered by the crystallization of the more thermodynamically stable $Bi_2Fe_4O_9$ phase at higher temperatures.

References
[1] A.K. Singh, S.D. Kaushik, B. Kumar, P.K. Mishra, A. Venimadhav, V. Siruguri, S. Patnaik, Appl. Phys. Lett. 92 (2008) 132910.
[2] N.I. Zakharchenko, Kinet. Catal. 43 (2002) 95.
[3] Q. Zhang, W. Gong, J. Wang, X. Ning, Z. Wang, X. Zhao, W. Ren, Z. Zhang, J. Phys. Chem. C 115 (2011) 25241.
[4] S-U. Weber, T.M. Gesing, J. Röder, F.J. Litterst, R.X. Fischer, K.-D. Becker, Int. J. Mat. Res. 103 (2012) 430.

SPE-P03

Temperature-dependent X-ray diffraction and Raman spectroscopy of $|Na_8(MnO_4)_2|[AlSiO_4]_6$

H. Petersen[1], L. Robben[1], T. M. Gesing[1]
[1]*Universität Bremen, Chemische Kristallographie fester Stoffe, Institut für Anorganische Chemie, Leobener Straße /NW2, D-28359 Bremen, Germany, Germany*

We report on the thermal expansion, structural development and intracage reactions of $|Na_8(MnO_4)_2|[AlSiO_4]_6$ from 300 K to 1140 K. The title compound was synthesized hydrothermally at 353 K using $NaAlO_2$, Na_2SiO_3, NaOH and $NaMnO_4$ with a molar ratio of 1 : 1 : 33 : 2.6 and investigated by temperature-dependent (TD) X-ray powder diffraction and TD Raman spectroscopy. TD XRD data were collected on a Philips X'Pert MPD diffractometer (Panalytical, Almelo, Netherlands), equipped with the high-temperature chamber HTK1200N (Anton Paar, Graz, Austria), using $CuK\alpha_{1,2}$ radiation. Rietveld refinements of the TD X-ray diffraction data show an increase of the lattice parameter *a* up to 680 K. Above this temperature *a* slightly increases nearly linear, indicating a phase transition from $P\bar{4}3n$ to the fully expanded sodalite structure in space group $Pm\bar{3}n$. At 900 K the upcoming of an additional reflection around 10° 2θ (Cu K_α) could be observed, indicating a symmetry reduction to space group *P*23 which is driven by template ordering effects. This ordering effect could be correlated with the reduction of the permanganate ion (MnO_4^-) to the manganate ion (MnO_4^{2-}) to occur around 873 K assumed by Brenchley and Weller [1] from TD IR investigations. The TD Raman data were collected from 123 K to 525 K on a LabRAM Aramis (Horiba Ltd, Kyoto, Japan) spectrometer using an excitation wavelength of 533 nm. Only the resonance excitation of the permanganate ion could be seen in the spectra. In the temperature range from 123 K to 300 K the respective harmonic frequency (υ_0) decreases linearly with υ_0 = 846.32(9) cm^{-1} - 2.62(5) $10^{-2}{\cdot}T$ cm^{-1}K. Above 300 K υ_0 = 840.7(4) cm^{-1} - 0.65(12) $10^{-2}{\cdot}T$ cm^{-1}K. For crystalline $NaMnO_4$ a value of υ_0 = 845.5(5) cm^{-1} was observed [2] which is close to υ_0 = 843.0(1) cm^{-1} evaluated at 123 K. At room-temperature the sodalite υ_0 value of 838.5(8) cm^{-1} matches well to the respective value determined for an aqueous $NaMnO_4$ solution (υ_0 = 839.6(4) cm^{-1}). This behavior could be interpreted as a vibration blocking caused by the framework-template interactions at low-temperatures. Above room-temperature the behavior of υ_0 could be interpreted as template-framework interactions similar to permanganate in an aqueous solution hydration shell.

Fig. 1: Harmonic frequencies (υ_0) of $|Na_8(MnO_4)_2|[AlSiO_4]_6$, an aqueous $NaMnO_4$ solution and crystalline $NaMnO_4$ determined by TD-Raman spectroscopy

[1] M. E. Brenchley and M. T. Weller, *Zeolites* 14 (1994) 682- 686.
[2] W. Kiefer and H. J. Bernstein, *Mol. Phys.* 5 (1972) 835-851.

Figure 1

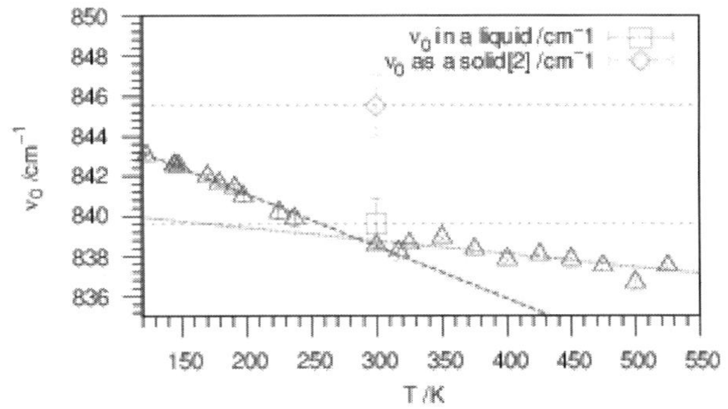

SPE-P04

First Experiments with the New Multiplex-system at the Three-Axes-Spectrometer PUMA

G. Eckold[1], O. Sobolev[1], R. Hoffmann[1]
[1]*Georg-August-Universität Göttingen, Institut für Physikalische Chemie, Göttingen, Germany*

Inelastic neutron scattering is one of the most powerful tools for the determination of chemical bond strengths and or magnetic interactions in condensed matter. Single crystal investigations using three axes spectrometers, in particular, provide detailed information about excitation spectra, eigenvectors, lifetimes or anharmonic effects that might be associated with lattice instabilities. The unprecedented flexibility of a three axis spectrometer allows focusing on special aspects of the phenomena under consideration. Being a point-by-point technique, however, this type of spectroscopy requires still rather long beam time even if modern neutron optical devices are used.

In order to improve the performance of this method a new type of multiplex technique has been realized and successfully commissioned at the PUMA spectrometer at FRM II which is one of the leading instruments worldwide. Consisting of eleven analyser-detector channels which can be configured individually, this technique is especially useful for kinetic experiments where a single excitation spectrum is recorded as a function of time without the need to move the spectrometer. On a time-scale of seconds an entire spectrum can be recorded thus allowing users to monitor changes during fast kinetic processes in single shot experiments without the need for stroboscopic techniques. Moreover, the multianalyser system provides an efficient and rapid tool for mapping excitations in (***Q***,*w*)-space without relaxing the resolution power. The results of pilot experiments demonstrate the performance of this new technique and a user-friendly software is presented which assists users during their experiments.

SPE-P05

NMR spectroscopy on magnetically aligned powder samples

A. Korthaus[1], B. Mausolf[1], F. Haarmann[1]
[1]*Institut für Anorganische Chemie / RWTH Aachen University, Aachen, Germany*

In contrast to its wide range of application and the variety of existing compounds, chemical bonding in intermetallic phases is not yet well understood. Due to the growing interest in intermetallic compounds for industrial applications and the desire for target-oriented synthesis routes, the knowledge of structure-bonding-property relationships is mandatory. The electric field gradient (EFG), accessible by solid state nuclear magnetic resonance (NMR) spectroscopy and quantum mechanical (QM) calculations, is very sensitive to the environment of a nucleus. The NMR line-shape of quadrupolar nuclei located at sites of low symmetry is predominately affected by each site's 2nd rank EFG tensor and its orientation towards the external magnetic field.

Wien2k is used to calculate the relevant quadrupolar NMR parameters as well as EFG tensor orientations. Regular powder samples contain randomly distributed EFG main directions towards the external field. As a consequence such spectra usually span a wide frequency range. Since NMR spectra are the superposition of every site's signal contribution weighted according to its relative abundance in the crystal structure, interpretation of such spectra becomes more challenging with an increasing number of different atom positions. Due to the random crystallite orientation there is in fact no way to conclude the absolute EFG main directions in the crystal structure from such spectra.

Today we would like to report on an approach to investigate chemical bonding in anisotropically electron-conducting materials by solid state NMR spectroscopy. Crystallites of the above mentioned materials align inside external magnetic fields striving to minimize the Lorentz Force. This is accomplished when the crystallites' direction of highest Fermi velocity is parallel to the external field [1]. The crystallite orientation can be fixed by dispersing the sample in a two-component adhesive, letting it set inside the magnet. Orientation-dependent wideline and frequency-sweep NMR experiments on aligned powder samples are performed using an automatic tuning and matching goniometer (ATMG) system [2].

Advantages of this method are a larger NMR data basis for signal line-shape analysis as well as the possibility of identifying atoms located at individual sites by their characteristic orientation-dependent line-shape development. Today we would like to put emphasis on the simulation of orientation-dependent NMR spectra of aligned powder samples based on QM calculations.

[1] F. Haarmann, in: Enzyclopedia of Magnetic Resonance (Eds.: R. K. Harris, R. E. Wasylishen), John Wiley & Sons, Ltd, Chichester, 2011.
[2] O. Pecher, F. Haarmann. *Z. Kristallogr.* Suppl. 32, 24 (2012).
[3] O. Pecher. Dissertation, RWTH Aachen University (2013).

Fig.1: Regular (left) and aligned (right) powder sample. Shaded areas indicate highest Fermi velocity [3]

Figure 1

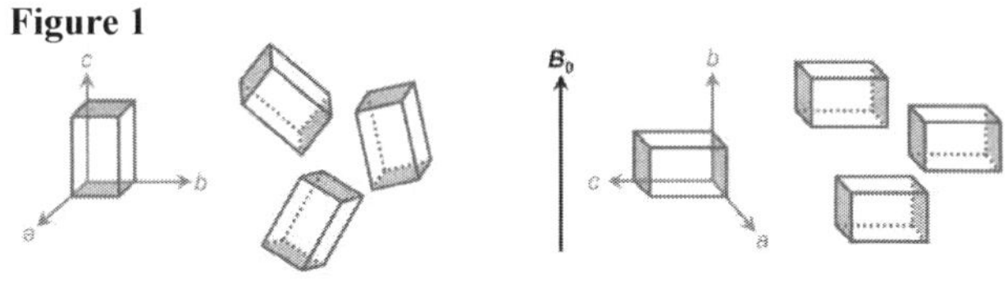

SPE-P06

Determination of local atomic arrangements in Ba($M_{1-x}M'_x$)$_4$ with (M,M' = Al, Ga, In) using quantum mechanical calculations and solid-state-NMR-spectroscopy.

V. Peters[1], B. Mausolf[1], F. Haarmann[1]
[1]*Instutut für Anorganische Chemie/RWTH Aachen University, Aachen, Germany*

NMR Spectroscopy for structure determination of organic molecules is widely used. The isotropic NMR signal shifts vary in certain ranges for specific functional groups. Due to large frequency distributions of the NMR signals in solid-state-NMR-Spectroscopy, slightly different local environments will most likely show overlapping signals. Therefore, distinguishing between different environments is challenging. In solid-state-NMR-Spectroscopy the quadrupolar coupling can be used to determine between different environments. Contrary to the isotropic NMR signal shift the relation between structure and the quadrupolar coupling parameter C_Q is rather ambiguous. Modelling different possible environments and calculating their quadrupolar coupling parameters is essential for further NMR-signal-lineshape-analysis. This method allows the determination of the local atomic arrangements [1].

In the present work we studied the intermetallic compounds Ba($M_{1-x}M'_x$)$_4$ with (M,M' = Al, Ga, In). They crystallize in the $BaAl_4$ - structure (*I*4/*mmm*), where Ba occupies the corners and the center of the unit cell. The other atoms build a cage-like structure around the Ba-atoms. These "cages" contain two different crystallographic sites. The first site is tetrahedron-like coordinated (4b) and the coordination polyhedron of the other site is quadratic pyramidal (5b). Previous studies on Al@$SrGa_4$ showed that Al-atoms tend to occupy the 4b position while the Ga-Atoms occupy the 5b position [2]. In $Ba(Al_{0.975}Ga_{0.025})_4$ Ga occupies the 5b Position while Ga-atoms in $Ba(Ga_{0.025}In_{0.975})_4$ occupy both sites. Further analysis reveal that the Ga-NMR signal of $Ba(Al_{0.975}Ga_{0.025})_4$ comprises of two signals. The first signal results from isolated Ga-atoms (5b1) and the other signal from two Ga-atoms along the diagonal in the *a*-*b*-plane (5b2)(Fig.1). Using the signal integral the frequency of occurrence of these arrangements in the sample can be determined. For higher Ga contents the frequency of occurrence of 5b2 arrangements rises. We expect to gain insight into preferential site occupation in solid state materials by applying this strategy to other samples of Ba($M_{1-x}M'_x$)$_4$.This may lead to new ways to improve material properties.

[1] F.Haarmann, K.Koch, P.Jeglič,O.Pecher, H.Rosner, Yu.Grin. NMR spectroscopy of Intermetallic Compounds: An Experimental and Theoretical Approach to Local Atomic Arrangements in Binary Gallides. *Chem. Eur. J.*17(27),7560-7568, 2011. [2] O. Pecher, F. Haarmann. Looking into Intermetallic Phases.*Nachrichten aus der Chemie*. 61(10), 1017-1021, 2013.
Fig 1: Orientation dependent ^{71}Ga-NMR signals of the Ga(5b1) and Ga(5b2)position in $Ba(Al_{0.975}Ga_{0.025})_4$

Figure 1

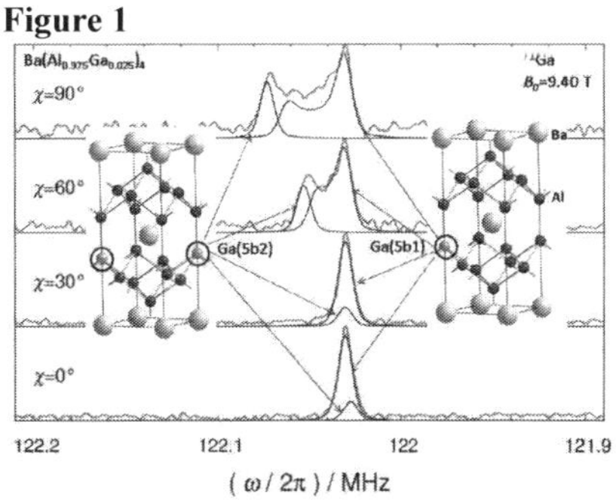

SPE-P07

Time-dependent XPS studies on $Li_{0.4}WO_3$ bronze stored at different conditions

N. Lefeld[1], A. Schaefer[2], M. M. Murshed[1], M. S. Rahman[1], M. Bäumer[1], T. M. Gesing[1]
[1]*Universität Bremen, Chemische Kristallographie fester Stoffe, Institut für Anorganische Chemie, Bremen, Germany*
[2]*Universität Bremen, Institut für angewandte und physikalische Chemie, Bremen, Germany*

Alkali metal tungsten bronzes (A_xWO_3: A = alkali metal, $0 < x < 1$) have been studied for years because of their remarkable electrical [1-2] and optical properties [1, 3]. Li_xWO_3 bronzes have drawn particular attention for their structural pitfalls and associated peculiarities in crystal-chemical properties. Recently, it has been shown that perovskite type $Li_{0.4}WO_3$ slowly transformed from a body centered cubic (*I*m-3, a = 744.9(2) pm) into a tetragonal (*P*4/nmm: a = 520.5(2) pm, c = 383.0(3) pm) phase when exposed to atmospheric conditions [4]. To get more insights of this transformation caused by aging, time-dependent investigations were performed on $Li_{0.4}WO_3$ powder samples kept under ultra-high vacuum (UHV), dry oxygen and open atmospheric conditions using x-ray photoelectron spectroscopy (XPS). The XPS spectrum of the as-synthesized samples exhibits two doublets of the W-4f orbital originating from different final states [5] induced by the interaction between the generated core-hole and the conduction band, which is populated by electrons donated by lithium. Samples exposed to open atmospheric conditions for a given time showed a shift of both W-4f doublets with a decrease of the binding energy separation between the doublets. Moreover, their relative intensity successively changed upon aging, indicating a loss of conduction band electrons. An additional signal evolved on the high binding-energy side in the O-1s spectrum in that given period, which we assume to attribute to the presence of OH/CO_3 species at the surface. These could be a reason for an observed decrease of conductivity compared to that of the parent $Li_{0.4}WO_3$. The sample stored at dry oxygen conditions showed a comparable behavior to that observed for open atmospheric storage, however, not as fast and with no indication for OH/CO_3 formation. Under UHV conditions we observed a reverse behavior in the W-4f signal, which is not fully understood yet, as the binding energy separation between the doublets increased and the relative intensity changed in an opposite way. Again, OH/CO_3 formation was not apparent. At this stage we suggest that moisture plays a role for the associated aging. The results shed light on the electronic changes involved in the structural transformation [4]. However, the question is still open whether the structural transformation is electronically driven or a result of a thermodynamic disequilibrium of the initial structure.

References

[1] P. G. Dickens, M. S. Whittingham, Q. Rev. Chem. Soc. 22 (1968) 30-44. [2] M. J. Sienko, T. B. N. Troung, J. Am. Chem. Soc. 83 (1961) 3939-3943. [3] M. Green, Z. Hussain, J. Appl. Phys. 74 (1993) 3451-3458. [4] M. S. Rahman, M. M. Murshed, Th. M. Gesing, Z. Kristallogr. 229 (2014) 797-805. [5] J. Chazalviel, M. Campagna, G. K. Wertheim, H. R. Shanks, Phys. Rev. B 16 (1977) 697-705.

SPE-P08

Of substitution and doping: Spatial and electronic structure in iron pnictides

M. Merz[1], P. Schweiss[1], P. Nagel[1], M.- J. Huang[1], R. Eder[1], T. Wolf[1], H. von Loehneysen[1], S. Schuppler[1]
[1]*Karlsruher Institut für Technologie, Institut für Festkörperphysik, Eggenstein-Leopoldshafen, Germany*

A highly intriguing aspect in iron-pnictide superconductors is the composition-dependent electronic structure, in particular the question if and how charge carriers are introduced to the system upon substitution of Ba by alkali metals or of Fe by other transition metals, *TM*. We report on a systematic study of spatial structure and electronic states by x-ray diffraction and x-ray absorption, performed on a large number of compositions in the $(Ba,K)(Fe,TM)_2As_2$ family of compounds. The coherent combination of detailed structural information with an in-depth analysis of the electronic structure allows us to disentangle very sensitively "doping" effects from "substitutional" effects. This balance between substitution and doping turns out to be crucial for an understanding of magnetism and superconductivity in iron pnictides.

SPE-P09

Infrared spectroscopy as a complementary tool for the structure analysis of hydrous layer silicates

B. Marler[1], S. Grabowski[1], H. Gies[1]
[1]*Ruhr University Bochum, Dept. of Geology, Mineralogy and Geophysics, Bochum, Germany*

Hydrous Layer Silicates (HLSs) like the mineral Kanemite, $Na_4[H_4Si_8O_{20}] \cdot 12\ H_2O$ or synthetic RUB-36, $[N(CH_3)_2(C_2H_5)_2]_4[H_4Si_{36}O_{76}]$ consist of complex silicate layers which intercalate cations of low charge density (e.g. $[Na(H_2O)_6]^+$ or $[N(CH_3)_2(C_2H_5)_2]^+$). HLSs have important applications as adsorbants and can be used as precursors to produce microporous and mesoporous framework materials. However, the structure analysis and in particular the structure solution of HLSs is often very difficult because of the limited crystallinity of the materials. Powder diffraction diagrams typically contain sharp as well as broad peaks and only very low peak intensities are observed at diffraction angles larger than ca. 35° 2theta (Fig. 1). Also, data mining and the comparison with related structures usually fail to reveal a first structure model since the powder XRD diagrams of structurally even closely related HLSs are very different. This precludes the direct recognition of a structural relationship.

IR spectroscopy, however, is a very useful tool to support the structure analysis of a HLS with unknown structure. In contrast to many other silicates the fingerprint region (ca. 400 - 1500 cm^{-1}) within the IR spectrum of a HLSs gives a clear indication on the layer type of the material. Absorption bands of the org. cations and the water molecules being present in the structure are usually very weak in the fingerprint region and can be neglected. As soon as the topology of the silicate layer is recognized a first structure model can be generated which is sufficient to solve and refine the structure of the HLS based on powder diffraction data.

ATR-FTIR spectra of RUB-36 ($Pn2_1a$), RUB-20 ($P2_1/m$) and PREFER (I2), all containing thefer type silicate layer are compared to RUB-52 possessing a layer of different topology (Fig. 2). The upper three spectra exhibit similar features even though the materials are different concerning space group symmetry, unit cell dimensions and chemical composition (type of cation, water content).

It is assumed that the weak interactions between silicate layer and cations of low charge density have only a minor impact on the structure of the silicate layer leading to similar IR spectra for all HLSs with identical layer types. This is in contrast to layer silicates containing cations of higher charge density: E.g., the IR spectra of hydrated H-Apophyllite, $H_8Si_8O_{20} \cdot 4\ H_2O$, and Apophyllite, $KCa_4[Si_8O_{20}(F,OH)] \cdot 8\ H_2O$, with small cations ($K^+$, Ca^{2+}) differ considerably.

Figure 1

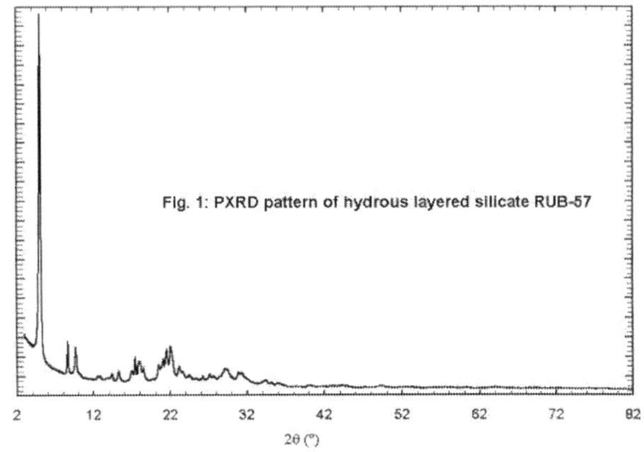

Figure 2

Fig. 2: FTIR Spectra

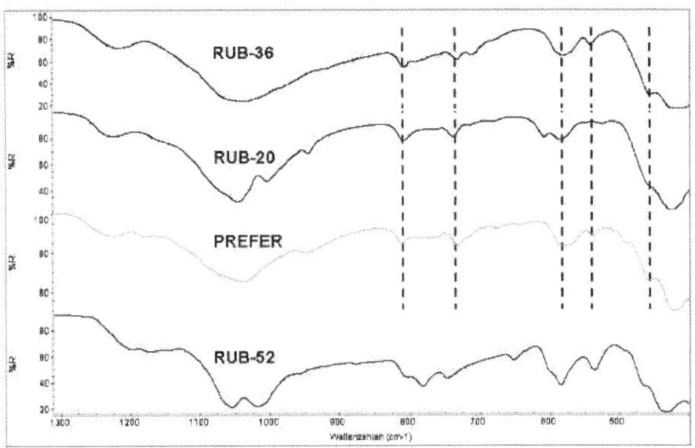

SCE - Structural chemistry at extreme conditions

SCE-P01
Trigonal bipyramidal coordination of Zn^{2+} in $Zn(NH_3)_2F_2$ and $Zn(NH_3)_3F_2$

T. Richter[1], N. S. A. Alt[2,3], E. Schlücker[3], R. Niewa[1]
[1]*Universität Stuttgart, Institut für Anorganische Chemie, Stuttgart, Germany*
[2]*Universität Stuttgart, Stuttgart, Germany*
[3]*Friedrich-Alexander-Universität Erlangen-Nürnberg, Lehrstuhl für Prozessmaschinen und Anlagentechnik, Erlangen, Germany*

We present two possible intermediates in a conceivable ammonothermal synthesis of Zn_3N_2, namely $Zn(NH_3)_2F_2$ ($P2_1/m$, a = 7.5335(4) Å, b = 4.2828(1) Å, c = 12.0201(2) Å, β = 108.26(6) °, V = 368.3(1) Å^3, Z = 4) and $Zn(NH_3)_3F_2$ ($P2_1/c$, a = 7.709(2) Å, b = 10.876(2) Å, c = 11.481(2) Å, β = 90.0(3) °, V = 962.2(3) Å^3, Z = 1).

Both compounds were obtained during ammonothermal transport reactions from NH_4F and Zn in supercritical ammonia at 570 °C and 2.9 kbar in the cold zone of the autoclave. We expect Zn_3N_2 (a promising semiconductor with a narrow band gap of 0.9 - 1.2 eV [1]) to form under similar conditions. Therefore the presented ammoniates of zinc fluoride represent possible intermediates in a conceivable ammonothermal synthesis of Zn_3N_2. Rarely pentacoordinated complexes are stable enough to be isolated. We present two new compounds manifesting a trigonal bipyramidal coordination on Zn^{2+}, namely $Zn(NH_3)_2F_2$ and $Zn(NH_3)_3F_2$. While tetrahedral (e.g. in Zn_3N_2 [2]) and octahedral environments (e.g. in ZnF_2 [3]) are well known for Zn^{2+}, only a few examples for trigonal bipyramidal coordinations are known in general and for Zn^{2+} in particular. Previously, a bipyramidal surrounding of five NH_3 ligands coordinating Cu^{2+} has been reported in $[Cu(NH_3)_5]F_2 \cdot NH_3$ [4]. Like $Zn(NH_3)_2Cl_2$ [5] both new compounds contain NH_3 and halide ligands at Zn. $Zn(NH_3)_3F_2$ contains isolated trigonal bipyramidal $[Zn(NH_3)_3F_2]$ units whereas in $Zn(NH_3)_2F_2$ axial F^- connect the trigonal bipyramids resulting in linear 1-D chains ($^1_\infty[Zn(NH_3)_2F_{2/2}F]$, see figure 1). In $Zn(NH_3)_2F_2$ and $Zn(NH_3)_3F_2$ all NH_3 ligands are in equatorial position and F^- in axial position of the trigonal bipyramidal surrounding of Zn. In $Zn(NH_3)_2F_2$ it is completed by a further F^- ligand in equatorial position. These arrangements correlate with the observation that for trigonal bipyramidal coordinations stronger σ donors prefer the equatorial position for d^0 - d^4 and d^{10}, with NH_3 being the stronger ligand [6]. Raman spectra were recorded and the bands assigned to ν-, δ- and ρ-vibrations (see figure 2).

Figure 1: a) Trigonal bipyramidal surrounding of Zn^{2+} a) in $Zn(NH_3)_3F_2$ and b) in $Zn(NH_3)_2F_2$ resulting in 1-D chains ($^1_\infty[Zn(NH_3)_2F_{2/2}F]$).

Figure 2: Raman spectra of $Zn(NH_3)_3F_2$ and $Zn(NH_3)_2F_2$.

[1] S.-H. Yoo, A. Walsh, D. O. Scanlon, A. Soon, *RSC Adv.* 2014, *4*, 3306.
[2] D. E. Partin, D. J. Williams, M. O'Keeffe, *J. Solid State Chem.* 1997, *132*, 56-59.
[3] J. W. Stout, S. A. Reed, *J. Am. Ceram. Soc.* 1954, *76*, 5279-5281.
[4] P. Woidy, A. J. Karttunen, M. Widenmeyer, R. Niewa, F. Kraus, *Chem. Eur. J.* 2014, submitted.
[5] C. H. MacGillavry, J. M. Bijvoet, *Z. Kristallogr.* 1936, *94*, 249 - 255.
[6] A. R. Rossi, R. Hoffmann, *Inorg. Chem.* **1975**, *14*, 365-374.

Figure 1

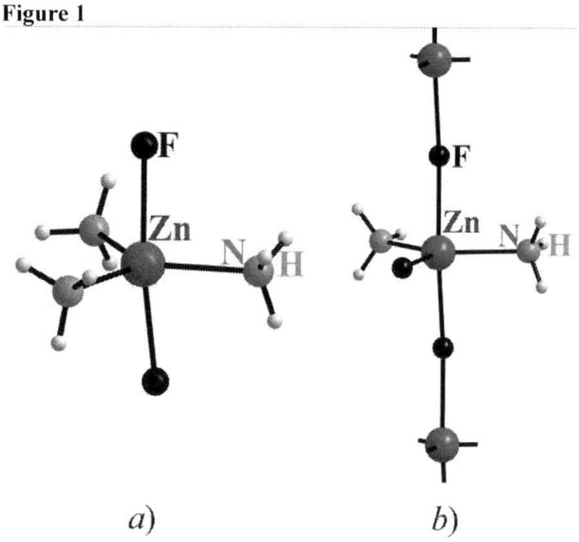

Figure 2

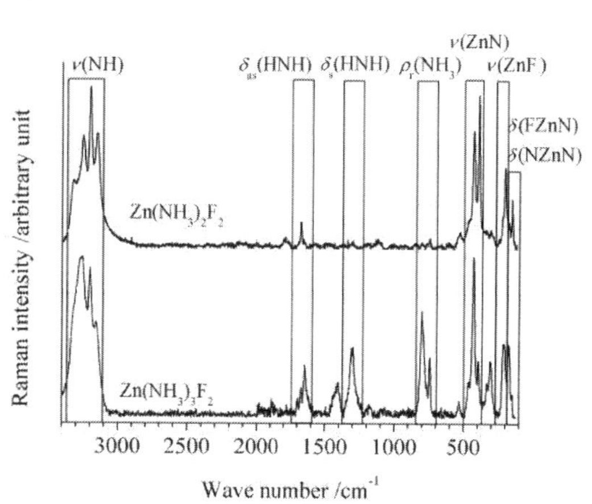

SCE-P02
TEV - a tool for the determination and visualization of the thermal expansion tensor from diffraction data

V. Kahlenberg[1], T. Langreiter[1]
[1]*University of Innsbruck, Institute of Mineralogy & Petrography, Innsbruck, Austria*

TEV (Thermal Expansion Visualizing) is a user-friendly program for the calculation of the thermal expansion tensor α_{ij} from diffraction data. Unit cell parameters determined from temperature dependent data collections can be provided as input. An intuitive graphical user interface enables fitting of the evolution of individual lattice parameters to polynomials up to fifth order. Alternatively, polynomial representations obtained from other fitting programs or from the literature can be entered.

The polynomials and their derivatives are employed for the calculation of the tensor components of α_{ij} in the infinitesimal temperature limit using the approach of Paufler & Weber (1999). The tensor components, eigenvalues, eigenvectors as well as their angles with the crystallographic axes can be evaluated for individual temperatures or within a temperature interval. Values of the tensor in specific directions parallel to either [uvw]'s of the crystal lattice or vectors (hkl) of reciprocal space can be calculated. Finally, the 3-D representation surface for the second rank tensor and pre- or user-defined 2-D sections can be plotted and saved in a bitmap format. An example for the representation surface of a triclinic crystal is given in Figure 1.

TEV is written in JAVA. The distribution contains an EXE-file for Windows users and a system independent JAR-file for running the software under Linux and Mac OS X.

Paufler, P. & Weber, T. (1999) On the determination of linear thermal expansion coefficients of triclinic crystals using X-ray diffraction. Eur. J. Mineral. 11, 721-730.

Figure 1

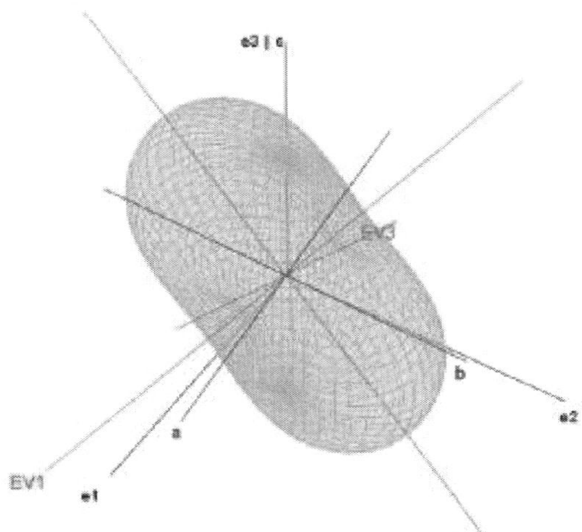

SCE-P03

Crystalline hydrates of an amphiphilic ionic liquid and its influence on drug polymorphism

S. Saouane[1], F. P. A. Fabbiani[1]
[1]*Georg-August-Universtaet Goettingen, GZG - Abt. Kristallographie, Göttingen, Germany*

Ionic liquids are versatile molten salts with a melting point below the boiling point of water. Thanks to some of their numerous physical and chemical properties, notably negligible vapour pressure, ionic liquids are gaining interest as green solvents.[1] More recently, the use of ionic liquids was extended to pharmaceutical research where known drug molecules are redesigned as active ionic liquids[2] or the polymorphic behaviour of drug compounds is investigated using ionic liquids as crystallisation media.[3] Following our successful elucidation of the complex crystal polymorphism in $[C_4mim][PF_6]$,[4] we extend our investigations of imidazolium-based ionic liquids by presenting herein the rich solid-state behaviour of 1-decyl-3-methylimidazolium chloride ($[C_{10}mim][Cl]$), a highly hygroscopic and amphiphilic ionic liquid which forms several hydrates at both low-temperature and high-pressure conditions. We have also been investigating the use of $[C_{10}mim][Cl]$ aqueous solutions as high-pressure crystallisation media for investigating the polymorphism of drug molecules, such as paracetamol.

(1) Wilkes, J. S. A Short History of Ionic Liquids—from Molten Salts to Neoteric Solvents. *Green Chem.* 2002, *4*, 73-80.

(2) Hough, W. L.; Smiglak, M.; Rodríguez, H.; Swatloski, R. P.; Spear, S. K.; Daly, D. T.; Pernak, J.; Grisel, J. E.; Carliss, R. D.; Soutullo, M. D.; et al. The Third Evolution of Ionic Liquids: Active Pharmaceutical Ingredients. *New J. Chem.* 2007, *31*, 1429.

(3) An, J.-H.; Kim, J.-M.; Chang, S.-M.; Kim, W.-S. Application of Ionic Liquid to Polymorphic Design of Pharmaceutical Ingredients. *Cryst. Growth Des.* 2010, *10*, 3044-3050.

(4) Saouane, S.; Norman, S. E.; Hardacre, C.; Fabbiani, F. P. a. Pinning down the Solid-State Polymorphism of the Ionic Liquid [bmim][PF6]. *Chem. Sci.* 2013, *4*, 1270.

SCE-P04

Effects of missing data on single-crystal refinement of organic structures at high pressure

J. Holstein[1], R. Granero-García [1], F. Fabbiani[1]
[1]*Georg-August Universität Göttingen, Crystallography, Göttingen, Germany*

Organic compounds have a tendency to crystallise in low-symmetry space groups.[1] This can have severe implications for accurate and precise high-pressure studies of these compounds. The diamond anvil cell poses a significant geometrical restriction in the measurement of X-ray data in transmission mode and consequently parts of the Ewald sphere are inaccessible.[2] Together with limited resolution this often results in systematically missing data and a low data to parameter ratio. As in the case of protein structure determination, the introduction of additional observations may be required in the structure refinement process to ensure a reliable molecular model. Typically, restraints for bonding geometries and atomic displacements would be implemented. The systematically missing data in reciprocal space can also introduce systematic errors to the refined model parameters, most notably in the shape of the anisotropic displacement parameters (APDs). In this work we would like to shed light on these and other effects caused by systematically missing data. For this purpose we employ the ECD program[3] to simulate high-pressure data by systematically cutting out cones of reflections in reciprocal space from complete single-crystal datasets. We then test different refinement techniques on these pruned datasets and compare the results with the model refined against the complete datasets. We also present a refinement methodology adapted from macromolecular techniques that foresees the use of R(free)[4] and R(complete)[5] to assess model bias in the refinement of high-pressure structures.

[1] F. H. Allen, Acta Cryst., B58, 380-388, 2002.
[2] R. Miletich, D. R. Allan and W. F. Kuhs, High-Temperature and High-Pressure Crystal Chemistry, 41, 445 - 519, 2000.
[3] R. Granero-Gracía and S. Parsons, Erase Crystallographic Data (ECD), The University of Edinburgh, 2011.
[4] A.T. Brünger, Nature, 355, 472-475, 1992.
[5] A. T. Brünger, Methods in Enzymology, 277, 366 - 396, 1997.

SCE-P05

Structural transformations of the silica clathrate chibaite

K. Scheidl[1], R. Miletich[1], T. Yagi[2], K. Momma[3]
[1]*University of Vienna, Vienna, Austria*
[2]*University of Tokyo, Tokyo, Japan*
[3]*National Museum of Nature and Science, Tsukuba, Japan*

Silica clathrates are characterized by an open SiO_2-framework and cage-like voids, analogical to gas hydrates. Chibaite represents a naturally analogue to synthetic silica clathrate MTN [1]. Besides melanophlogite, it is the only naturally formed silica clathrate found so far. Due to the natural formation in marine sediments, hydrocarbons (CH_4, C_2H_6, C_3H_8 and i-C_4H_{10}) are intercalated in the cages [2] and thus chibaite currently is considered as a potential carbon reservoir and a further sink within the global carbon cycle similar to the role of gas clathrates [3,4,5]. Chibaite crystalizes in cubic type-II gas hydrate structure (Fd3 symmetry for the SiO_2-framework). The high-pressure behavior of chibaite was investigated by means of *in situ* Raman-spectroscopy and X-ray diffraction in a diamond anvil cell. Due to the open framework of chibaite host structure guest molecules are supposed to have a strong influence on the high-pressure behavior. In order to study the influence of the guest molecules pristine as well as heated samples were used. Elasticity and phase transitions of pristine and heated chibaite were studied up to 10 GPa. The Raman-spectra for the experiments using a methanol-ethanol mixture as pressure medium show two structural transitions with distinct spectral changes indicating the break of the cubic symmetry at 1.3 and 3.1 for the pristine (Figure 1), and at 1.4 and 3.7 for the heated chibaite sample (Figure 2). The spectra of the pristine and the heated sample differ due to the lower amount of volatile guest molecules inside the cages of the treated sample. Within the applied pressure range both samples do not show any evidence of pressure-induced amorphization.

Figure 1: Raman spectra of pristine chibaite

Figure 2: Raman spectra of heated chibaite

References
[1] Schlenker J L et al.; Zeitschrift für Kristallographie, 167 (1984), 73-82.
[2] Momma K et al.; Nature 2, (2011), 1-7.
[3] Lu, H et al.; Nature, 445 (2007), 303-306.
[4] Kvenvolden K A; Organic Geochemistry, 23 (1995), 997-1008.
[5] Davidson, D W et al.; Geochimica et Cosmochimica Acta, 50 (1986), 619-623.

Figure 1

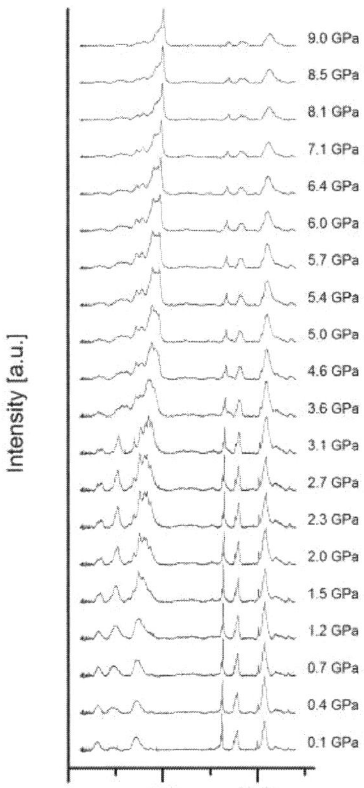

Figure 2

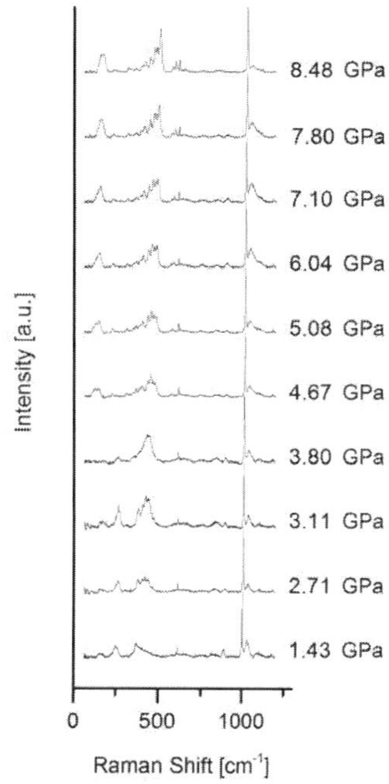

SCE-P06

Melamine at high pressures and high temperatures

N. Rademacher[1], L. Bayarjargal[1], W. Morgenroth[1], B. Winkler[1]
[1]*Goethe University, Institute of Geosciences, Frankfurt am Main, Germany*

Chemical reactions at high pressures and temperatures are of interest because extreme conditions can lead to reaction pathways which result in novel compounds with unusual structural features and interesting properties [1]. Reactions of molecules from the chemical system C-H-N-O are of particular interest for example in astrophysics, condensed-matter physics or chemistry [2]. This study presents high pressure and high temperature experiments of melamine (C3H6N6). Melamine has previously been studied at high pressures and high temperatures by means of diffraction and spectroscopy with the aim of synthesising g-C3N4 and superhard C3N4 [e.g. 3,4]. Yu et al. [3] reported on the chemical reaction of melamine at around 5 GPa and temperatures of up to 900 °C. The authors found that depending on the conditions the products can be carbon nitrides or triazine polymers. Fang et al. [3] studied the pyrolysis of melamine at 22 GPa and up to 2000 °C and found that diamond was formed at this conditions. High pressure phase transitions have also been studied [e.g. 5,6]. Odake et al. [5] reported on the Raman spectroscopy of melamine up to 60 GPa and did not find indications for phase transformations. In contrast Galley et al. [6] studied melamine up to 36 GPa and found three phase transitions at 1 to 2 GPa, 7 to 9 GPa and around 16 GPa.
We have performed diamond anvil cell (DAC) experiments at pressures up to 50 GPa in order to complement previously published data and to study the stability field of melamine. The experiments were performed with neon as pressure transmitting medium. We have followed the structural changes during pressure increase by means of micro-Raman spectroscopy. Our Raman spectra show phase transitions between 0 and 3 GPa, between 7 and 10 GPa and at around 20 GPa. Above 36 GPa we observed strong fluorescence of the sample which resulted in a very high background. At around 47 GPa the fluorescence decreased and stronger Raman bands could be observed again. The experiments at high pressures and temperatures up to 600 K are still ongoing.

The authors gratefully acknowledge financial support from the DFG (project RA 2585/1-1) and the BMBF (project 05K13RF1).

Figure 1: Micro-Raman spectra of melamine up to 50 GPa (the spectra are not background corrected).

[1] R. Bini, M. Ceppatelli, M. Citroni and V. Schettino, Chemical Physics 398, 262 (2012).
[2] F. Datchi and G. Weck, Z. Kristallogr. 229, 135 (2014).
[3] D. L. Yu, J. L. He, Z. Y. Liu, B. Xu, D. C. Li and Y. J. Tian, J. Mater. Sci. 43, 689 (2008).
[4] L.-M. Fang, X.-P. Chen, H. Ohfuji, T. Irifune, G.-A. Sun, B. Chen and S.-M. Peng, Chinese Physics C 37, 088002 (2013).
[5] S. Odake, P. V. Zinin and L.-C. Ming, High Pressure Res. 33, 392 (2013).
[6] M. Galley, M. Pravica and Z. Liu, High Pressure Res. 33, 40 (2013).

Figure 1

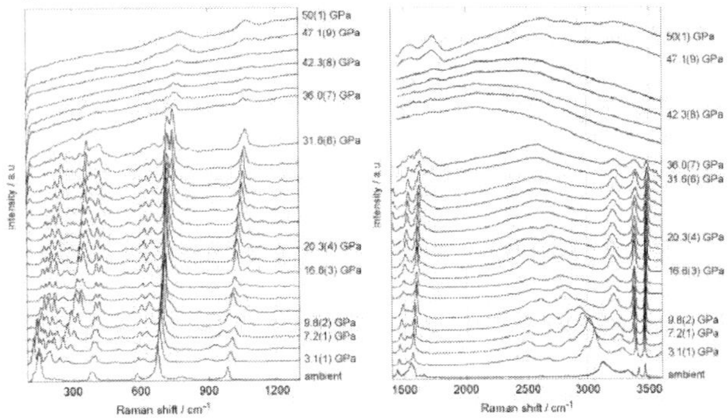

SCE-P07

Shock-wave treatment of kaolinite under extreme conditions

T. Schlothauer[1], K. Keller[1], B. Erica[1], H. Gerhard[1], E. Kroke[1]
[1]*TU Bergakademie Freiberg, Institut für Mineralogie, Freiberg, Germany*

Question

Commonly shock-wave techniques were used to recognize the behavior of materials under intense dynamic loading, the synthesis of new ultra hard high pressure phases for technical applications and for special manufacturing processes (e.g. explosive welding). In the case of fluid rich natural mineral phases like carbonates or clay minerals this method was used for experimental investigations of large impact events (Chicxulub-event at the cretaceous-tertiary boundary). For investigations of the deep earth interior this group of minerals is in present time an object of great interest. This concerns for example the "deep carbon cycle" or the behavior of hydratized silicates in the time of subduction processes. Because especially the carbonates are very sensitive against shock-wave treatment ("degassing along the release path") new sample recovery systems will be required.

Methods

For the shock-wave treatment of fluid rich mineral phases some requirements should be fulfilled: avoiding of Mach-effects, avoiding of shock-wave reflections and avoiding of adiabatic decompression along the release path. Especially the adiabatic decompression into vacuoles near the sample or into gaps between sample and sample holder falsifies the experimental results. Despite the occurrence of phase transformation in the high pressure range the degassing along the release path will result in a complete decay of the obtained high pressure phases. For this reason the salt based shock-wave synthesis without adiabatic decompression was developed.

Results

For first methodical tests of this method four experiments were performed. In two cases the impedance method was used and for two experiments the reflection method with Cu-encapsuled W-reflectors was selected. For the changing of temperature without changing of other experimental parameters (porosity, container material, kind of halogenides) the buffer-plate-method between flyer plate and container top was used.The kaolinite bearing samples were treated with pressures in the first set of experiments from 12-20 GPa and the second set with pressures of approx. 120 GPa. The rocksalt-sample ratio in every case was >9:1. In all cases the sample recovery rate was 100%, the samples were available as compact discs with significant residual moisture at pressures up to 20 GPa, indicating the absence of degassing along the release path. The NMR- and XRD-analyses show clearly the existence of OH-bearing phases. Thermogravometric measurements indicate an increasing mass-loss of the HP-phase with decreasing relative temperatures (K/GPa) at comparable pressures.

Conclusions

Despite the fact that the method "or the encapsulated shock-wave treatment of fluid bearing natural phases" was new developed in November 2014 in Freiberg it shows great possibilities also for the treatment of carbonates under deep mantle conditions.

SCE-P08

Structural Properties of Nickel Dimethylglyoxime at High Pressure. Single-Crystal X-ray Diffraction and DFT Studies.

I. Bruce-Smith[1,2], B. Zakharov[3,1], J. Stare[4], E. Boldyreva[3,1], C. Pulham[2]
[1]*Novisibirsk State University, Pirogova Str. 2, 630090, Russia, Russian Federation*
[2]*School of Chemistry and Centre for Science at Extreme Conditions, University of Edinburgh, Edinburgh, West Mains Road, EH9 3JJ, United Kingdom*
[3]*Institute of Solid State Chemistry and Mechanochemistry SB RAS, Novosibirsk, Russian Federation*
[4]*National Institute of Chemistry, Ljubljana, Hajdrihova 19, SI- 1000, Slovenia*

Structural changes in nickel dimethylglyoxime ($Ni(dmg)_2$) were followed by single-crystal X-ray diffraction in a diamond-anvil cell (DAC) at pressures up to 5.1 GPa, i.e. in the pressure range through the major color change point (2 GPa), but before the phase transition at 7.4 GPa.

High quality single crystals of $Ni(dmg)_2$ have been grown and single-crystal X-ray diffraction data recorded up to pressures of 5 GPa. There were no detectable phase changes in this region, with the color change being attributed to the influence of pressure on the Ni-Ni distance in the chains formed by the molecules. Significant average compression (~ 4% / GPa) was observed, with anisotropic, but continuous and monotonic lattice strain. The maximum compression was observed for the direction perpendicular to planar layers of $Ni(dmg)_2$ and thus corresponds to decreasing the shortest contacts between nickel cations. Compression within the layers was not so pronounced as the compression between the layers. The strong anisotropy of structural strain results from an interplay of Ni-Ni interactions and the steric effects imposed by the repulsion of the methyl groups. As the use of complementary experimental and theoretical methods offers several advantages in the structural studies of new compounds, the experimental diffraction study was assisted by periodic DFT calculations, yielding improved refinement of hydrogen atoms and important information about the structure and dynamics of the short hydrogen bond. A peculiar feature is the significant coupling between methyl torsions and the proton potential, resulting in an asymmetric, single-well hydrogen bond. Due to the pressure-driven steric repulsion the proton exhibits the tendency to be located nearer to the donor site at higher pressures.

This high-pressure study complements previous work on organic molecular crystals, for which coupling of methyl-group dynamics and adjacent O-H...O and N-H...O hydrogen bonds was analyzed as a function of temperature. It also show a new effect of pressure on the short O-H...O bond, as compared with previously reported symmetrization, or, on the contrary, a complete ionization of the two components resulting from a shift of the donor H-atom to the acceptor.

LIG - Young Crystallograhers Lightning Session Posters

LIG-P01

Biochemical and structural studies on reaction mechanism of the acetyl-CoA synthesis

Y. Dendra[1], S. Götzl[1], J.- H. Jeoung[1], H. Dobbek[1]
[1]*Humboldt Universität zu Berlin, Institut für Biologie, Berlin, Germany*

Acetyl-CoA synthase (ACS) is one of the key enzymes of the Wood-Ljundahl pathway of anaerobic CO_2 fixation and is abundant in a number of anaerobic bacteria (sulfate reducers, methanogens, acetogens, etc) (1). ACS catalyzes the formation of acetyl-CoA by condensing the methyl group ($-CH_3$) of a methylated corrinoid iron-sulfur protein (CoFeSP) with CO and coenzyme A (1, 2). The active site of the ACS, denoted as A-cluster, contains a Fe_4S_4 cubane bridged to a nickel-nickel binuclear site. In acetogenic bacteria, ACS forms a tight tetrameric complex together with CO Dehydrogenase (CODH). The first crystal structure of **ACS/CODH complex** from *Moorella thermoacetica* was solved at 2.5 Å resolution by Doukov et. al (1). The 2.2 Å resolution structure of the monomeric ACS from hydrogenogenic bacterium *Carboxydothermus hydrogenoformans* was obtained by Svetlitchnyi in 2004 (3). In methanogenic archaeae, ACS together with CODH, CoFeSP and methyltransferase are a part of a 2.4 MDa **acetyl-CoA decarbonylase/synthase (ACDS) complex** (4). No crystal structures of the methanogenic ACS alone or as part of the ACDS complex has been reported so far.

Despite nearly three decades of extensive spectroscopic and biochemical studies and several high-resolution structures of the ACS, many questions concerning the mechanism of acetyl-CoA formation remain open. Here we present the current state of knowledge on reaction mechanism of acetyl-CoA formation.

(1) Doukov et al., Science 298, 567 (2002);
(2) Ragsdale et al., Chem. Rev. 96, 2515 (1996);
(3) Svetlitchnyi et al., PNAS 101, 2 (2004);
(4) Terlesky et al., J. Bacteriol 168, 3 (1986).

LIG-P02

Epitaxial growth of pulsed laser deposited Ge-Sb-Te thin films on (111) oriented substrates

E. Thelander[1], U. Ross[1], J. Gerlach[1], A. Lotnyk[1], B. Rauschenbach[1]
[1]*Institut für Oberflächenmodifzierung e.V., Leipzig, Germany*

Phase change materials based on the material system Ge-Sb-Te (GST) have been widely used for optical storage for decades and are gaining more interest as candidates for next generation non-volatile electronic memories. Recently, it was shown that memory devices with an ordered fiber textured structure possesses improved switching characteristics in terms of speed and stability [1]. Even more highly ordered Ge-Sb-Te-films in form of epitaxial layers have been achieved with MBE [2], however, with severe limitations regarding deposition rate. Pulsed laser deposition (PLD) employs a compound target and offers high deposition rate rendering it more interesting from an industrial point of view. Furthermore, it has been successfully used for the deposition of epitaxial GST films on cubic (100) oriented substrates[3].

In this contribution, PLD was employed to deposit epitaxial GST-layers on the (111) plane of BaF_2 and Si single crystal substrates with deposition rates between 2-200 nm/min[4]. X-ray diffraction measurements show a process temperature window for epitaxial growth between roughly 100°C and 295°C. No polycrystalline growth is observed for lower temperatures, i.e. a direct epitaxial growth is obtained, whereas higher temperatures lead to strong desorption of the film constituents. The films are of hexagonal structure with lattice parameters consistent with existing structure models. X-ray pole figure measurements reveal that the films grow with one single out-of-plane crystal orientation, but rotational twin domains are present. The out-of-plane epitaxial relationship is determined to be $Ge_2Sb_2Te_5(0001)$ || $BaF_2(111)$, whereas the in-plane relationship is characterized by two directions i.e. $Ge_2Sb_2Te_5$[-12-10] || BaF_2[1-10] and $Ge_2Sb_2Te_5$[1-210] || BaF_2[1-10] as revealed by pole figure measurements. For films deposited on Si, the epitaxial relationships are completely analogue. Aberration-corrected high-resolution scanning transmission electron microscopy was used to resolve the local atomic structure, including vacancy layer distribution, and confirm the hexagonal structure of the films.

References:

1. R. E. Simpson et al., *Interfacial phase-change memory,* Nat. Nano **6,** 501 (2011).
2. F. Katmis et al., *Insight into the Growth and Control of Single-Crystal Layers of Ge-Sb-Te Phase-Change Material,* Cryst. Growth Des. 11, 4606 (2011).
3. E. Thelander et al. , *Epitaxial growth of Ge-Sb-Te films on KCl by high deposition rate pulsed laser deposition,* J. Appl. Phys. 115, 213504 (2014).
4. E. Thelander et al., *Low temperature epitaxy of Ge-Sb-Te films on $BaF_2(111)$ by pulsed laser deposition,* In Press at Appl. Phys. Lett. (2014).

Figure caption

Pole figure measurement of GST films on $BaF_2(111)$ deposited at different substrate temperature (a). Intensity (cps) is shown on a logarithmic scale and applies for all pole figures. In (b), (c) and (d), STEM images of epitaxial GST films deposited at 180°C (b and c) and 295°C (d) are shown.

Figure 1

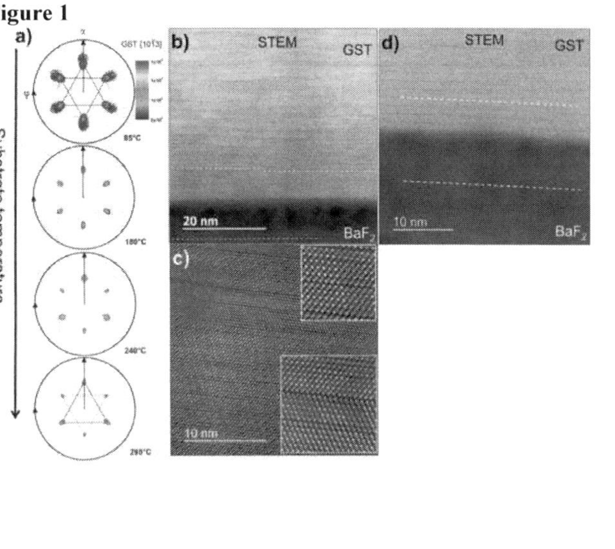

LIG-P03

Real time structural dynamics and entropy insights into the early unfolding kinetics of ubiquitin - rapid mixing micro-channel based time-resolved small angle X-ray scattering at synchrotrons

R. Jain[1], M. Petri[1], S. Sonnenkalb[1], S. Becker[1], C. Griesinger[1], A. Menzel[2], T. P. Burg[1], S. Techert[1,3,4]
[1]*Max Planck Institute of Biophysical Chemistry, Structural Dynamics of Biochemical Systems, Goettingen, Germany*
[2]*Paul Scherrer Institut, Villigen, Switzerland*
[3]*Georg-August-University, Goettingen, Germany*
[4]*Deutsches Elektronen-Synchrotron (DESY), Hamburg, Germany*

The overall size and shape of a macromolecule as a function of time, salt, or denaturant concentration can be determined with small-angle X-ray scattering (SAXS). Time-resolved SAXS (TR-SAXS) is a state of the art technique for studying protein folding-unfolding *in vitro* at all time-scales - from slow to very fast. It requires the setup of novel experimental schemes at pulsed, high flux X-ray synchrotrons. Doing so, TR-SAXS provides unique information about global structure of transient intermediates.

An experimental set-up with low protein consumption is essential for the extensive application of TR-SAXS. Together with the workgroup of T. Burg, we set up a newly developed 20-microchannel microfluidic continuous-flow mixer and combined that with SAXS at high-flux X-ray synchrotrons. With this experimental setup, I will demonstrate the kinetic ubiquitin unfolding dynamics after the rapid mixing of folded ubiquitin with unfolding agent Guanidinium-HCl. It required only ~ 40 nanoliters of protein sample per scattering image. This new platform might increase the use of TR-SAXS, TR-WAXS and neutron scattering experiments for studying milliseconds dynamics. The potential research field includes protein folding, protein misfolding, aggregation in amyloidogenic diseases, function of intrinsically disordered proteins and various protein-ligand interactions.

With the above mentioned experimental set-up, the structural dynamics investigation of ubiquitin unfolding will be presented on the diffusion limited time scales. Time-dependent variation in the X-ray scattering intensity, size and shape of ubiquitin after the complete mixing of ubiquitin and Gdn-HCl were monitored with the Fourier transformation of collected X-ray scattering signals during unfolding. The summarized results yielded a "real-time millisecond ubiquitin unfolding movie", and I will finally present that. These structural intermediate changes will be discussed as a sequential multi-state ubiquitin unfolding model along with detailed kinetic analysis with Boltzmann-fitting, time-dependent dispersive kinetics and JMAK model. The entropy and enthalpy contributions derived with the dispersive kinetics were used to describe the driving forces contributing to the ubiquitin unfolding funnel

References:

Jain et al., X-ray scattering experiments with high flux X-ray source coupled rapid mixing microchannel device...... European Physical Journal E, Sep; 36(9):109.

Real-time small-angle X-ray scattering reveals entropy-driven multistate ubiquitin unfolding reaction. (Jain et al., Submission in Nature Communications)

Visualization of millisecond ubiquitin unfolding kinetics in a microchannel device by TR-SAXS. (Jain et al., under preparation).

A glimpse of substrate induced conformational changes in human Guanylate kinase, (Jain et al., under preparation).

Fig. 1: Scheme of the TR-SAXS set-up.

Figure 1

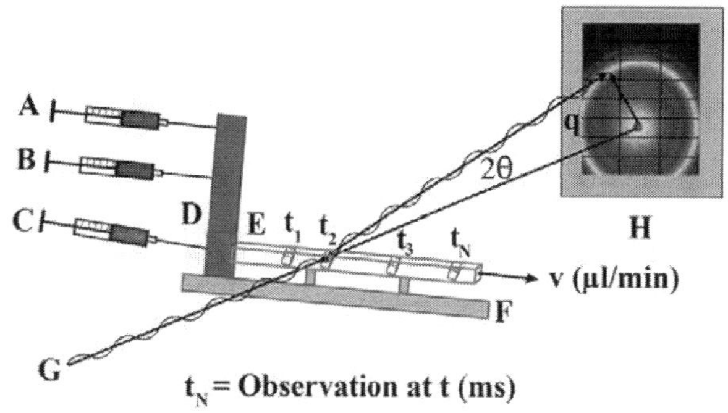

LIG-P04

Hexanuclear Metallacryptates: Cu(α-amino-carboxylate)$_2$ as Building Units for Self Assembly.

K. Lamberts[1], U. Englert[1]
[1]*RWTH Aachen, Inorganic Chemistry, Aachen, Germany*

Crown ethers and cryptands are organic molecules, introduced in the late 1960s [1], with the ability to selectively coordinate specific cations. Metalla-crowns [2] and metallacryptands [3] are their analogues in coordination-chemistry. However, their synthesis is not straightforward, as coordination chemistry is strongly dependent on weak interactions and self assembly. Cu(α-amino-carboxylate)$_2$ has proven a suitable building unit for the formation of hexanuclear metallacryptates: The square planar units occupy the equatorial positions of a distorted octahedron and bridge the axial Cu(II) cations. Perchlorate anions stabilize the four Cu(α-amino-carboxylate)$_2$-lamellas by connecting them over Cu---O Jahn-Teller contacts.

In the center of the octahedral unit the carboxylate-O form a coordination pocket in almost square prismatic geometry. Crystals with incorporated Na(I) cations were reported for glycine, L-proline, L-hydroxyproline, L-alanine, L-threonine and 1-amino-1-cyclopropane [4].

We recently discovered that such Metallacryptates are also capable of coordinating the transition metal cation Ag(I) (Fig 1). Structural and chemical properties, trends and chemical variations will be presented.

Fig. 1: Hexanuclear metallacryptate cation from Cu(II), perchlorate and L-proline with incorporated Ag(I).

[1] C.J. Pedersen, *Journal of the American Chemical Society*, 1967, 89, 7017-7036.
[2] *e.g.*: a) M.S. Lah and V.L. Pecoraro, *Comments on Inorganic Chemistry*, 1990, 11, 59-84. b) F.C.J.M. van Veggel, W. Verboom and D.N. Reinhoudt, *Chemical Reviews*, 1994, 94, 279-299.
[3] *e.g.*: a) M.S. Lah, B.R. Gibney, D.L. Tierney, J.E. Penner-Hahn and V.L. Pecoraro, *Journal of the American Chemical Society*, 1993, 115, 5857-5858. b) B.R. Gibney, A.J. Stemmler, S. Pilotek, J.W. Kampf and V.L. Pecoraro, *Inorganic Chemistry*, 1993, 32, 6008-6015.
[4] a) S. Hu, W. Du, J. Dal, L. Wu, C. Cui, Z. Fu and X. Wu, *Dalton Transactions*, 2001, 2963-2964. b) L.-Y. Wang, S. Igarashi,

Y. Yukawa, Y. Hoshino, O. Roubeau, G. Aromí and R.E.P. Winpenny, *Dalton Transactions*, 2003, 2318-2324. c) S.-M. Hu, S.-C. Xiang, J.-J. Zhang, T.-L. Sheng, R.-B. Fu and X.-T. Wu, *European Journal of Inorganic Chemistry*, 2008, 1141-1146. d) S.-C. Xiang, S.-M. Hu, J.-J. Zhang, X.-T. Wu and J.-Q. Li, *European Journal of Inorganic Chemistry*, 2005, 2706-2713. e) W. Ghattas, R. Ricoux, H. Korri-Youssoufi, R. Guillot, E. Rivière and J.-P. Mahy, *Dalton Transactions*, 2014, 7708-7711.

Figure 1

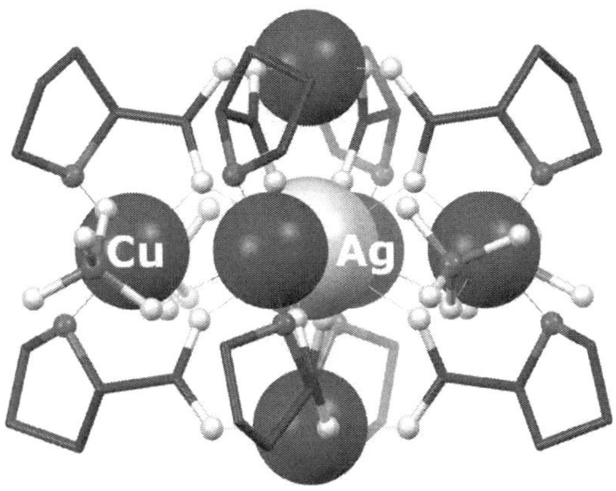

LIG-P05

Crystallization kinetics of zeolite Li-A(BW) in dependence of the aluminosilicate educt and the reaction temperature and time

D. Zeibig[1], J.- C. Buhl[1]
[1]*Leibniz Universität Hannover, Institut für Mineralogie, Hannover, Germany*

Zeolites of ABW structure type like Li-A(BW) (composition: $Li_4(H_2O)_4[Al_4Si_4O_{16}]$ are microporous materials with a channel system of dimensions of 3.8 x 3.4 Å, formed of 8-rings [1]. According to the pore wide and the manyfold substitution possibilities of T-atoms and cations in this structure the A(BW) zeolites are interesting for future applications. Energy efficient crystallization procedures would thus become important. The aim of the present work was to optimize the synthesis, first performed by Barrer and White [2] and modified by Norby et al. [3] to test a complete conversion at economical conditions using different aluminosilicate educts, low temperatures and short reaction times.

Therefore the crystallization of Li-A(BW) has been examined by testing three different alumosilicate educts (Na-LTA as in [3], Na-13X and Na-LSX) in dependence of the synthesis temperature and synthesis time. The educts have been dissolved in LiCl-solution and heated under hydrothermal conditions at 200 °C, 150 °C and 100 °C for 72 h, 36 h and 18 h. All products have been analyzed using XRD. Furthermore, for some selected samples FTIR, SEM/EDX and quantitative Rietveld-refinements were carried out.

The results show that 200 °C and 72 h are suitable conditions for all starting materials to obtain pure phase zeolite Li-A(BW). 200 °C and 36 h are still adequate for Na-LTA and Na-LSX, whereas Na-13X does not crystallize completely to Li-A(BW). A further decrease of synthesis time down to 18 h leads to an incomplete conversion of Na-LSX. Na-LTA is the only reactant which produces Li-A(BW) at 200 °C and 18 h.

Even at lower temperature of 150°C the starting material Na-LTA reveals as suitable source for the crystallization of the desired product for both times (72 h and 36 h). In contrast Na-13X and Na-LSX does not show any conversion, even at 72 h. The synthesis temperature of 100 °C is generally inappropriate, because the conversion to the zeolite Li-A(BW) could not be observed for any of the three educt compounds.

Summarizing the results it could be shown that using Na-LTA as starting material is the best method for synthesizing the desired Li-A(BW) even for lower temperatures and times.

Figure 1: Conversion of the different solid educts into zeolite Li-A(BW) in dependence of the reaction time at 150°C and 200°C (results from quantitative Rietveld refinements).

Figure 2: SEM image of zeolite Li-A(BW) crystals obtained from zeolite LTA at 150°C and 72 h reaction time in 1,6 M-LiCl solution.

References:
[1] Baerlocher, Ch, Meier, W. M., Olson D. H.: "Atlas of Zeolite Framework Types". 5th ed., Elsevier, Amsterdam, New York, 2001.
[2] Barrer, R. M., White, E. A. D.:J.: "The Hydrothermal Chemistry of Silicates". Part 1. Synthetic Lithium Aluminosilicates. Chem. Soc. (1951) 1267 - 1278.
[3] Norby, P., Norlund Christensen, A., Krogh Andersen, I. G.: "Hydrothermal preparation of zeolite Li-A(BW), $LiAlSiO_4 \cdot H_2O$, and structure determination from powder diffraction data by direct methods". Acta Chem. Scand. A40 (1986) 500-506.

Figure 1

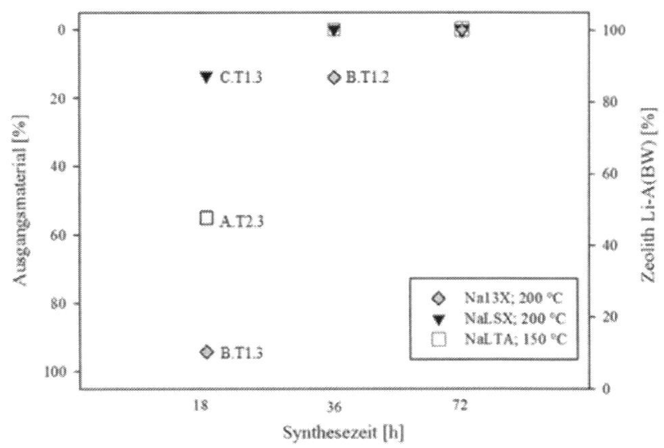

Figure 2

LIG-P06

Anisotropic Displacement Parameters from Dispersion-Corrected DFT and their Experimental Validation by X-ray Diffraction

A. Wang[1], U. Englert[1]
[1]*RWTH Aachen University, Institute of Organic Chemistry, Aachen, Germany*

Single crystal X-ray diffraction experiments have been conducted for the monoclinic form of pentachloropyridine[1] as a function of temperature. For the anisotropic displacement parameters (ADPs) of the five chloride atoms in the molecule experimental values have been refined based on these intensity data, and theoretical values based on dispersion-corrected density-functional theory (DFT) have been obtained by phonon computations using the PHONOPY[2] software. Dispersion correction represents an important aspect in these calculations: With the PBE+TS method[3], excellent agreement for the calculated and experimentally observed thermal motion in the crystal is found at 100, 150 and 200 K. The agreement can be conveniently visualised by thermal ellipsoids as shown in the figure. Alternative validation approaches are discussed on the poster. Our method promises reliable predictions for thermal motion in molecular crystals.

Figure: Comparion betweeen ADPs from experiment (Exp) and the most suitable theoretical method among those tested (PBE+TS).

[1] J. George, A. Wang, V. L. Deringer, R. M. Wang, R. Dronskowski and U. Englert, *CrystEngComm*, preparation.

[2]A. Togo, F. Oba and I. Tanaka, *Phys. Rev. B: Condens. Matter. Mater. Phys.*, 2008, **78**, 134106.

[3]J. P. Perdew, K. Burker, M. Ernzerhof. *Phys, Rev. Lett*, 1996, **77**, 3865-3868; A. Tkatchenko and M. Scheffler, *Phys. Rev. Lett.*, 2009, **102**, 073005.

Figure 1

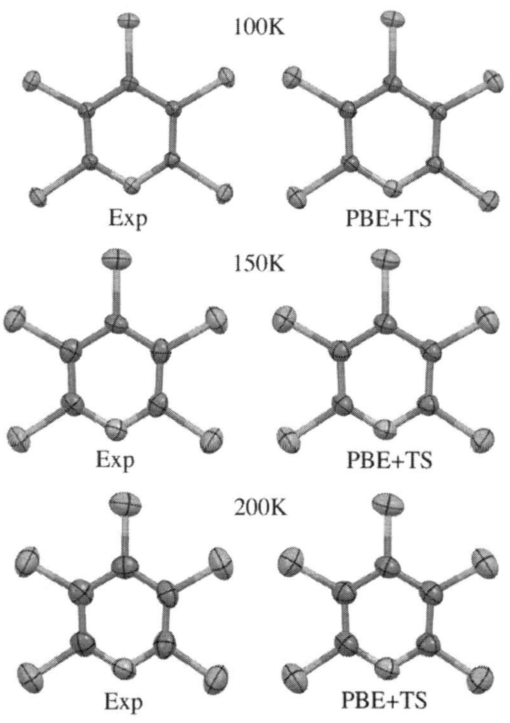

LIG-P07

Isoreticular Solvo- and Polymorphs: Cu(I) in a Stable Bimetallic Coordination Network

F. Steuber[1], U. Englert[1]
[1]*RWTH Aachen, Institute of Inorganic Chemistry, Aachen, Germany*

Organic-inorganic hybrid materials such as coordination polymers have gained tremendous interest in basic research as well as applied materials science. The ditopic ligand 3-cyano-2,4-pentanedione (HacacCN) has proven to be a suitable candidate for a stepwise bimetallic network construction [1, 2]. Isoreticular interpenetrated diamond networks have been synthesised from Sc(III) and Y(III) in combination with Ag(I). A similar structure from La(III) has been reported previously [3]. The preferred self-assembly of the motif allowed for the synthesis of an isomorphous Cu(I)-Y(III) coordination network. Structural relationships between five novel bimetallic networks will be discussed.

Figure 1: Structural motif found in presented compounds.

[1] A. D. Burrows, K. Cassar, M. F. Mahon and J. E. Warren, *Dalton Trans.*, 2007, 2499-2509.
[2] C. Merkens, N. Becker, K. Lamberts, U. Englert, *Dalton Trans.*, 2012, **41**, 8594-8599.
[3] C. Merkens, U. Englert, *Dalton Trans.*, 2012, **41**, 4664-4673.

Figure 1

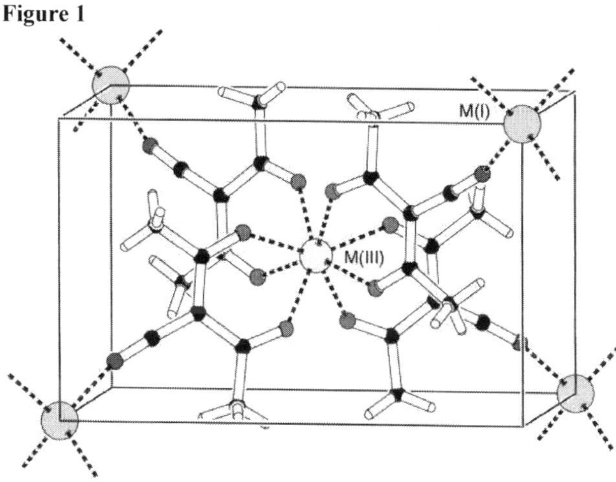

LIG-P08

Do we know the Pyroelectric Coefficient of Barium Titanate?

H. Stöcker[1], B. Störr[1], S. Jachalke[1], E. Mehner[1], J. Hanzig[1], M. Sonntag[1], R. Schmid[1], T. Leisegang[2], D. C. Meyer[1]
[1]*TU Bergakademie Freiberg, Institut für Experimentelle Physik, Freiberg, Germany*
[2]*Fraunhofer-Technologiezentrum Halbleitermaterialien, Freiberg, Germany*

Barium titanate is a ferroelectric material widely used in capacitors because of its high permittivity and as a chemically and mechanically stable pyro- and piezoelectric. For utilization in pyroelectric devices, e. g. for infrared detection or waste heat harvesting, a high pyroelectric coefficient is often desired. Literature data for barium titanate compounds vary between 4.8 $\mu C/m^2K$ and 13 C/m^2K, depending on doping, synthesis procedure, and measurement method. Hence, the pyroelectric coefficient of barium titanate seems to span more than six orders of magnitude, which renders predictions on the efficiency of devices nearly impossible.

Here, we try to give an overview on the factors influencing the pyroelectric coefficient of barium titanate to finally extract meaningful values of this fundamental material property. First, a short overview on pyroelectric measurement methods and their application criteria is given. This allows identifying reliable literature values measured with an appropriate method for further discussion. Since barium titanate can be produced in many different forms, e. g. thin films, ceramics or single crystals, polarization and measurement of the samples have to be conducted accordingly. Depending on crystallinity and morphology, different pyroelectric coefficients and Curie temperatures are reported in literature. Finally, doping of barium titanate with calcium, cobalt, iron, nickel, strontium, tin or zirconium is found to decrease the Curie temperature. Shifting the Curie temperature in turn changes the pyroelectric coefficient at a fixed temperature. Doping with calcium or nickel leads to an increased pyroelectric coefficient, whereas the substitution with strontium or tin does not affect the pyroelectric coefficient at the Curie temperature.

LIG-P09

CO Dehydrogenase II - Digging channels for small molecules

L. Domnik[1], J.- H. Jeoung[1], J. Fesseler[1], H. Dobbek[1]
[1]*Humboldt Universität zu Berlin, Berlin, Germany*

Rising of CO_2 levels in the earth's atmosphere demands the development of energy sources that are CO_2 neutral. Therefore understanding how microorganisms utilize C1-compounds will be of crucial importance in the future. The carbon monoxide dehydrogenase II (CODHII) of the thermophilic bacterium *Carboxydothermus hydrogenoformans* catalyses the reduction of CO_2 to CO and water using two protons and two electrons with a turnover rate of 15 s^{-1}.

In order to guarantee an efficient supply of substrates and release of products controlled routes are needed within the enzyme [1, 2, 3]. Biochemical and structural studies on mutants have been applied to illuminate the putative proton and water transfer pathways. Exploiting the ability of xenon to bind hydrophobic protein cavities allowed determination of gas channels in CODHII.

References:
[1] Jeoung JH and Dobbek H (2007) Science 318(5855):1461-4

[2] Kim EJ, Feng J, Bramlett MR, Lindahl PA (2004) Biochemistry 43(19):5728-34

[3] Doukov TI, Blasiak LC, Seravalli J, Ragsdale SW, Drennan CL (2008) Biochemistry 47(11): 3474-3483.

LIG-P10

Identifying new inorganic Na-ion conductors using data-mining and geometric crystallography - the Voronoi-Dirichlet approach

F. Meutzner[1], T. Nestler[1], M. Wolfram[1], N. A. Kabanova[2], T. Leisegang[3], V. A. Blatov[2], D. C. Meyer[1]
[1]*TU Bergakademie Freiberg, Institut für Experimentelle Physik, Freiberg, Germany*
[2]*Samara State University, Samara, Russian Federation*
[3]*Fraunhofer Technologiezentrum Halbleitermaterialien, Freiberg, Germany*

Due to its high achievable energy densities, stationary electrochemical energy storage devices are most important for the stabilisation of the electric grid, compensating for the capricious nature of renewable energies. Sodium is one of the most used candidates in this field and applied in stationary Na-S and Zebra cells for example in Japan. These are high-temperature cells in which Na and S react with each other, mediated by a solid Na-ion conductor such as ß-aluminate. In order to achieve reasonable conductivities, high temperatures have to be applied leaving Na and S in their liquid state. For energetic and environmental reasons, however, it would be best to lower temperatures. Therefore, new solid electrolytes have to be identified.

Crystallographic Databases, such as the inorganic crystal structure database (ICSD), store large amounts of measured structures and thus offer knowledge which is partially not systematised. Parameters of related structures can be "*data-mined*" to find their most common values or familiar structure motives. On the other hand, these databases can be used in an inductive way, using this newly created information to find compounds which show the same behaviour but have not been considered for their potential function, yet.

The Voronoi-Dirichlet tessellation offers a possibility to partition space - in this case a crystal structure. The polyhedra created by this routine depend solely on the crystal structure's geometry and metrics offering readily available parameters, such as volumes, vertex-coordinates, *etc*. These can be used to describe structures very easily and fast. These parameters also have crystallographic meanings that can be translated into functional properties as well. Through this, structure-immanent voids and channels between these voids can be identified. They are important factors in compounds showing ionic conductivity. Because of its simplicity, the partition can be easily applied on huge databases. This combination of data-mining and Voronoi-Dirichlet tessellation is used in this work to identify compounds which are new potential ionic conductors or intercalation hosts for Na-ions.

This work is funded by the German BMBF (CryPhysConcept: 03EK3029A) and the German BMU (BaSta: 0325563B).

LIG-P11

Synthesis and local structure of $Cu_xZn_{2-x}TiO_4$ spinel unveiled by Raman spectroscopy

M. He[1], J. Ruiz-Fuertes[1], T. Bernert[1], B. Winkler[1], V. L. Vinograd[2]
[1]*Goethe-Universitaet Frankfurt, Frankfurt am Main, Germany*
[2]*Forschungszentrum Juelich GmbH, Juelich, Germany*

The local structure of solid solutions is essential for understanding their physical properties. Since their structural investigations are challenging, up to now there are just a few studies addressing the local symmetry of inverse spinel-solid solutions, especially when they are affected by a Jahn-Teller distortion [1]. Here we present the solid-state synthesis of the $Cu_xZn_{2-x}TiO_4$ solid solution and the investigation of the influence of the incorporation of Cu^{2+} on the local structure of technologically important Zn_2TiO_4 [2]. X-ray diffraction shows no phase change accross the solid solution. The analysis of the lineshape and frequency of the Raman active modes reveal the incorporation of some Cu^{2+} in tetrahedral sites up to x = 0.6. The observed ordering scheme is in agreement with our *ab initio* calculations with pseudo-random generated structures, where an octahedral over tetrahedral site preference energy for Cu^{2+} of only ~-6 kJ/mol has been computed [3]. With increasing Cu^{2+} content, a non-linear change of the unit-cell volume is observed (Figure 1). The excess volume is associated to an increase of the distortion of the coordination polyhedra which is maximum for x = 0.6.

Fig. 1: (a) Lattice parameter a of $Cu_xZn_{2-x}TiO_4$ as a function of Cu concentration and (b) the oxygen thermal displacement parameter

U_{iso}.

M.H. acknowledges the China Scholarship Council for a PhD grant and J.R.-F. thanks the Alexander von Humboldt Foundation for a postdoctoral fellowship.

[1] A. Le Nestour, M. Goudon, G. Villeneuve, R. Andrissen, and A. Demourges, Inorg. Chem. **46**, 2645 (2007).
[2] T. Esaka, T. Ikebe, and M. Kamata, Solid State Ionics **76**, 237 (1995).

[3] J. Ruiz-Fuertes, T. Bernert, M. He, B. Winkler, V. L. Vinograd, and V. Milman Appl. Phys. Lett. **105**, 071911 (2014).

Figure 1

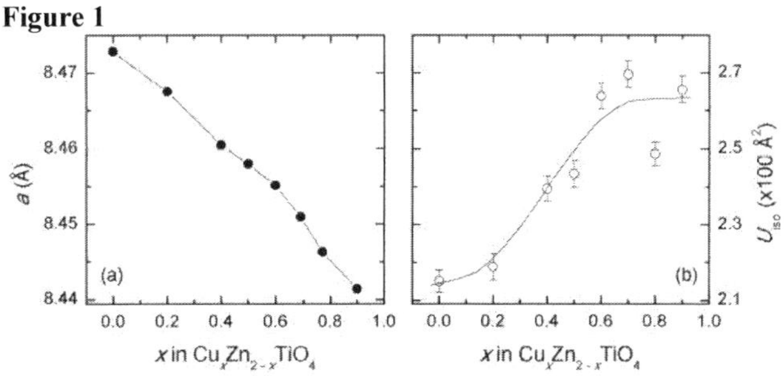

LIG-P12

High resolution polarized Raman scattering study on multiferroic $MnWO_4$

F. Ziegler[1], H. Gibhardt[1], J. Leist[1], P. Becker[2], L. Bohatý[2], G. Eckold[1]
[1]*Georg-August-Universität Göttingen, Institut für Physikalische Chemie, Göttingen, Germany*
[2]*Universität zu Köln, Institut für Kristallographie, Köln, Germany*

Multiferroic materials with magnetic and electric ordering within a single phase gained attention during the last years due to their potential for industrial applications. Materials with a cycloidal spin-arrangement are of particular interest because the electrical polarisation is directly caused by the complex magnetic order. Hence, both ordering phenomena are tightly coupled in this category of compounds. One example is $MnWO_4$, which passes successively three antiferromagnetic phase transitions on cooling. The phase AF2 between 12.5 K and 7 K exhibits a cycloidal spin arrangement with an incommensurate magnetic propagation vector. This ordering induces a spontaneous polarisation by the inverse Dzyaloshinski-Moriya-interaction. As a consequence, the interaction between spin and polarisation could possibly lead to a change in the spectrum of lattice vibrations.

Due to the tiny polarization, it was believed up to now that this effect was too small to be observed experimentally. Using high resolution polarized Raman scattering, however, we were able to detect clear signatures of this magnetoelectric interaction. The Raman spectra were taken from a $MnWO_4$ single crystal in the temperature range from 100 K to 6 K covering the three magnetic phase transitions. A careful analysis of the experimental data allowed the precise determination of both, the wave numbers and intrinsic linewidths.

Furthermore, the effect of an electric field applied along the polar axis was investigated for each of the 18 Raman active modes. Most of the phonons show a significant shift in wave number below the magnetic ordering temperature that cannot be explained merely on the basis of lattice dynamics, but should be attributed to spin-phonon-interaction. The most striking effect is observed for the A_g mode with the highest frequency (about 884 cm^{-1}), which corresponds to an oxygen stretching vibration and exhibits a pronounced softening on cooling.

LIG-P13

New $Mn_{23}Pt_{12}Ga_{11}$ phase, synthesis, crystal structure and its complex phase reaction scheme

B. Jamiyansuren[1], W. Carrillo-Cabrera[1], G. Kreiner[1], C. Felser[1]
[1]*Max-Planck-Institut für Chemische Physik fester Stoffe, Anorganische Chemie, Dresden, Germany*

Mn_2-based Heusler compounds have attracted much interest as magnetic materials because of high spin polarization at the Fermi level. In addition, tetragonal compounds can exhibit large magnetocrystalline anisotropy [1,2]. Recently the tetragonal Heusler compound Mn_2PtGa has been reported to show a large exchange-bias effect and a first order ferrimagnetic to ferromagnetic transition [3,4]. Here, we report on the crystal structure and the heterogeneous equilibria of the new intermetallic compound $Mn_{23}Pt_{12}Ga_{11}$, which forms close to the composition Mn_2PtGa. Bulk samples were prepared by inductively melting the elements in Al_2O_3 crucibles in Ar atmosphere. The resulting polycrystalline ingots were placed in Al_2O_3 crucibles. These crucibles were then encapsulated in cleaned and weld-sealed Ta containers which in turn were jacketed in fused silica ampoules. The samples were then annealed for several days at elevated temperatures and subsequently quenched in water. The samples were characterized by chemical bulk analysis (ICP-OES), metallographic examination, EDXS, WDXS, thermal analysis and by powder X-ray diffraction. The crystal structure of $Mn_{23}Pt_{12}Ga_{11}$ was solved using electron diffraction data (ADT3D, [5]) obtained by manual electron diffraction tomography as described in [6]. The structural model was validated by powder diffraction data (Rietveld technique, Jana2006, [7]). $Mn_{23}Pt_{12}Ga_{11}$ crystallizes in a new crystal structure type with 46 atoms per unit cell ($P3m1$, a = 8.649(1) Å, c = 10.848(1) Å) and shows a complex phase reaction scheme, which includes a four phase reaction of class II. The title compound exhibits heterogeneous equilibria to MnPtGa (BeZrSi type, [8]), to a structurally closely related monoclinic phase with slightly more Ga content, to a phase of $TiAl_3$ structure type close to Mn_3Pt and to a Mn-rich Heusler compound. The crystal structure of $Mn_{23}Pt_{12}Ga_{11}$ contains structural entities as found in NiAs type based MnPtGa as well as in Heusler structures.

[1] T. Graf, C. Felser, S.S.P. Parkin, Prog. Solid State Chem. 39 (2011) 1.
[2] G. Kreiner, A. Kalache, S. Hausdorf, V. Alijani, J.-F. Qian, G. Shan, U. Burkhardt, S. Ouardi, C. Felser, Z. Anorg. Allg. Chem. 640 (2014) 738.
[3] A.K. Nayak, M. Nicklas, S. Chadov, C. Shekar, Y. Skourski, J. Winterlik, C. Felser, Phys. Rev. Lett. 110 (2013) 127204.
[4] A.K. Nayak, M. Nicklas, C. Shekar, C. Felser, J. Appl. Phys. 110 (2013) 17E308.
[5] U. Kolb, T. Gorelik, C. Kübel, M.T. Otten, D. Hubert, Ultramicroscopy, 107 (2007) 507.
[6] J. Fan, W. Carrillo-Cabrera, L. Akselrud, I. Antonyshyn, L. Chen, Y. Grin, Inorg. Chem. 52 (2013) 11067.
[7] V. Petricek, M. Dusek, L. Palatinus, Jana2006. Structure Determination Software Programs. Institute of Physics, Praha, Czech Republic, 2006.
[8] P. Villars, K. Cenzual, Pearsons's Crystal Data - Crystal Structure Data Base for Inorganic Compounds , Release 2012/2013, ASM International, Materials Park, Ohio, USA.

LIG-P14

Structural chemistry of the SrO-CuO-As_2O_5 ternary system: crystal structure of $Sr_2Cu_7(AsO_4)_6$

S. Gerger[1], T. Djordjevic[1]
[1]*University of Vienna, Wien, Austria*

A new arsenate, $Sr_2Cu_7(AsO_4)_6$ was synthesised under low-temperature hydrothermal conditions during detailed structural investigations of the compounds from the MO-CuO-As_2O_5-H_2O system (M = Cd^{2+}, Sr^{2+}). $Sr_2Cu_7(AsO_4)_6$ crystallised as transparent, light-blue crystals with lengths up to 0.11 mm from a mixture of $Sr(OH)_2 \cdot 8H_2O$, $Cu(OH)_2$, As_2O_5 and distilled water (Teflon-lined steel autoclaves, 493 K, 7 d). Besides $SrCuAs_2O_7$ [1], it represents the second structurally characterized compound in the ternary SrO-CuO-As_2O_5 system.

The crystal structure of $Sr_2Cu_7(AsO_4)_6$ was refined from the single-crystal X-ray diffraction data (293 K, Mo*Ka*, $2q_{max}$ = 70°) starting from the atomic coordinates of the isotypic $Pb_2Cu_7(AsO_4)_6$ [2] (space group P-1, a = 5.1413(10), b = 8.3536(17), c = 11.173(2) Å, a = 90.52(3), β = 90.62(3), g = 92.76(3)°, V = 479.26(17) Å^3, Z = 1). The refinement (182 free parameters) yielded $R_1(F)$ = 0.0472 for 4481 observed reflections with F_o^2 ³ $4\sigma(F_o^2)$ [3].

The crystal structure of $Sr_2Cu_7(AsO_4)_6$ is built up from three crystallographically unique CuO_4 squares and two unique CuO_5 trigonal pyramids, three distinct AsO_4 tetrahedra and one SrO_8 polyhedron. CuO_x polyhedra are linked in two different kinds of chains: Cu_3O_{10} chains running along a-axis are formed by corner connection of one $Cu(3)O_4$ square and four different CuO_5 polyhedra. Cu_4O_{12} chains are parallel to [110] and are formed by edge and corner sharing of $Cu(1)O_4$ and $Cu(2)O_4$ squares and the $Cu(4)O_5$ polyhera. These chains are interconnected by AsO_4 tetrahedra thus forming a 3D-network, where in channels parallel to the a-axis slightly distorted SrO_8 tetragonal antiprisms are situated. SrO_8 polyhedra are interconnected to dimers *via* one O-O edge by a centre of symmetry. Both $Cu(4)O_5$ and $Cu(5)O_5$ polyhedra are trigonal dipyramids, which is a rare coordination figure for divalent copper atoms. However, as already observed in $Pb_2Cu_7(AsO_4)_6$ [2], the trigonal dipyramids are notably regular concerning Cu-O bond lengths and O-Cu-O bond angles: Cu-O bond legths 1.91 to 2.10 Å and O_a-Cu-O_a angles (O_a is apical ligand) are just somewhat bent and amount 174.8(2) and 173.7(2)° in $Cu(4)O_5$ and $Cu(5)O_5$ polyhedra, respectively. The O_e-Cu-O_a in equatorial plane of the polyhedra deviate only 21° from ideal value; the largest is 131.7° and the smallest is 101.7°.

Financial support of the Austrian Science Foundation (FWF) (Grant V203-N19) is gratefully acknowledged.

[1] Chen, T.-C &Wang, S.-L. (1996) J. Solid State Chem., 121, 350-355.
[2] Effenberger, H. (1995) J. Solid State Chem., 114, 413-419.
[3] Sheldrick, G.M. (2008) Acta Crystallogr., A64, 112-122.

LIG-P15

Crystal branching phenomena in natural and gel grown calcites

F. Wiethoff[1], J. Schreuer[1]
[1]*Ruhr-University Bochum, Geology, Mineralogy & Geophysics, 44780, Germany*

The morphological features of calcite crystals building up marine cements and speleothems are not easy to address due to the space filling crystallization based on competitive growth. Especially the calcite morphology depends strongly on the physico-chemical conditions during crystallization with the molar Mg/Ca ratio as the main parameter controlling the morphology [1]. Several studies pointed out that Mg^{2+} bearing solutions cause a larger growth vector parallel the c-axis resulting in elongated calcite crystals [2]. Therefore, measuring the aspect ratio in thin sections is one of the first approaches of sediment-petrologists to assess the presence of Mg^{2+} in the parental solution on a qualitative basis. Other features of such fabrics, like systematic undulous extinction and curved cleavage planes and/or crystal faces (Fiure 1a) occur quite common but can't yet employed as a tool to reveal further information due to the poorly understood underlying processes [3].

For this reason we grew calcite crystals in gels with different Mg/Ca ratios of the parental solutions. As a result of the spatially separated sites of crystallization in the porous gel media we observed plenty of different crystal morphologies some of them exhibiting subparallel branched plates with macroscopically curved adjacent faces (Figure 1b). In comparison to natural samples we found clear similarities in thin sections regarding the curved intracrystalline marks (Figure 1a), the undulous (divergent-) extinction behaviour, and in the morphology and branching phenomena of some natural, freely grown calcites.

On an atomistic level branching in calcites can be attributed to the incorporation of Mg^{2+} into the calcite lattice resulting in accumulation of internal stress-fields which can be reduced by developing slightly tilted sub-domains. Our observations suggest that crystal branching is a common feature during calcite crystallization in climate-research relevant archives and in principle influences the distribution and partitioning of trace elements [4] and isotopes through face specific branching phenomena.

Figure 1: (a)Thin-section of a marine calcitic cement showing curved intracrystalline marks (linear polarizedlight).(b)In gel grown synthetic calcite (Mg/Ca [mol/mol] = 0.5) with strong crystal branching phenomena resulting in subparallel crystal units.

References:
[1] Given, R.K. and Wilkinson, B.H. Kinetic control of morphology, composition, and mineralogy of abiotic sedimentary carbonates. J. Sediment. Petrol. **55** (1985) 109-119
[2] Folk, R.L. The natural history of crystalline calcium carbonate: Effect of magnesium content and salinity. J. Sediment. Petrol. **44** (1974) 40-53
[3] Richter, D. K., Neuser, R. D., Schreuer, J., Gies, H. and Immenhauser, A.: Radiaxial-fibrous calcites: A new look at an old problem. Sediment. Geol. **239** (2011) 23-36
[4] Rimstidt, J.D., Balog, A. and Webb, J. Distribution of trace elements between carbonate minerals and aqueous solutions. Geochim. Cosmochim Ac. **62** (1998) 1851-1863

Figure 1

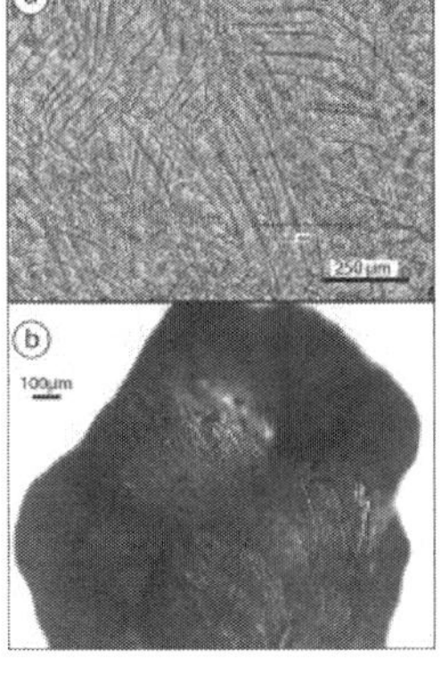

LIG-P16

Structure and Dynamics of Substituted η^3-Cycloheptatrienide-Pd Complexes

C. Jandl[1], A. Pöthig[1]
[1]*Technische Universität München, Catalysis Research Center, Garching b. München, Germany*

We report the structural characterisation of the first monomeric allyl complexes of Pd dihalides. Single crystal X-ray diffraction studies were essential not only in the identification of the monomeric structure, as both mass spectrometry and IR/Raman spectroscopy were ambiguous, but also in the further characterisation: Our approach combined crystallography and computational chemistry to gain insights into the rapid dynamic of the complexes and besides the global minimum isomer, suitable derivatisation also allowed us to characterise one structure as a snapshot of an intermediate in the fluctuation process (Figure 1). Additionally the compounds feature strong intramolecular H-bonds the exact conformation of which could only be determined *via* X-ray diffraction and in accord with DFT calculations was found to be highly solvent-dependent. This combination of molecular crystallography and computational chemistry proved to be a very valuable tool in investigations of fluxional systems that are traditionally seen as problematic for X-ray studies.[1]

Figure 1. Fluxional coordination of a substituted cycloheptatrienide ligand towards a Pd-dihalide moiety.

[1] Christian Jandl, Karl Öfele, Fritz E. Kühn, Wolfgang A. Herrmann, and Alexander Pöthig*, *Organometallics*, 2014, *33 (22)*, 6398. DOI: 10.1021/om500738d .

Figure 1

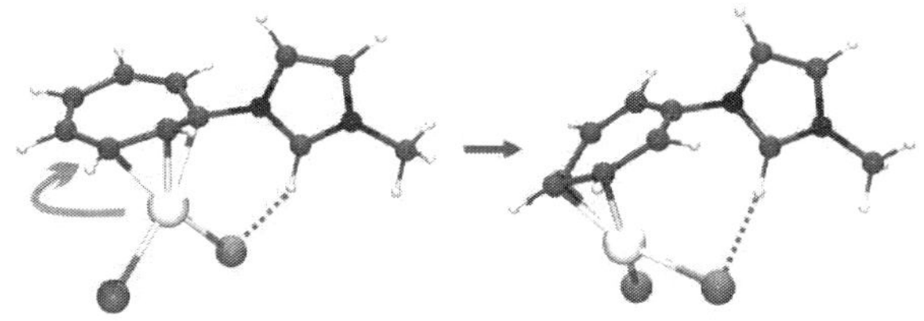

LIG-P17

Synthesis, X-Ray Structure, Infrared study and anti-corrosion activity of 1,6-diammoniumhexandihydrogenodiphosphate dihydrate

A. Zaraq [1], A. Kheireddine [2], A. El Bouari[1], S. Belaaouad [2]
[1]*Laboratory physics and chemistry of applied materials - University of Hassan II Casablanca , Casablanca, Morocco*
[2]*Laboratory chemistry and physics of materials - University Hassan II Casablanca, Casablanca, Morocco*

The crystal structure and IR investigation of 1,6-diammoniumhexandihydrogenodiphosphate dehydrate $(C_6H_{18}N_2)H_2P_2O_7.2H_2O$ (denoted DHDP) synthesized newly are explored through measurements of X-ray four-circle diffraction. This system is triclinic with space group P-1and the lattice constants are a=9.1744(4) Å, b=13.6977(5) Å, c=13.6977(5) Å, α=115.514(4)°, β=111.940(4)°, γ=91.186(3)° and V=1501.72(10)Å^3. The main feature of the atomic arrangement of DHDP is the existence of infinite lines parallel to the c direction of the unit cell, made up of $H_2P_2O_7$ groups. These phosphoric lines are organized to create with the help of the organic cation, large rectangular channels in which water molecules are located. The compound was tested as corrosion inhibitors for mild steel in 0.1 M nitric acid medium using potentiodynamic polarization technique.

LIG-P18

High resolution structure of apo-furin reveals competitive and allosteric inhibitor mechanisms

S. O. Dahms[1], M. E. Than[1]
[1]*Leibniz Institute for Age Research - Fritz Lipmann Institute (FLI), Protein Crystallography, Jena, Germany*

Furin is a subtilisin-like endoproteinase belonging to the pro-protein convertase (PC) family. PCs are required for the activation and maturation of many secreted proteins, including peptide hormones, growth factors, matrix metalloproteases, blood coagulation factors, regulators of the cholesterol metabolism, bacterial toxins and viral capsid proteins. Therefore furin and other PCs are highly interesting targets for the treatment of many diseases, e.g. cancer as well as viral- and bacterial infections (Seidah & Prat, 2012).

Development of furin inhibitors for pharmacological use requires a detailed structural understanding of the mechanisms involved in substrate-binding, catalysis and inhibition (Dahms et al, 2014; Henrich et al, 2003). Here we present the structure of apo-furin at 1.8 Å resolution and different inhibitor-bound structures. Our crystals in absence of any inhibitor showed the protease catalytically active in the crystalline state. These crystals proved to be highly suited to investigate the mechanism of competitive and allosteric inhibitors by soaking. Allosteric mechanisms involve the reorientation of the catalytic residues and/or the structural changes proximal to the catalytic center. Crystals of apo-furin enable the investigation of inhibitory compounds with low and very low affinities, the use of novel inhibitory mechanisms and to target novel surface patches of the enzyme, e.g. in crystallography based fragment screening approaches. Thus, our work provides novel possibilities for structure-based lead discovery and the design of next-generation furin inhibitors.

Dahms SO, Hardes K, Becker GL, Steinmetzer T, Brandstetter H, Than ME (2014) X-ray structures of human furin in complex with competitive inhibitors. *ACS chemical biology* **9:** 1113-1118

Henrich S, Cameron A, Bourenkov GP, Kiefersauer R, Huber R, Lindberg I, Bode W, Than ME (2003) The crystal structure of the proprotein processing proteinase furin explains its stringent specificity. *Nature structural biology* **10:** 520-526

Seidah NG, Prat A (2012) The biology and therapeutic targeting of the proprotein convertases. *Nature reviews* **11:** 367-383

LIG-P19

In-situ X-ray studies of GaAs Nanowire growth onto silicon(111) substrate

S. M. Mostafavi Kashani[1], P. Schroth[1], H. Letzguss [2], M. Koehl[2], S. Bauer [3], T. Baumbach[2], U. Pietsch[1]
[1]*University of Siegen, Department of Physics, Siegen, Germany*
[2]*Karlsruhe Institute of Technology (KIT) , Institute for Photon Science and Sychrotron Radiation (IPS)/ANKA, Karlsruhe, Germany*
[3]*Karlsruhe Institute of Technology, Synchrotron Facility ANKA, Karlsruhe, Germany*

GaAs nanowires (NWs) are promising candidates in optoelectronic applications e.g., low power-consumption light emitting diodes (LEDs) and laser diods, and solar cells with high efficiency. Naturally, NWs contain random structural defects and polytypism which may degrade electrical and optoelectrical properties of NW devices. In-situ x-ray studies of NW growth can play a key role to reveal understandings in order to control the NW structures.

We report on monitoring the evolution of NW polytypism during growth using high intensity x-rays provided at NANO beamline of ANKA synchrotron (KIT). The growth of GaAs NWs onto silicon (111) substrate was performed using a Portable Molecular Beam Epitaxy (PMBE) chamber. Before the in-situ experiment, the growth parameters such as gallium and arsenic flux, and the substrate temperature have been calibrated by ex-situ growth runs. Inspected by SEM, it turned out that at a substrate temperature of 600 °C, a V/III ratio of 2.55, and a growth time duration of 1 hour, the epitaxially grown NWs have an average length of ~1 μm, a diameter of ~40 nm, and a number density of 0.76 μm^{-2}.

During the in-situ experiment, we inspected the (220), (311), and (103) Bragg reflection which were sensitive for zinc-blende (ZB), twin ZB, and wutzite (WZ) phases, respectively, using time resolution of 5 minutes. NWs have been grown preferentially in ZB phase with a small content of WZ phase.

LIG-P20

Symmetry-mode analysis of the commensurately modulated superstructure of tetragonal tungsten bronze $KMnCrF_6$

C. Drathen [1]
[1]ESRF - The European Synchrotron, Grenoble, France

Multiferroic materials showing coupling of the different order parameters (ferroelectric, ferromagnetic, ferroelastic) are interesting not only from a fundamental perspective, but also from a technological point of view, e.g. for to the development of new storage technologies. However, the coexistence of (ferro)magnetism and ferroelectricity is considered a rare phenomenon. Whilst this may be true for perovskite oxides, where empty *d*-shells favor the off-centering of ions but counteract magnetism, this intrinsic limitation can be avoided by moving to different structure types, and/or away from oxides. An example of non-perovskite, non-oxide multiferroic systems are the tetragonal tungsten bronze (TTB) fluorides $K_xM^{2+}{}_xM^{3+}{}_{1-x}F_3$ (x = 0.4 - 0.6), which show coexistence of electric and magnetic ordering [1]. Here we present a detailed structural study of the TTB fluoride $K_xMn_xCr_{1-x}F_3$ (x = 0.5). $K_{0.5}Mn_{0.5}Cr_{0.5}F_3$ has been previously described as tetragonal $P4_2bc$ and orders ferrimagnetically below T ~ 30 K [2]. We used high-resolution powder diffraction techniques to reinvestigate the crystal structure as a function of temperature. Our results reveal a structural distortion to monoclinic symmetry (*P*2) at room temperature, with a 2√2x2√2x1 supercell as compared to the tetragonal cell, containing 400 and 15 independent atoms, respectively. To handle the high number of parameters, we use the symmetry mode description in a combined neutron- and X-ray refinement. This approach leads us to a model with only ~ 60 variables describing the fractional coordinates of all atoms and allows us to identify those modes that are driving the structural transition to tetragonal symmetry at high temperatures. Although structurally subtle, this distortion may indicate a ferroelectric state, similar to K_xFeF_3, where ferroelectricity is observed only in the orthorhombic phase.

References:

[1] Scott, J.F., Blinc, R., *J. Phys. Condens. Matter*, **2011**, *23*, 113202.,
[2] Banks E., Shone, M. Hong, Y.S., Williamson, R.F., Boo, W.O.J., *Inorg Chem.* **1982**, *21* (11) 3894.

LIG-P21

Spinophilin restricts Neurexin signaling to protect from excessive seeding of new active zones

J. H. Driller[1], K. G. Muhammad[2,3], S. Reddy[2,3], U. Rey[3], M. A. Böhme[3], C. Hollmann[3], N. Ramesh[2], H. Depner[2,3], J. Lützkendorf[2,3], T. Matkovic[2], D. Bergeron[2], C. Quentin[2,3], J. Schmoranzer[4], F. Goettfert[5], M. Holt[6], M. C. Wahl[1], S. W. Hell[5], A. Walter[2,4], B. Loll[1], S. J. Sigrist[2,3]
[1]*Freie Universität Berlin, Biochemie, Berlin, Germany*
[2]*Freie Universität Berlin, Biologie, Berlin, Germany*
[3]*Charité, NeuroCure, Berlin, Germany*
[4]*Leibniz Institut für Molekulare Pharmakologie, Berlin, Germany*
[5]*Max Planck Institute, Department of NanoBiophotonics, Göttingen, Germany*
[6]*VIB Center for the Biology of Disease, Leuven, Belgium*

Cytoplasmic active zone scaffolds, composed of several conserved multi-domain proteins, decorate the presynaptic membrane. While scaffold sizes were suggested to influence synaptic vesicle release, core mechanisms to assemble functional synapses remain to be characterized. Assembly and maturation of synapses depend on trans-synaptic Neurexin/Neuroligin (Nrx/Nlg) signaling, which is promoted by the regulator Syd-1 at *Drosophila* neuromuscular synapses.Here, we show that presynaptic Spinophilin, dependent on its PDZ domain by which it binds Nrx with micromolar affinity, attenuates Nrx/Nlg signaling and antagonizes Syd-1.Loss of Spinophilin resulted in the formation of excess but atypically small active zone scaffolds. We verified these interactions *in vitro* via pull-down assays and characterized them by ITC measurements. We also determined a high resolution crystal structure of the PDZ domain of Spinophilin bound to a peptide derived of the last 10 C-terminal amino acids of Nrx-1.

LIG-P22

Crystallographic analysis of the ClpP1/2 heterocomplex from Listeria monocytogenes

M.- T. Vielberg[1], M. Dahmen[2], S. A. Sieber[2], M. Groll[1]
[1]*Department of Chemistry, Technische Universität München, Chair of Biochemistry, Garching, Germany*
[2]*Department of Chemistry, Technische Universität München, Chair of Organic Chemistry, Garching, Germany*

The virulence of many obligate as well as facultative pathogens is mediated by the caseinolytic protease P (ClpP). This multimeric serine protease was shown to degrade small peptides independently, whereas digestion of proteins requires the interaction with an AAA+-chaperone, as for example ClpX or ClpA. Listeria monocytogenes encodes not only one, but two isoforms of ClpP. Prior investigations revealed that LmClpP2 alone forms a catalytically active, tetradecameric complex. In contrast, LmClpP1 was found to exist as a heptameric ring in vitro that does not exhibit any enzymatic activity. However, labeling studies with LmClpP1-specific inhibitors proved that both isoforms interact to build up a functional degradation machinery in vivo. Recently, we could determine the crystal structure of the LmClpP1/2 heterocomplex at 2.8 Å resolution (PDB: 4RYF)[1]. It preserves all characteristic features of active ClpP proteins which is particularly reflected in the correct alignment of all catalytic centers. The substitution of Asp172 with an asparagine in LmClpP1 leads to replacement of the highly conserved triad by a functional dyad. The reactivity of this assembly is strengthened by the polarizing properties of the mutated residue. Further differences between LmClpP1 and LmClpP2 were found in both the N-terminal region and the S1-substrate pocket. The latter may be useful for isoform-specific drug design in the future.

[1] Maria Dahmen*, Marie-Theres Vielberg*, Michael Groll, Stephan A. Sieber: Structure and mechanism of the caseinolytic protease ClpP1/2 heterocomplex from Listeria monocytogenes, Angew. Chem. Int. Ed. 2015, in press